Student Solution
for Gustafson

COLLEGE
ALGEBRA
5TH EDITION

Michael G. Welden
Rock Valley College

Brooks/Cole Publishing Company
Pacific Grove, California

I(T)P™ The trademark ITP is used under license.

Brooks/Cole Publishing Company
A Division of Wadsworth, Inc.

Printed in the United States of America

10 9 8 7 6 5 4 3 2 1

ISBN 0-534-20883-5

Sponsoring Editor: Audra C. Silverie
Editorial Assistant: Carol Ann Benedict
Production Coordinator: Tessa A. McGlasson
Cover Design: Roy R. Neuhaus
Cover Photo: Ed Young
Printing and Binding: Malloy Lithographing, Inc.

Preface

This manual contains detailed solutions to the odd-numbered exercises of the text *College Algebra*, 5th edition, by R. David Gustafson and Peter D. Frisk. This manual can help you better understand the concepts presented in the textbook, if it is used correctly. However, if you do not use it correctly, you may actually find it more difficult to perform similar exercises on quizzes or tests.

In order to benefit most from the textbook exercises, you should attempt to work each exercise on your own first. You may not immediately succeed. If not, you should try to solve the problem again, possibly by using a slightly different method or approach. Answers to odd-numbered exercises can be compared with those in the back of your textbook to determine if you have obtained the correct solution.

If you have not solved the exercise correctly, or if you have been unable to solve the exercise after several honest attempts, you should then refer to this manual. It would be best to look only at the beginning of the solution to the exercise for help in starting to solve it. If this does not help, analyze the solution in the manual carefully. DO NOT SIMPLY COPY DOWN THE SOLUTION. After you have studied the solution, close this manual and try to solve the exercise on your own. You learn very little, if anything, from only copying solutions. Be sure to solve additional similar exercises without referring to this manual.

Many of the exercises can be solved in more than one way, but it is not feasible to list all solutions in this manual. If you obtain the correct answer, but have solved the exercise in a different way, your work is likely correct. Also, some problems have been solved using methods slightly different from those covered in the textbook. This is done to show you other ways of solving the problems.

I would like to thank the members of the mathematics faculty of Rock Valley College for their assistance and Audra Silverie of Brooks/Cole for her support. This book is dedicated to John, who helps me to realize that mathematics cannot describe everything in life.

May your study of this material be successful and rewarding.

Michael G. Welden

CONTENTS

Exercise 1.1 (page 11)

1. $A \subseteq C$, since every element of set A is also an element of set C.

3. $7 \in B$, since 7 is in the set $B = \{2, 4, 6, 7\}$.

5. $\emptyset \subseteq B$, since the empty set $(\emptyset)$ is a subset of every set.

7. $B \subseteq B$, since every set is a subset of itself.

9. $A \cap B = \{2, 3, 5, 7\} \cap \{1, 3, 5, 7, 9\} = \{3, 5, 7\}$.
   The intersection is the set of elements which are members of **both** sets.

11. $B \cup C = \{1, 3, 5, 7, 9\} \cup \{2, 4, 6, 8, 10\} = \{1, 2, 3, 4, 5, 6, 7, 8, 9, 10\}$.
    The union is the set of elements which are members of **either** set.

13. $\begin{aligned}(A \cap \emptyset) \cup B &= \emptyset \cup B \\ &= B \\ &= \{1, 3, 5, 7, 9\}\end{aligned}$ The empty set has no elements in common with A. There are no elements in the empty set which are not already listed in **B**.

15. $\begin{aligned}(A \cap B) \cup C &= (\{2, 3, 5, 7\} \cap \{1, 3, 5, 7, 9\}) \cup \{2, 4, 6, 8, 10\}) \\ &= \{3, 5, 7\} \cup \{2, 4, 6, 8, 10\} \\ &= \{2, 3, 4, 5, 6, 7, 8, 10\}\end{aligned}$

17. 10, 12, 14, 15, 16, 18   "Between" means greater than 9 and less than 19, so 9 is not included.

19. 2   (Any other even number is composite.)

21. Infinite

23. Finite

25. Finite

27. 53

29. False. 27 has factors other than 1 and itself (3 and 9).

31. True. Every composite number is also an integer.

33. False. 0 can be written as $\frac{0}{2}$.

35. True. An integer n can be written $\frac{n}{1}$.

37. False. 1 is neither prime nor composite.

39. False. 2 and 3 are prime numbers, but $2 \times 3 = 6$, which is composite.

41. False. 2 is a prime number, but $2^2 = 4$, which is composite.

43. False. 4 and 6 are composite numbers, but $4 + 6 = 10$, which is composite.

45. False. 2 is an even prime.

47. True.

49. True.

51. False. $3 + 5 + 6 = 14$.

1

53. True. Every rational number is also a real number.

55. False. All integers are rational numbers (see **#35**), so no irrational number can be an integer.

57. False. The set of real numbers is the union of the sets of rational and irrational numbers, so any real number must be in one of those sets.

59. True. The intersection of the set of real numbers and the set of rational numbers is the set of rational numbers, which is the set of decimals that either terminate or repeat.

61.
20 22 24 26 28 30

63.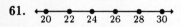
11 13   17 19

65.
1

67. There are no numbers which are both even and odd, so there is no graph.

69.
3   10

71.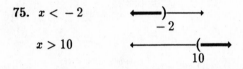
−4   0

73. $3 < x$

$x \le 7$

$3 < x$ and $x \le 7$

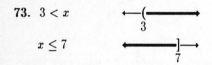
3   7

75. $x < -2$

$x > 10$

$x < -2$ or $x > 10$

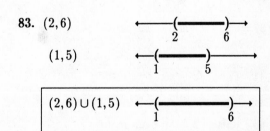
−2   10

77. $x \ge 0$

$x \le 0$

$x \ge 0$ and $x \le 0$

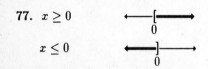
0

79.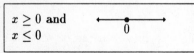
−2   3

81.
4   8

83. $(2,6)$

$(1,5)$

$(2,6) \cup (1,5)$

2   6
1   5
1   6

2

85. $[-2,2)$

$[1,2]$

$[-2,2) \cap [1,2]$

87. $(-5,-3]$

$[-3,-2)$

$(-5,-3) \cup [-3,-2)$

89. $x \neq 0$

$(-3,2]$

$\{x \mid x \neq 0\} \cap (-3,2]$

91. $-(-12) = 12$

93. $4 - (-2) = 4 + 2 = 6$

95. $(-12)(-3) = 36$

97. $\dfrac{-36}{12} = -3$

99. $|-7| = 7$

101. $|7| - |-6| = 7 - 6 = 1$

103. $|\sqrt{2} - 1| = \sqrt{2} - 1$
Note that $\sqrt{2} > 1$, so $\sqrt{2} - 1 > 0$.

105. $|1 - \pi| = -(1 - \pi) = \pi - 1$
Note that $\pi > 1$, so $1 - \pi < 0$.

107. $\begin{aligned} d &= |x - y| \\ &= |-5 - (-2)| \\ &= |-5 + 2| = |-3| = 3 \end{aligned}$

109. Transitive

111. Reflexive, transitive

113. Reflexive, symmetric, transitive

115. Reflexive, symmetric, transitive

117. Commutative property of addition

119. Transitive property of equality

121. Reflexive property of equality

123. Commutative property of addition

125. Symmetric property of equality

127. Associative property of multiplication

Exercise 1.2 (page 20)1. $13^2 = 13 \cdot 13 = 169$

3. $-5^2 = -1 \cdot 5^2 = -1 \cdot 5 \cdot 5 = -25$

5. $4x^3 = 4xxx$

7. $(-5x)^4 = (-5)^4 x^4 = 625xxxx$

9. $7xxx = 7x^3$

11. $(-x)(-x) = (-x)^2 = x^2$

13. $(3t)(3t)(-3t) = -27t^3$

15. $xxxyy = x^3 y^2$

17. $x^2 x^3 = x^{2+3} = x^5$

19. $(z^2)^3 = z^{2 \cdot 3} = z^6$

21. $(y^5 y^2)^3 = (y^{5+2})^3 = (y^7)^3 = y^{7 \cdot 3} = y^{21}$

23. $(z^2)^3 (z^4)^5 = z^{2 \cdot 3} z^{4 \cdot 5} = z^6 z^{20} = z^{6+20} = z^{26}$

25. $(3x^3) = 3^3 x^3 = 27x^3$

27. $(x^2 y)^3 = (x^2)^3 y^3 = x^{2 \cdot 3} y^3 = x^6 y^3$

29. $\left(\dfrac{a^2}{b}\right)^3 = \dfrac{(a^2)^3}{b^3} = \dfrac{a^{2 \cdot 3}}{b^3} = \dfrac{a^6}{b^3}$

31. $(-x)^0 = 1$

33. $(4x)^0 = 1$

35. $z^{-4} = \dfrac{1}{z^4}$

37. $y^{-2} y^{-3} = y^{-2+(-3)} = y^{-5} = \dfrac{1}{y^5}$

39. $(x^3 x^{-4})^{-2} = (x^{-1})^{-2} = x^{-1(-2)} = x^2$

41. $\dfrac{x^7}{x^3} = x^{7-3} = x^4$

43. $\dfrac{a^{21}}{a^{17}} = a^{21-17} = a^4$

45. $\dfrac{(x^2)^2}{x^2 x} = \dfrac{x^4}{x^3} = x^{4-3} = x^1 = x$

47. $\left(\dfrac{m^3}{n^2}\right)^3 = \dfrac{(m^3)^3}{(n^2)^3} = \dfrac{m^{3 \cdot 3}}{n^{2 \cdot 3}} = \dfrac{m^9}{n^6}$

49. $\dfrac{(a^3)^{-3}}{aa^2} = \dfrac{a^{-9}}{a^3} = a^{-9-3} = a^{-12} = \dfrac{1}{a^{12}}$

51. $\left(\dfrac{a^{-3}}{b^{-1}}\right)^{-4} = \dfrac{(a^{-3})^{-4}}{(b^{-1})^{-4}} = \dfrac{a^{12}}{b^4}$

53. $\left(\dfrac{r^4 r^{-6}}{r^3 r^{-3}}\right)^2 = \left(\dfrac{r^{-2}}{r^0}\right)^2 = \left(\dfrac{r^{-2}}{1}\right)^2 = \dfrac{(r^{-2})^2}{1^2} = r^{-4} = \dfrac{1}{r^4}$

55. $\left(\dfrac{x^5 y^{-2}}{x^{-3} y^2}\right)^4 = \left(x^{5-(-3)} y^{-2-(2)}\right)^4 = (x^8 y^{-4})^4 = (x^8)^4 (y^{-4})^4 = x^{32} y^{-16} = \dfrac{x^{32}}{y^{16}}$

4

57.
$$\left(\frac{5x^{-3}y^{-2}}{3x^2y^{-2}}\right)^{-2} = \left(\frac{5}{3}x^{-3-(2)}y^{-2-(-2)}\right)^{-2} = \left(\frac{5}{3}x^{-5}y^0\right)^{-2}$$
$$= \left(\frac{5}{3x^5}\right)^{-2}$$
$$= \left(\frac{3x^5}{5}\right)^2 = \frac{9x^{10}}{25}$$

59. $\left(\dfrac{x^2+3x+y}{2x^5-y}\right)^0 = 1$

61.
$$\left(\frac{3x^5y^{-3}}{6x^{-5}y^3}\right)^{-2} = \left(\frac{3}{6}x^{5-(-5)}y^{-3-(3)}\right)^{-2} = \left(\frac{1}{2}x^{10}y^{-6}\right)^{-2} = \left(\frac{x^{10}}{2y^6}\right)^{-2}$$
$$= \left(\frac{2y^6}{x^{10}}\right)^2$$
$$= \frac{4y^{12}}{x^{20}}$$

63. $(-x^2y^x)^5 = (-1)^5(x^2)^5(y^x)^5 = -x^{10}y^{5x}$

65. $-(x^2y^3)^{4xy} = -(x^2)^{4xy}(y^3)^{4xy} = -x^{8xy}y^{12xy}$

67. $x^{m+1}x^{3-m} = x^{m+1+3-m} = x^4$

69. $(x^{2n}x^{3n})^n = (x^{2n+3n})^n = (x^{5n})^n = x^{5n^2}$

71. $\dfrac{x^{3m+5}}{x^{3m}} = x^{3m+5-3m} = x^5$

73.
$$\frac{(8^{-2}z^{-3}y)^{-1}}{(5y^2z^{-2})^3(5yz^{-2})^{-1}} = \frac{8^2z^3y^{-1}}{(5^3y^6z^{-6})(5^{-1}y^{-1}z^2)} = \frac{64z^3y^{-1}}{5^2y^5z^{-4}}$$
$$= \frac{64}{25}z^{3-(-4)}y^{-1-(5)}$$
$$= \frac{64}{25}z^7y^{-6}$$
$$= \frac{64z^7}{25y^6}$$

75.
$$\left[\frac{(m^{-2}n^{-1}p^3)^{-2}}{(mn^2)^{-3}(p^{-3})^4}\right]^{-2} = \left[\frac{m^4n^2p^{-6}}{m^{-3}n^{-6}p^{-12}}\right]^{-2}$$
$$= \left[m^{4-(-3)}n^{2-(-6)}p^{-6-(-12)}\right]^{-2}$$
$$= [m^7n^8p^6]^{-2}$$
$$= m^{-14}n^{-16}p^{-12}$$
$$= \frac{1}{m^{14}n^{16}p^{12}}$$

77. $x^2 = (-2)^2 = 4$ **79.** $x^3 = (-2)^3 = -8$

81. $(-xz)^3 = [(-1)(-2)(3)]^3 = 6^3 = 216$

83. $\dfrac{-(x^2z^3)}{z^2-y^2} = \dfrac{-(-2)^2(3)^3}{(3)^2-(0)^2} = \dfrac{-4\cdot 27}{9-0} = \dfrac{-108}{9} = -12$

85. $5x^2 - 3y^3z = 5(-2)^2 - 3(0)^3(3) = 5(4) - 0 = 20$

87. $3x^3y^7z^{15} = 3(-2)^3(0)^7(3)^{15} = 0$

89.
$$\frac{-3x^{-3}z^{-2}}{6x^2z^{-3}} = -\frac{3}{6}x^{-3-2}z^{-2-(-3)} = -\frac{1}{2}x^{-5}z = -\frac{z}{2x^5} = -\frac{3}{2(-2)^5}$$
$$= -\frac{3}{2(-32)}$$
$$= -\frac{3}{-64}$$
$$= \frac{3}{64}$$

91. $x^z z^x y^z = (-2)^3(3)^{-2}(0)^3 = 0$ **93.** $372{,}000 = 3.72 \times 10^5$

95. $-177{,}000{,}000 = -1.77 \times 10^8$ **97.** $0.007 = 7.0 \times 10^{-3}$

99. $-0.000000693 = -6.93 \times 10^{-7}$

101. one trillion $= 1{,}000{,}000{,}000{,}000 = 1.0 \times 10^{12}$

103. one trillionth $= 0.000000000001 = 1.0 \times 10^{-12}$

105. $99.7 \times 10^{-4} = 0.00997 = 9.97 \times 10^{-3}$

107. $9.37 \times 10^5 = 937{,}000$ **109.** $2.21 \times 10^{-5} = 0.0000221$

6

111. $0.00032 \times 10^4 = 3.2$

113. $-3.2 \times 10^{-3} = -0.0032$

115.

$$\frac{(65,000)(45,000)}{250,000} = \frac{(6.5 \times 10^4)(4.5 \times 10^4)}{2.5 \times 10^5} = \frac{(6.5)(4.5)(10^4)(10^4)}{2.5 \times 10^5}$$

$$= \frac{29.25 \times 10^8}{2.5 \times 10^5}$$

$$= \frac{29.25}{2.5} \times \frac{10^8}{10^5}$$

$$= 11.7 \times 10^3$$

$$= 1.17 \times 10^4$$

117.

$$\frac{(0.00000035)(170,000)}{0.00000085} = \frac{(3.5 \times 10^{-7})(1.7 \times 10^5)}{8.5 \times 10^{-7}} = \frac{(3.5)(1.7)(10^{-7})(10^5)}{8.5 \times 10^{-7}}$$

$$= \frac{5.95 \times 10^{-2}}{8.5 \times 10^{-7}}$$

$$= \frac{5.95}{8.5} \times \frac{10^{-2}}{10^{-7}}$$

$$= 0.7 \times 10^5$$

$$= 7.0 \times 10^4$$

119.

$$3.31 \times 10^4 \, \frac{cm}{sec} \cdot \frac{1 \, m}{100 \, cm} \cdot \frac{60 \, sec}{min} = \frac{3.31 \times 10^4 \cdot 6 \times 10^1}{1 \times 10^2} \, \frac{m}{min} = \frac{19.86 \times 10^5}{1 \times 10^2} \, \frac{m}{min}$$

$$= 19.86 \times 10^3 \, \frac{m}{min}$$

$$= 1.986 \times 10^4 \, \frac{m}{min}$$

121. To find the mass of one billion protons, multiply the mass of one proton by one billion (1,000,000,000).

1,000,000,000 × 0.00000000000000000000000167248

$$= 1 \times 10^9 \times 1.67248 \times 10^{-24} = 1.67248 \times 10^{-15} \text{ grams}$$

For exercises #123 – 127, keystrokes for the TI-81 calculator are given. The problems may be done on other calculators using slightly different keystrokes. Refer to your calculator's owner's manual. Numbers, letters, symbols, etc. in brackets [ ] represent buttons to be pushed on the TI-81. Numbers in set brackets {} represent output from the calculator.

123. [4] [.] [5] [7] [∧] [0] [ENTER]   {1}

125. $[\,4\,]\,[\,.\,]\,[\,1\,]\,[\,x^2\,]\,[\,+\,]\,[\,5\,]\,[\,.\,]\,[\,2\,]\,[\,x^2\,]\,[\,\text{ENTER}\,]$ $\{43.85\}$

   $[\,(\,]\,[\,4\,]\,[\,.\,]\,[\,1\,]\,[\,+\,]\,[\,5\,]\,[\,.\,]\,[\,2\,]\,[\,)\,]\,[\,x^2\,]\,[\,\text{ENTER}\,]$ $\{86.49\}$

127. $[\,3\,]\,[\,.\,]\,[\,2\,]\,[\,\wedge\,]\,[\,(-)\,]\,[\,3\,]\,[\,\times\,]\,[\,3\,]\,[\,.\,]\,[\,2\,]\,[\,\wedge\,]\,[\,7\,]\,[\,\text{ENTER}\,]$

   $\{104.8576\}$

   $[\,3\,]\,[\,.\,]\,[\,2\,]\,[\,\wedge\,]\,[\,4\,]\,[\,\text{ENTER}\,]$ $\{104.8576\}$

Exercise 1.3   (page 32)

1. $9^{1/2} = 3$

3. $-16^{1/4} = -2$

5. $\left(\dfrac{1}{25}\right)^{1/2} = \dfrac{1^{1/2}}{25^{1/2}} = \dfrac{1}{5}$

7. $(81r^8)^{1/4} = 81^{1/4}(r^8)^{1/4} = 3r^2$

9. $(-64p^6)^{1/3} = (-64)^{1/3}(p^6)^{1/3} = -4p^2$

11. $\left(\dfrac{16a^4}{25b^2}\right)^{1/2} = \dfrac{16^{1/2}(a^4)^{1/2}}{25^{1/2}(b^2)^{1/2}} = \dfrac{4a^2}{5b}$

13. $(16a^2)^{1/2} = 16^{1/2}(a^2)^{1/2} = 4\,|\,a\,|$

15. $(16a^{12})^{1/4} = 16^{1/4}(a^{12})^{1/4} = 2\,|\,a^3\,|$

17. $(-32a^5)^{1/5} = (-32)^{1/5}(a^5)^{1/5} = -2a$

19. $(-216b^6)^{1/3} = (-216)^{1/3}(b^6)^{1/3} = -6b^2$

21. $4^{3/2} = (4^{1/2})^3 = 2^3 = 8$

23. $-16^{3/4} = -(16^{1/4})^3 = -2^3 = -8$

25. $-1000^{2/3} = -(1000^{1/3})^2 = -10^2 = -100$

27. $64^{-1/2} = \left(\dfrac{1}{64}\right)^{1/2} = \dfrac{1^{1/2}}{64^{1/2}} = \dfrac{1}{8}$

29. $64^{-3/2} = \left(\dfrac{1}{64}\right)^{3/2} = \dfrac{1^{3/2}}{64^{3/2}} = \dfrac{1}{(64^{1/2})^3} = \dfrac{1}{8^3} = \dfrac{1}{512}$

31. $-9^{-3/2} = -\left(\dfrac{1}{9}\right)^{3/2} = -\dfrac{1^{3/2}}{9^{3/2}} = -\dfrac{1}{(9^{1/2})^3} = -\dfrac{1}{3^3} = -\dfrac{1}{27}$

33. $\left(\dfrac{4}{9}\right)^{5/2} = \left(\left(\dfrac{4}{9}\right)^{1/2}\right)^5 = \left(\dfrac{2}{3}\right)^5 = \dfrac{2^5}{3^5} = \dfrac{32}{243}$

8

35. $\left(-\dfrac{27}{64}\right)^{-2/3} = \left(-\dfrac{64}{27}\right)^{2/3} = \left(\left(-\dfrac{64}{27}\right)^{1/3}\right)^{2} = \left(-\dfrac{4}{3}\right)^{2} = \dfrac{16}{9}$

37. $(100s^4)^{1/2} = 100^{1/2}(s^4)^{1/2} = 10s^2$

39. $(32y^{10}z^5)^{-1/5} = 32^{-1/5}y^{-2}z^{-1} = \dfrac{1}{32^{1/5}y^2z} = \dfrac{1}{2y^2z}$

41. $(x^{10}y^5)^{3/5} = (x^{10})^{3/5}(y^5)^{3/5} = x^6y^3$

43. $(r^8s^{16})^{-3/4} = r^{8(-3/4)}s^{16(-3/4)} = r^{-6}s^{-12} = \dfrac{1}{r^6s^{12}}$

45. $\left(-\dfrac{8a^6}{125b^9}\right)^{2/3} = \left(\left(-\dfrac{8a^6}{125b^9}\right)^{1/3}\right)^{2} = \left(-\dfrac{2a^2}{5b^3}\right)^{2} = \dfrac{4a^4}{25b^6}$

47. $\left(\dfrac{27r^6}{1000s^{12}}\right)^{-2/3} = \left(\dfrac{1000s^{12}}{27r^6}\right)^{2/3} = \left(\left(\dfrac{1000s^{12}}{27r^6}\right)^{1/3}\right)^{2} = \left(\dfrac{10s^4}{3r^2}\right)^{2} = \dfrac{100s^8}{9r^4}$

49. $\left(\dfrac{a^8c^{24}}{b^{20}d^0}\right)^{0.25} = \dfrac{a^{8(0.25)}c^{24(0.25)}}{b^{20(0.25)}\cdot 1} = \dfrac{a^2c^6}{b^5}$

51. $\dfrac{a^{2/5}a^{4/5}}{a^{1/5}} = \dfrac{a^{6/5}}{a^{1/5}} = a^{5/5} = a^1 = a$

53. $\dfrac{(16a^{-6}s^2)^{1/4}}{a^{3/2}s^{-1/2}} = \dfrac{16^{1/4}a^{-3/2}s^{1/2}}{a^{3/2}s^{-1/2}} = 2a^{-6/2}s^{2/2} = 2a^{-3}s = \dfrac{2s}{a^3}$

55. $\dfrac{b^{x/3}b^{x/4}}{b^{x/8}} = \dfrac{b^{8x/24}b^{6x/24}}{b^{3x/24}} = \dfrac{b^{14x/24}}{b^{3x/24}} = b^{11x/24}$

57. $\sqrt{49} = 7$

59. $\sqrt[3]{125} = 5$

61. $-\sqrt[4]{81} = -3$

63. $\sqrt[5]{-\dfrac{32}{100,000}} = -\dfrac{2}{10} = -\dfrac{1}{5}$

65. $\sqrt{36x^2} = \sqrt{36}\sqrt{x^2} = 6x$

67. $\sqrt{x^2y^4} = \sqrt{x^2}\sqrt{y^4} = xy^2$

69. $\sqrt[3]{8y^3} = \sqrt[3]{8}\,\sqrt[3]{y^3} = 2y$

71. $\sqrt[4]{\dfrac{x^4y^8}{z^{12}}} = \dfrac{\sqrt[4]{x^4}\,\sqrt[4]{y^8}}{\sqrt[4]{z^{12}}} = \dfrac{xy^2}{z^3}$

73. $\sqrt{4x^2} = \sqrt{4}\sqrt{x^2} = 2\,|x|$

75. $-\sqrt{16b^4} = -\sqrt{16}\sqrt{b^4} = -4\,|b^2| = -4b^2$

77. $\sqrt[3]{27x^3} = \sqrt[3]{27}\ \sqrt[3]{x^3} = 3x$ **79.** $\sqrt[6]{x^6 y^{12}} = |x|\,|y^2| = |x|\,y^2$

81. $\sqrt{8} - \sqrt{2} = \sqrt{4\cdot 2} - \sqrt{2} = \sqrt{4}\sqrt{2} - \sqrt{2} = 2\sqrt{2} - \sqrt{2} = \sqrt{2}$

83. $\sqrt{200x^2} + \sqrt{98x^2} = \sqrt{100x^2 \cdot 2} + \sqrt{49x^2 \cdot 2} = 10x\sqrt{2} + 7x\sqrt{2} = 17x\sqrt{2}$

85. $2\sqrt{48y^5} - 3y\sqrt{12y^3} = 2\sqrt{16y^4 \cdot 3y} - 3y\sqrt{4y^2 \cdot 3y}$

$$= 2(4y^2)\sqrt{3y} - 3y(2y)\sqrt{3y}$$

$$= 8y^2\sqrt{3y} - 6y^2\sqrt{3y} = 2y^2\sqrt{3y}$$

87. $2\ \sqrt[3]{81} + 3\ \sqrt[3]{24} = 2\ \sqrt[3]{27\cdot 3} + 3\ \sqrt[3]{8\cdot 3} = 2(3)\ \sqrt[3]{3} + 3(2)\ \sqrt[3]{3}$

$$= 6\ \sqrt[3]{3} + 6\ \sqrt[3]{3}$$

$$= 12\ \sqrt[3]{3}$$

89. $\sqrt[4]{768z^5} + \sqrt[4]{48z^5} = \sqrt[4]{256z^4 \cdot 3z} + \sqrt[4]{16z^4 \cdot 3z} = 4z\ \sqrt[4]{3z} + 2z\ \sqrt[4]{3z}$

$$= 6z\ \sqrt[4]{3z}$$

91. $\sqrt{8x^2 y} - x\sqrt{2y} + \sqrt{50x^2 y} = \sqrt{4x^2 \cdot 2y} - x\sqrt{2y} + \sqrt{25x^2 \cdot 2y}$

$$= 2x\sqrt{2y} - x\sqrt{2y} + 5x\sqrt{2y}$$

$$= 6x\sqrt{2y}$$

93. $\sqrt[3]{16xy^4} + y\ \sqrt[3]{2xy} - \sqrt[3]{54xy^4} = \sqrt[3]{8y^3 \cdot 2xy} + y\ \sqrt[3]{2xy} - \sqrt[3]{27y^3 \cdot 2xy}$

$$= 2y\ \sqrt[3]{2xy} + y\ \sqrt[3]{2xy} - 3y\ \sqrt[3]{2xy}$$

$$= 0$$

95. $\dfrac{3x\sqrt{2x} - 2\sqrt{2x^3}}{\sqrt{18x} - \sqrt{2x}} = \dfrac{3x\sqrt{2x} - 2\sqrt{x^2 \cdot 2x}}{\sqrt{9\cdot 2x} - \sqrt{2x}} = \dfrac{3x\sqrt{2x} - 2x\sqrt{2x}}{3\sqrt{2x} - \sqrt{2x}} = \dfrac{x\sqrt{2x}}{2\sqrt{2x}} = \dfrac{x}{2}$

97. $\dfrac{3}{\sqrt{3}} = \dfrac{3}{\sqrt{3}} \cdot \dfrac{\sqrt{3}}{\sqrt{3}} = \dfrac{3\sqrt{3}}{3} = \sqrt{3}$ **99.** $\dfrac{2}{\sqrt{x}} = \dfrac{2}{\sqrt{x}} \cdot \dfrac{\sqrt{x}}{\sqrt{x}} = \dfrac{2\sqrt{x}}{x}$

101. $\dfrac{2}{\sqrt[3]{2}} = \dfrac{2}{\sqrt[3]{2}} \cdot \dfrac{\sqrt[3]{4}}{\sqrt[3]{4}} = \dfrac{2\ \sqrt[3]{4}}{\sqrt[3]{8}} = \dfrac{2\ \sqrt[3]{4}}{2} = \sqrt[3]{4}$

103. $\dfrac{5a}{\sqrt[3]{25a}} = \dfrac{5a}{\sqrt[3]{25a}} \cdot \dfrac{\sqrt[3]{5a^2}}{\sqrt[3]{5a^2}} = \dfrac{5a\ \sqrt[3]{5a^2}}{\sqrt[3]{125a^3}} = \dfrac{5a\ \sqrt[3]{5a^2}}{5a} = \sqrt[3]{5a^2}$

105. $\dfrac{2b}{\sqrt[4]{3a^2}} = \dfrac{2b}{\sqrt[4]{3a^2}} \cdot \dfrac{\sqrt[4]{27a^2}}{\sqrt[4]{27a^2}} = \dfrac{2b\sqrt[4]{27a^2}}{\sqrt[4]{81a^4}} = \dfrac{2b\sqrt[4]{27a^2}}{3a}$

107. $\sqrt{\dfrac{x}{2y}} = \dfrac{\sqrt{x}}{\sqrt{2y}} \cdot \dfrac{\sqrt{2y}}{\sqrt{2y}} = \dfrac{\sqrt{2xy}}{2y}$

109. $\sqrt[3]{\dfrac{2u^4}{9v}} = \dfrac{\sqrt[3]{2u^4}}{\sqrt[3]{9v}} = \dfrac{u\sqrt[3]{2u}}{\sqrt[3]{9v}} \cdot \dfrac{\sqrt[3]{3v^2}}{\sqrt[3]{3v^2}} = \dfrac{u\sqrt[3]{6uv^2}}{\sqrt[3]{27v^3}} = \dfrac{u\sqrt[3]{6uv^2}}{3v}$

111. $\sqrt{\dfrac{1}{3}} - \sqrt{\dfrac{1}{27}} = \dfrac{1}{\sqrt{3}} \cdot \dfrac{\sqrt{3}}{\sqrt{3}} - \dfrac{1}{\sqrt{27}} \cdot \dfrac{\sqrt{3}}{\sqrt{3}} = \dfrac{\sqrt{3}}{3} - \dfrac{\sqrt{3}}{9} = \dfrac{3\sqrt{3}}{9} - \dfrac{\sqrt{3}}{9} = \dfrac{2\sqrt{3}}{9}$

113. $\sqrt{\dfrac{x}{8}} - \sqrt{\dfrac{x}{2}} + \sqrt{\dfrac{x}{32}} = \dfrac{\sqrt{x}}{\sqrt{8}} \cdot \dfrac{\sqrt{8}}{\sqrt{8}} - \dfrac{\sqrt{x}}{\sqrt{2}} \cdot \dfrac{\sqrt{2}}{\sqrt{2}} + \dfrac{\sqrt{x}}{\sqrt{32}} \cdot \dfrac{\sqrt{32}}{\sqrt{32}} = \dfrac{\sqrt{8x}}{8} - \dfrac{\sqrt{2x}}{2} + \dfrac{\sqrt{32x}}{32}$

$$= \dfrac{2\sqrt{2x}}{8} - \dfrac{\sqrt{2x}}{2} + \dfrac{4\sqrt{2x}}{32}$$

$$= \dfrac{\sqrt{2x}}{4} - \dfrac{\sqrt{2x}}{2} + \dfrac{\sqrt{2x}}{8}$$

$$= \dfrac{2\sqrt{2x}}{8} - \dfrac{4\sqrt{2x}}{8} + \dfrac{\sqrt{2x}}{8}$$

$$= -\dfrac{\sqrt{2x}}{8}$$

115. $\dfrac{\sqrt[3]{x^6 y^3}}{\sqrt{9x^2 y}} = \dfrac{x^2 y}{3x\sqrt{y}} \cdot \dfrac{\sqrt{y}}{\sqrt{y}} = \dfrac{x^2 y \sqrt{y}}{3xy} = \dfrac{x\sqrt{y}}{3}$

117. $\dfrac{\sqrt[4]{81x^{-6}y^8}}{\sqrt[4]{x^{-2}y^4}} = \sqrt[4]{\dfrac{81x^{-6}y^8}{x^{-2}y^4}} = \sqrt[4]{\dfrac{81y^4}{x^4}} = \dfrac{3y}{x}$

119. $\dfrac{\sqrt{x^{12}y^{12}}}{\sqrt[5]{32x^6 y^{11}}} = \dfrac{x^6 y^6}{2xy^2 \sqrt[5]{xy}} \cdot \dfrac{\sqrt[5]{x^4 y^4}}{\sqrt[5]{x^4 y^4}} = \dfrac{x^6 y^6 \sqrt[5]{x^4 y^4}}{2xy^2 \sqrt[5]{x^5 y^5}} = \dfrac{x^6 y^6 \sqrt[5]{x^4 y^4}}{2x^2 y^3}$

$$= \dfrac{x^4 y^3 \sqrt[5]{x^4 y^4}}{2}$$

121. $\sqrt[4]{9} = 9^{1/4} = (3^2)^{1/4} = 3^{2/4} = 3^{1/2} = \sqrt{3}$

123. $\sqrt[10]{16x^6} = (16x^6)^{1/10} = \left[(4x^3)^2\right]^{1/10} = (4x^3)^{1/5} = \sqrt[5]{4x^3}$

125. $\sqrt{2}\ \sqrt[3]{2} = 2^{1/2} \cdot 2^{1/3} = 2^{3/6} \cdot 2^{2/6} = \sqrt[6]{2^3} \cdot \sqrt[6]{2^2} = \sqrt[6]{8} \cdot \sqrt[6]{4} = \sqrt[6]{32}$

127.

$$\frac{\sqrt[4]{3}}{\sqrt{2}} = \frac{\sqrt[4]{3}}{\sqrt{2}} \cdot \frac{\sqrt{2}}{\sqrt{2}} = \frac{3^{1/4} \cdot 2^{1/2}}{2} = \frac{3^{2/8} \cdot 2^{4/8}}{2} = \frac{\sqrt[8]{3^2} \cdot \sqrt[8]{2^4}}{2} = \frac{\sqrt[8]{9} \cdot \sqrt[8]{16}}{2}$$

$$= \frac{\sqrt[8]{144}}{2}$$

$$= \frac{(12^2)^{1/8}}{2}$$

$$= \frac{\sqrt[4]{12}}{2}$$

129. If x is unrestricted, then $\sqrt{x^2} = |x|$, so the question becomes "When is $|x| = x$?" This is true is and only if $x \geq 0$. So when $x \geq 0$, $\sqrt{x^2} = x$.

131. If x is unrestricted, then $\sqrt[4]{x^4} = |x|$, so the question becomes "When is $|x| = -x$?" This is true if and only if $x \leq 0$. So when $x \leq 0$, $\sqrt[4]{x^4} = -x$.

133.

$$\left(\frac{x}{y}\right)^{-m/n} = \left(\frac{y}{x}\right)^{m/n} \qquad \text{from the theorem on page 17}$$

$$= \left(\left(\frac{y}{x}\right)^m\right)^{1/n} \qquad \text{from the rule on page 24}$$

$$= \sqrt[n]{\left(\frac{y}{x}\right)^m} \qquad \text{from the definition of radical notation on page 26}$$

$$= \sqrt[n]{\frac{y^m}{x^m}} \qquad \text{from the power rules of exponents on page 15}$$

For exercises #135 − 139, keystrokes for the TI-81 calculator are given. The problems may be done on other calculators using slightly different keystrokes. Refer to your calculator's owner's manual. Numbers, letters, symbols, etc. in brackets [ ] represent buttons to be pushed on the TI-81. Numbers in set brackets {} represent output from the calculator.

135. [( ] [3] [.] [5] [×] [1] [.] [2] [ )] [∧] [1] [.] [4] [ENTER]

{7.456740105}

[( ] [3] [.] [5] [∧] [1] [.] [4] [ )] [×] [( ] [1] [.] [2] [∧] [1] [.]

[4] [ )] [ENTER]   {7.456740105}

137. $[\,(\,]\,[\,3\,]\,[\,.\,]\,[\,5\,]\,[\,\wedge\,]\,[\,1\,]\,[\,.\,]\,[\,2\,]\,[\,)\,]\,[\,\wedge\,]\,[\,1\,]\,[\,.\,]\,[\,4\,]\,[\,\text{ENTER}\,]$

$$\{8.204163646\}$$

$[\,(\,]\,[\,3\,]\,[\,.\,]\,[\,5\,]\,[\,\wedge\,]\,[\,1\,]\,[\,.\,]\,[\,4\,]\,[\,)\,]\,[\,\wedge\,]\,[\,1\,]\,[\,.\,]\,[\,2\,]\,[\,\text{ENTER}\,]$

$$\{8.204163646\}$$

139. $[\,(\,]\,[\,3\,]\,[\,.\,]\,[\,5\,]\,[\,\wedge\,]\,[\,(\,]\,[\,1\,]\,[\,.\,]\,[\,2\,]\,[\,x^{-1}\,]\,[\,)\,]\,[\,)\,]\,[\,\wedge\,]\,[\,(\,]\,[\,3\,]\,[\,.\,]$
$[\,5\,]\,[\,x^{-1}\,]\,[\,)\,]\,[\,\text{ENTER}\,]$ $\qquad\{1.347534865\}$

$[\,3\,]\,[\,.\,]\,[\,5\,]\,[\,\wedge\,]\,[\,(\,]\,[\,(\,]\,[\,3\,]\,[\,.\,]\,[\,5\,]\,[\,\times\,]\,[\,1\,]\,[\,.\,]\,[\,2\,]\,[\,)\,]\,[\,x^{-1}\,]\,[\,)\,]$
$[\,\text{ENTER}\,]$ $\qquad\{1.347534865\}$

Exercise 1.4 (page 43)

1. 2nd degree trinomial

3. Not a polynomial

5. 3rd degree binomial

7. 0th degree monomial

9. Monomial with no degree

11. Not a polynomial

13. $(x^3 - 3x^2) + (5x^3 - 8x) = x^3 - 3x^2 + 5x^3 - 8x = 6x^3 - 3x^2 - 8x$

15. $(y^5 + 2y^3 + 7) - (y^5 - 2y^3 - 7) = y^5 + 2y^3 + 7 - y^5 + 2y^3 + 7 = 4y^3 + 14$

17. $\begin{aligned} 2(x^2 + 3x - 1) - 3(x^2 + 2x - 4) + 4 &= 2x^2 + 6x - 2 - 3x^2 - 6x + 12 + 4 \\ &= -x^2 + 14 \end{aligned}$

19. $\begin{aligned} 8(t^2 - 2t + 5) + 4(t^2 - 3t + 2) - 6(2t^2 - 8) \\ &= 8t^2 - 16t + 40 + 4t^2 - 12t + 8 - 12t^2 + 48 \\ &= -28t + 96 \end{aligned}$

21. $y(y^2 - 1) - y^2(y + 2) - y(2y - 2) = y^3 - y - y^3 - 2y^2 - 2y^2 + 2y = -4y^2 + y$

23. $x(x^2 - 1) - x^2(x + 2) - x(2x - 2) = x^3 - x - x^3 - 2x^2 - 2x^2 + 2x = -4x^2 + x$

25. $\begin{aligned} xy(x - 4y) - y(x^2 + 3xy) + xy(2x + 3y) \\ &= x^2y - 4xy^2 - x^2y - 3xy^2 + 2x^2y + 3xy^2 \\ &= 2x^2y - 4xy^2 \end{aligned}$

27. $2x^2y^3(4xy^4) = 2(4)x^2xy^3y^4 = 8x^3y^7$

29. $-3m^2n(2mn^2)\left(-\dfrac{mn}{12}\right) = -3(2)\left(-\dfrac{1}{12}\right)m^2mmnn^2n = \dfrac{1}{2}m^4n^4 = \dfrac{m^4n^4}{2}$

31. $-4rs(r^2 + s^2) = -4rs(r^2) - 4rs(s^2) = -4r^3s - 4rs^3$

13

33. $6ab^2c(2ac + 3bc^2 - 4ab^2c) = 6ab^2c(2ac) + 6ab^2c(3bc^2) + 6ab^2c(-4ab^2c)$
$$= 12a^2b^2c^2 + 18ab^3c^3 - 24a^2b^4c^2$$

35. $(a+2)(a+2) = a^2 + 2a + 2a + 2^2 = a^2 + 4a + 4$

37. $(a-6)^2 = (a-6)(a-6) = a^2 - 6a - 6a + 36 = a^2 - 12a + 36$

39. $(x+4)(x-4) = x^2 - 4x + 4x - 4^2 = x^2 - 16$

41. $(a-2b)^2 = (a-2b)(a-2b) = a^2 - 2ab - 2ab + (-2b)^2 = a^2 - 4ab + 4b^2$

43. $(3m+4n)(3m-4n) = 9m^2 - 12mn + 12mn - 16n^2 = 9m^2 - 16n^2$

45. $(x-3)(x+5) = x^2 + 5x - 3x - 15 = x^2 + 2x - 15$

47. $(u+2)(3u-2) = 3u^2 - 2u + 6u - 4 = 3u^2 + 4u - 4$

49. $(2x+3)(3x-5) = 6x^2 - 10x + 9x - 15 = 6x^2 - x - 15$

51. $(5x-1)(2x+3) = 10x^2 + 15x - 2x - 3 = 10x^2 + 13x - 3$

53. $(3v-4)(4v+3) = 12v^2 + 9v - 16v - 12 = 12v^2 - 7v - 12$

55. $(2y-4x)(3y-2x) = 6y^2 - 4xy - 12xy + 8x^2 = 6y^2 - 16xy + 8x^2$

57. $(9x-y)(x^2-3y) = 9x^3 - 27xy - x^2y + 3y^2 = 9x^3 - x^2y - 27xy + 3y^2$

59. $(5z+2t)(z^2-t) = 5z^3 - 5zt + 2z^2t - 2t^2 = 5z^3 + 2z^2t - 5zt - 2t^2$

61. $(3x-1)^3 = (3x-1)(3x-1)(3x-1) = (9x^2 - 6x + 1)(3x-1)$
$$= 27x^3 - 9x^2 - 18x^2 + 6x + 3x - 1$$
$$= 27x^3 - 27x^2 + 9x - 1$$

63. $(xy^2-1)(x^2y+2) = x^3y^3 + 2xy^2 - x^2y - 2 = x^3y^3 - x^2y + 2xy^2 - 2$

65. $(3x+1)(2x^2+4x-3) = 6x^3 + 12x^2 - 9x + 2x^2 + 4x - 3 = 6x^3 + 14x^2 - 5x - 3$

67. $(3x+2y)(2x^2-3xy+4y^2) = 6x^3 - 9x^2y + 12xy^2 + 4x^2y - 6xy^2 + 8y^3$
$$= 6x^3 - 5x^2y + 6xy^2 + 8y^3$$

69. $(x^2+x+1)(x^2+x-1) = x^4 + x^3 - x^2 + x^3 + x^2 - x + x^2 + x - 1$
$$= x^4 + 2x^3 + x^2 - 1$$

71.
$$2y^n(3y^n - 2y^2 + y^{-n}) = 2y^n(3y^n) + 2y^n(-2y^2) + 2y^n(y^{-n})$$
$$= 6y^{2n} - 4y^{n+2} + 2y^0$$
$$= 6y^{2n} - 4^{n+2} + 2$$

73.
$$-5x^{2n}y^n(2x^{2n}y^{-n} + 3x^{-2n}y^n) = -5x^{2n}y^n(2x^{2n}y^{-n}) - 5x^{2n}y^n(3x^{-2n}y^n)$$
$$= -10x^{4n}y^0 - 15x^0y^{2n}$$
$$= -10x^{4n} - 15y^{2n}$$

75. $(x^n + 3)(x^n - 4) = x^{2n} - 4x^n + 3x^n - 12 = x^{2n} - x^n - 12$

77. $(2r^n - 7)(3r^n - 2) = 6r^{2n} - 4r^n - 21r^n + 14 = 6r^{2n} - 25r^n + 14$

79.
$$x^{1/2}(x^{1/2}y + xy^{1/2}) = x^{1/2}(x^{1/2}y) + x^{1/2}(xy^{1/2}) = xy + x^{1/2}(x^{2/2}y^{1/2})$$
$$= xy + x^{3/2}y^{1/2}$$

81. $(a^{1/2} + b^{1/2})(a^{1/2} - b^{1/2}) = a^{1/2}a^{1/2} - a^{1/2}b^{1/2} + a^{1/2}b^{1/2} - b^{1/2}b^{1/2} = a - b$

83. $\dfrac{2}{\sqrt{3}-1} = \dfrac{2}{(\sqrt{3}-1)} \cdot \dfrac{(\sqrt{3}+1)}{(\sqrt{3}+1)} = \dfrac{2(\sqrt{3}+1)}{3-1} = \dfrac{2(\sqrt{3}+1)}{2} = \sqrt{3}+1$

85.
$$\frac{3x}{\sqrt{7}+2} = \frac{3x}{(\sqrt{7}+2)} \cdot \frac{(\sqrt{7}-2)}{(\sqrt{7}-2)} = \frac{3x(\sqrt{7}-2)}{7-4} = \frac{3x(\sqrt{7}-2)}{3}$$
$$= x(\sqrt{7}-2)$$

87.
$$\frac{6}{\sqrt{5}-\sqrt{2}} = \frac{6}{(\sqrt{5}-\sqrt{2})} \cdot \frac{(\sqrt{5}+\sqrt{2})}{(\sqrt{5}+\sqrt{2})} = \frac{6(\sqrt{5}+\sqrt{2})}{5-2} = \frac{6(\sqrt{5}+\sqrt{2})}{3}$$
$$= 2(\sqrt{5}+\sqrt{2})$$

89. $\dfrac{x}{x-\sqrt{3}} = \dfrac{x}{(x-\sqrt{3})} \cdot \dfrac{(x+\sqrt{3})}{(x+\sqrt{3})} = \dfrac{x(x+\sqrt{3})}{x^2-3}$

91. $\dfrac{y+\sqrt{2}}{y-\sqrt{2}} = \dfrac{(y+\sqrt{2})}{(y-\sqrt{2})} \cdot \dfrac{(y+\sqrt{2})}{(y+\sqrt{2})} = \dfrac{(y+\sqrt{2})^2}{y^2-2}$

93. $\dfrac{\sqrt{5}-2}{\sqrt{5}+2} = \dfrac{(\sqrt{5}-2)}{(\sqrt{5}+2)} \cdot \dfrac{(\sqrt{5}-2)}{(\sqrt{5}-2)} = \dfrac{5-2\sqrt{5}-2\sqrt{5}+4}{5-4} = 9-4\sqrt{5}$

95.

$$\frac{\sqrt{2}-\sqrt{3}}{1-\sqrt{3}} = \frac{(\sqrt{2}-\sqrt{3})}{(1-\sqrt{3})} \cdot \frac{(1+\sqrt{3})}{(1+\sqrt{3})} = \frac{\sqrt{2}+\sqrt{6}-\sqrt{3}-3}{1-3}$$

$$= \frac{\sqrt{2}+\sqrt{6}-\sqrt{3}-3}{-2}$$

$$= -\frac{\sqrt{2}+\sqrt{6}-\sqrt{3}-3}{2}$$

$$= \frac{\sqrt{3}+3-\sqrt{2}-\sqrt{6}}{2}$$

97.

$$\frac{\sqrt{x}-\sqrt{y}}{\sqrt{x}+\sqrt{y}} = \frac{(\sqrt{x}-\sqrt{y})}{(\sqrt{x}+\sqrt{y})} \cdot \frac{(\sqrt{x}-\sqrt{y})}{(\sqrt{x}-\sqrt{y})} = \frac{x-\sqrt{xy}-\sqrt{xy}+y}{x-y} = \frac{x-2\sqrt{xy}+y}{x-y}$$

99.

$$\frac{\sqrt{x+3}-\sqrt{x}}{3} = \frac{(\sqrt{x+3}-\sqrt{x})}{3} \cdot \frac{(\sqrt{x+3}+\sqrt{x})}{(\sqrt{x+3}+\sqrt{x})} = \frac{x+3-x}{3(\sqrt{x+3}+\sqrt{x})}$$

$$= \frac{3}{3(\sqrt{x+3}+\sqrt{x})}$$

$$= \frac{1}{\sqrt{x+3}+\sqrt{x}}$$

101. $\dfrac{36a^2b^3}{18ab^6} = 2a^1b^{-3} = \dfrac{2a}{b^3}$

103. $\dfrac{16x^6y^4z^9}{-24x^9y^6z^0} = -\dfrac{2}{3}x^{-3}y^{-2}z^9 = -\dfrac{2z^9}{3x^3y^2}$

105. $\dfrac{5x^3y^2+15x^3y^4}{10x^2y^3} = \dfrac{5x^3y^2}{10x^2y^3} + \dfrac{15x^3y^4}{10x^2y^3} = \dfrac{1}{2}x^1y^{-1} + \dfrac{3}{2}x^1y^1 = \dfrac{x}{2y} + \dfrac{3xy}{2}$

107.

$$\frac{24x^5y^7-36x^2y^5+12xy}{60x^5y^4} = \frac{24x^5y^7}{60x^5y^4} - \frac{36x^2y^5}{60x^5y^4} + \frac{12xy}{60x^5y^4}$$

$$= \frac{2}{5}x^0y^3 - \frac{3}{5}x^{-3}y^1 + \frac{1}{5}x^{-4}y^{-3}$$

$$= \frac{2y^3}{5} - \frac{3y}{5x^3} + \frac{1}{5x^4y^3}$$

109.

$$
\begin{array}{r}
3x \quad + \; 2 \\
x+3 \,\overline{\big)\; 3x^2 \;+\; 11x \;+\; 6} \\
-(\,3x^2 \;+\; 9x) \\
\hline
2x \;+\; 6 \\
-(\quad 2x \;+\; 6) \\
\end{array}
$$

111.

$$
\begin{array}{r}
x \quad - 7 \quad + \dfrac{2}{2x-5}
\end{array}
$$

$2x-5\;)\;\overline{\;2x^2 \;-\; 19x \;+\; 37\;}$

$\underline{-(\,2x^2 \;-\; 5x\,)}$

$\qquad\qquad -14x \;+\; 37$

$\qquad\quad\underline{-(-\,14x \;+\; 35)}$

$\qquad\qquad\qquad\qquad 2$

113.

$$
x \quad - 3
$$

$x^2+x-1\;)\;\overline{\;x^3 \;-\; 2x^2 \;-\; 4x \;+\; 3\;}$

$\underline{-(\,x^3 \;+\; x^2 \;-\; x\,)}$

$\qquad\qquad -3x^2 \;-\; 3x \;+\; 3$

$\qquad\quad\underline{-(-\,3x^2 \;-\; 3x \;+\; 3)}$

115.

$$
x^2 \quad - 2 \quad + \dfrac{-x^2+5}{x^3-2}
$$

$x^3-2\;)\;\overline{\;x^5 \;+\; 0x^4 \;-\; 2x^3 \;-\; 3x^2 \;+\; 0x \;+\; 9\;}$

$\underline{-(\,x^5 \;+\; 0x^4 \;+\; 0x^3 \;-\; 2x^2\,)}$

$\qquad\qquad\qquad\qquad -2x^3 \;-\; x^2 \;+\; 0x \;+\; 9$

$\qquad\qquad\qquad\underline{-(-\,2x^3 \;+\; 0x^2 \;+\; 0x \;+\; 4)}$

$\qquad\qquad\qquad\qquad\qquad -x^2 \qquad\qquad +5$

117.

$$
x^4 \quad + 2x^3 \quad + 4x^2 \quad + 8x \quad + 16
$$

$x-2\;)\;\overline{\;x^5 \;+\; 0x^4 \;+\; 0x^3 \;+\; 0x^2 \;+\; 0x \;-\; 32\;}$

$\underline{-(\,x^5 \;-\; 2x^4\,)}$

$\qquad\qquad 2x^4 \;+\; 0x^3$

$\qquad\underline{-(\,2x^4 \;-\; 4x^3\,)}$

$\qquad\qquad\qquad 4x^3 \;+\; 0x^2$

$\qquad\qquad\underline{-(\,4x^3 \;-\; 8x^2\,)}$

$\qquad\qquad\qquad\qquad 8x^2 \;+\; 0x$

$\qquad\qquad\qquad\underline{-(\,8x^2 \;-\; 16x\,)}$

$\qquad\qquad\qquad\qquad\qquad 16x \;-\; 32$

$\qquad\qquad\qquad\qquad\underline{-(\,16x \;-\; 32)}$

$$\begin{array}{r} x^2 \quad + \quad 2 \quad + \dfrac{20}{x^2-2} \end{array}$$

119. x^2-2 $\overline{)\,\begin{array}{l} x^4 \;+\; 0x^3 \;+\; 0x^2 \;+\; 0x \;+\; 16 \\ \underline{-(\;\; x^4 \;+\; 0x^3 \;-\; 2x^2)} \\ \qquad\qquad\qquad\; 2x^2 \;+\; 0x \;+\; 16 \\ \qquad\qquad\;\; \underline{-(\;\; 2x^2 \;+\; 0x \;-\; 4)} \\ \qquad\qquad\qquad\qquad\qquad\qquad\; 20 \end{array}}$

$$\begin{array}{r} 6x^2 \quad + \quad x \quad - \quad 12 \end{array}$$

121. $6x^2+11x-10$ $\overline{)\,\begin{array}{l} 36x^4 \;+\; 72x^3 \;-\; 121x^2 \;-\; 142x \;+\; 120 \\ \underline{-(\;\; 36x^4 \;+\; 66x^3 \;-\; 60x^2)} \\ \qquad\qquad\quad 6x^3 \;-\; 61x^2 \;-\; 142x \\ \qquad\quad\;\; \underline{-(\;\; 6x^3 \;+\; 11x^2 \;-\; 10x)} \\ \qquad\qquad\qquad\qquad -\,72x^2 \;-\; 132x \;+\; 120 \\ \qquad\qquad\qquad\quad \underline{-(-\,72x^2 \;-\; 132x \;+\; 120)} \end{array}}$

$$\begin{array}{r} 3x^2 \quad - \quad x \quad + \quad 2 \end{array}$$

123. $4x^4+5x^3-3$ $\overline{)\,\begin{array}{l} 12x^6 \;+\; 11x^5 \;+\; 3x^4 \;+\; 10x^3 \;-\; 9x^2 \;+\; 3x \;-\; 6 \\ \underline{-(\;\; 12x^6 \;+\; 15x^5 \;+\; 0x^4 \;+\; 0x^3 \;-\; 9x^2)} \\ \qquad\qquad -\,4x^5 \;+\; 3x^4 \;+\; 10x^3 \;+\; 0x^2 \;+\; 3x \\ \qquad\;\; \underline{-(-\;\; 4x^5 \;-\; 5x^4 \;+\; 0x^3 \;+\; 0x^2 \;+\; 3x)} \\ \qquad\qquad\qquad\quad 8x^4 \;+\; 10x^3 \;+\; 0x^2 \;+\; 0x \;-\; 6 \\ \qquad\qquad\quad \underline{-(\;\; 8x^4 \;+\; 10x^3 \;+\; 0x^2 \;+\; 0x \;-\; 6)} \end{array}}$

125. $\quad (a+b+c)^2 = (a+b+c)(a+b+c)$
$$= a(a+b+c)+b(a+b+c)+c(a+b+c)$$
$$= a^2+ab+ac+ab+b^2+bc+ac+bc+c^2$$
$$= a^2+b^2+c^2+2ab+2ac+2bc$$

Exercise 1.5 (page 51)

1. $3x-6=3(x-2)$ 3. $8x^2+4x^3=4x^2(2+x)$

5. $7x^2y^2+14x^3y^2=7x^2y^2(1+2x)$

7. $3a^2bc+6ab^2c+9abc^2=3abc(a+2b+3c)$

18

9. $b(x+y) - a(x+y) = (b-a)(x+y)$

11. $\quad 4a+b-12a^2-3ab = 4a-12a^2+b-3ab = 4a(1-3a)+b(1-3a)$
$$= (4a+b)(1-3a)$$

13. $3x^3+3x^2-x-1 = 3x^2(x+1)-1(x+1) = (3x^2-1)(x+1)$

15. $\quad 2txy+2ctx-3ty-3ct = t(2xy+2cx-3y-3c) = t[2x(y+c)-3(y+c)]$
$$= t(2x-3)(y+c)$$

17. $ax+bx+ay+by+az+bz = x(a+b)+y(a+b)+z(a+b) = (x+y+z)(a+b)$

19. $\quad 6xc+yd+2dx+3cy = 6xc+3cy+2dx+yd = 3c(2x+y)+d(2x+y)$
$$= (3c+d)(2x+y)$$

21. $4x^2-9 = (2x)^2-3^2 = (2x+3)(2x-3)$

23. $4-9r^2 = 2^2-(3r)^2 = (2+3r)(2-3r)$

25. $(x+z)^2-25 = (x+z+5)(x+z-5)$

27. $25x^4+1 = (5x^2)^2+1^2$: PRIME (sum of squares)

29. $x^2-(y-z)^2 = [x+(y-z)][x-(y-z)] = (x+y-z)(x-y+z)$

31. $\quad (x-y)^2-(x+y)^2 = [(x-y)+(x+y)][(x-y)-(x+y)]$
$$= (x-y+x+y)(x-y-x-y)$$
$$= (2x)(-2y) = -4xy$$

33. $x^4-y^4 = (x^2)^2-(y^2)^2 = (x^2+y^2)(x^2-y^2) = (x^2+y^2)(x+y)(x-y)$

35. $x^8-64z^4 = (x^4)^2-(8z^2)^2 = (x^4+8z^2)(x^4-8z^2)$

37. $3x^2-12 = 3(x^2-4) = 3(x+2)(x-2)$

39. $18xy^2-8x = 2x(9y^2-4) = 2x(3y+2)(3y-2)$

41. $x^2+8x+16 = (x+4)(x+4) = (x+4)^2$

43. $b^2-10b+25 = (b-5)(b-5) = (b-5)^2$

45. $m^2+4mn+4n^2 = (m+2n)(m+2n) = (m+2n)^2$

47. $x^2+10x+21 = (x+3)(x+7)$ $\qquad$ **49.** $x^2-4x-12 = (x-6)(x+2)$

51. $x^2 - 2x + 15$: PRIME

53. $12x^2 - xy - 6y^2 = (4x - 3y)(3x + 2y)$

55. $24y^2 + 15 - 38y = 24y^2 - 38y + 15 = (4y - 3)(6y - 5)$

57. $-15 + 2a + 24a^2 = 24a^2 + 2a - 15 = (4y - 3)(6y - 5) = (4a - 3)(6a + 5)$

59. $6x^2 + 29xy + 35y^2 = (2x + 5y)(3x + 7y)$

61. $-35 - x + 6x^2 = 6x^2 - x - 35 = (2x - 5)(3x + 7)$

63. $12y^2 - 58y - 70 = 2(6y^2 - 29y - 35) = 2(6y - 35)(y + 1)$

65. $-6x^3 + 23x^2 + 35x = -x(6x^2 - 23x - 35) = -x(x - 5)(6x + 7)$

67. $6x^4 - 11x^3 - 35x^2 = x^2(6x^2 - 11x - 35) = x^2(2x - 7)(3x + 5)$

69. $-35 + 47x - 6x^2 = -6x^2 + 47x - 35 = -(6x^2 - 47x + 35) = -(6x - 5)(x - 7)$

71. $x^4 + 2x^2 - 15 = (x^2 + 5)(x^2 - 3)$

73. $2y^2z^2 - 16yz + 80 = 2(y^2z^2 - 8yz + 40)$ (cannot be factored further)

75. $a^{2n} - 2a^n - 3 = (a^n - 3)(a^n + 1)$ **77.** $6x^{2n} - 7x^n + 2 = (2x^n - 1)(3x^n - 2)$

79. $4x^{2n} - 9y^{2n} = (2x^n)^2 - (3y^n)^2$
$$= (2x^n + 3y^n)(2x^n - 3y^n)$$

81. $10y^{2n} - 11y^n - 6 = (2y^n - 3)(5y^n + 2)$

83. $8z^3 - 27 = (2z)^3 - 3^3 = (2z - 3)[(2z)^2 + (2z)(3) + 3^2] = (2z - 3)(4z^2 + 6z + 9)$

85. $2x^3 + 2000 = 2(x^3 + 1000) = 2(x^3 + 10^3) = 2(x + 10)(x^2 - 10x + 100)$

87. $(x + y)^3 - 64 = [(x + y) - 4][(x + y)^2 + (x + y)4 + 4^2]$
$$= (x + y - 4)(x^2 + 2xy + y^2 + 4x + 4y + 16)$$

89. $1 - (x + 1)^3 = [1 - (x + 1)][1^2 + 1(x + 1) + (x + 1)^2]$
$$= (1 - x - 1)(1 + x + 1 + x^2 + 2x + 1)$$
$$= -x(x^2 + 3x + 3)$$

91. $64a^6 - y^6 = (8a^3)^2 - (y^3)^2 = (8a^3 + y^3)(8a^3 - y^3)$
$$= [(2a + y)(4a^2 - 2ay + y^2)] \, [(2a - y)(4a^2 + 2ay + y^2)]$$

93. $a^3 - b^3 + a - b = (a^3 - b^3) + (a - b) = (a - b)(a^2 + ab + b^2) + 1(a - b)$
$$= (a - b)(a^2 + ab + b^2 + 1)$$

95. $64x^6 + y^6 = (4x^2)^3 + (y^2)^3 = (4x^2 + y^2)(16x^4 - 4x^2y^2 + y^4)$

97. $x^2 - 6x + 9 - 144y^2 = (x^2 - 6x + 9) - (144y^2) = (x - 3)^2 - (12y)^2$
$$= (x - 3 + 12y)(x - 3 - 12y)$$

99. $(a + b)^2 - 3(a + b) - 10 = [(a + b) - 5][(a + b) + 2] = (a + b - 5)(a + b + 2)$

101. $6(u + v)^2 + 11u + 4 + 11v = 6(u + v)^2 + 11(u + v) + 4$
$$= [2(u + v) + 1][3(u + v) + 4]$$
$$= (2u + 2v + 1)(3u + 3v + 4)$$

103. $x^6 + 7x^3 - 8 = (x^3 + 8)(x^3 - 1) = (x + 2)(x^2 - 2x + 4)(x - 1)(x^2 + x + 1)$

105. $a(c + d) + a + b(c + d) + b = a(c + d + 1) + b(c + d + 1) = (a + b)(c + d + 1)$

107. $x^4 + 3x^2 + 4 = x^4 + 3x^2 + x^2 + 4 - x^2 = (x^4 + 4x^2 + 4) - x^2$
$$= (x^2 + 2)^2 - x^2$$
$$= (x^2 + 2 + x)(x^2 + 2 - x)$$

109. $x^4 + 7x^2 + 16 = x^4 + 7x^2 + x^2 + 16 - x^2 = (x^4 + 8x^2 + 16) - x^2$
$$= (x^2 + 4)^2 - x^2$$
$$= (x^2 + 4 + x)(x^2 + 4 - x)$$

111. $4a^4 + 1 + 3a^2 = 4a^4 + 3a^2 + 1 = (4a^4 + 4a^2 + 1) - a^2$
$$= (2a^2 + 1)^2 - a^2$$
$$= (2a^2 + 1 + a)(2a^2 + 1 - a)$$

113. $2x^4 + 8 = 2(x^4 + 4) = 2(x^4 + 4x^2 + 4 - 4x^2)$
$$= 2[(x^2 + 2)^2 - 4x^2]$$
$$= 2(x^2 + 2 + 2x)(x^2 + 2 - 2x)$$

115. $3x + 2 = 2\left(\dfrac{3x}{2} + \dfrac{2}{2}\right) = 2\left(\dfrac{3}{2}x + 1\right)$

117. $x + x^{1/2} = x^{1/2}\left(\dfrac{x}{x^{1/2}} + \dfrac{x^{1/2}}{x^{1/2}}\right) = x^{1/2}\left(x^{1/2} + 1\right)$

119. $ab^{3/2} - a^{3/2}b = ab\left(\dfrac{ab^{3/2}}{ab} - \dfrac{a^{3/2}b}{ab}\right) = ab\left(b^{1/2} - a^{1/2}\right)$

121. $a^{n+2} + a^{n+3} = a^2\left(\dfrac{a^{n+2}}{a^2} + \dfrac{a^{n+3}}{a^2}\right) = a^2\left(a^n + a^{n+1}\right)$

Exercise 1.6   (page 58)

1. $\dfrac{8x}{3y} \overset{?}{=} \dfrac{16x}{6y}$

     $8x(6y) \overset{?}{=} 16x(3y)$

     $48xy = 48xy$

The fractions are equal.

3. $\dfrac{25xyz}{12ab^2c} \overset{?}{=} \dfrac{50a^2bc}{24xyz}$

     $25xyz(24xyz) \overset{?}{=} 50a^2bc(12ab^2c)$

     $600x^2y^2z^2 \neq 600a^3b^3c^2$

The fractions are not equal.

5. $\dfrac{4x}{7} \cdot \dfrac{2}{5a} = \dfrac{8x}{35a}$

7. $\dfrac{8m}{5n} \div \dfrac{3m}{10n} = \dfrac{8m}{5n} \cdot \dfrac{10n}{3m} = \dfrac{80mn}{15mn} = \dfrac{16}{3} = 5\dfrac{1}{3}$

9. $\dfrac{3z}{5c} + \dfrac{2z}{5c} = \dfrac{3z + 2z}{5c} = \dfrac{5z}{5c} = \dfrac{z}{c}$

11. $\dfrac{15x^2y}{7a^2b^3} - \dfrac{x^2y}{7a^2b^3} = \dfrac{15x^2y - x^2y}{7a^2b^3} = \dfrac{14x^2y}{7a^2b^3} = \dfrac{2x^2y}{a^2b^3}$

13. $\dfrac{2x - 4}{x^2 - 4} = \dfrac{2(x - 2)}{(x + 2)(x - 2)} = \dfrac{2}{x + 2}$

15. $\dfrac{25 - x^2}{x^2 + 10x + 25} = \dfrac{-(x^2 - 25)}{x^2 + 10x + 25} = -\dfrac{(x + 5)(x - 5)}{(x + 5)(x + 5)} = -\dfrac{x - 5}{x + 5} = \dfrac{-(x - 5)}{x + 5} = \dfrac{5 - x}{x + 5}$

17. $\dfrac{6x^3 + x^2 - 12x}{4x^3 + 4x^2 - 3x} = \dfrac{x(6x^2 + x - 12)}{x(4x^2 + 4x - 3)} = \dfrac{\cancel{x}(2x + 3)(3x - 4)}{\cancel{x}(2x + 3)(2x - 1)} = \dfrac{3x - 4}{2x - 1}$

19. $\dfrac{x^3 - 8}{x^2 + ax - 2x - 2a} = \dfrac{(x - 2)(x^2 + 2x + 4)}{x(x + a) - 2(x + a)} = \dfrac{(x - 2)(x^2 + 2x + 4)}{(x - 2)(x + a)} = \dfrac{x^2 + 2x + 4}{x + a}$

21. $\dfrac{x^2 - 1}{x} \cdot \dfrac{x^2}{x^2 + 2x + 1} = \dfrac{(x + 1)(x - 1)}{x} \cdot \dfrac{x^2}{(x + 1)(x + 1)} = \dfrac{(x + 1)(x - 1)\cancel{x}x}{\cancel{x}(x + 1)(x + 1)} = \dfrac{x(x - 1)}{x + 1}$

23. $\dfrac{3x^2 + 7x + 2}{x^2 + 2x} \cdot \dfrac{x^2 - x}{3x^2 + x} = \dfrac{(3x + 1)(x + 2)}{x(x + 2)} \cdot \dfrac{x(x - 1)}{x(3x + 1)} = \dfrac{(3x + 1)(x + 2) \cdot x(x - 1)}{x(x + 2) \cdot x(3x + 1)}$

         $= \dfrac{x - 1}{x}$

25. $\dfrac{x^2 + x}{x - 1} \cdot \dfrac{x^2 - 1}{x + 2} = \dfrac{x(x + 1)}{x - 1} \cdot \dfrac{(x + 1)(x - 1)}{x + 2} = \dfrac{x(x + 1)(x + 1)(x - 1)}{(x - 1)(x + 2)} = \dfrac{x(x + 1)^2}{x + 2}$

27. $\dfrac{2x^2 + 32}{8} \div \dfrac{x^2 + 16}{2} = \dfrac{2x^2 + 32}{8} \cdot \dfrac{2}{x^2 + 16} = \dfrac{2(x^2 + 16) \cdot 2}{8(x^2 + 16)} = \dfrac{4}{8} = \dfrac{1}{2}$

29.
$$\frac{z^2 + z - 20}{z^2 - 4} \div \frac{z^2 - 25}{z - 5} = \frac{z^2 + z - 20}{z^2 - 4} \cdot \frac{z - 5}{z^2 - 25} = \frac{(z+5)(z-4)(z-5)}{(z+2)(z-2)(z+5)(z-5)}$$
$$= \frac{z - 4}{(z+2)(z-2)}$$

31.
$$\frac{3x^2 + 5x - 2}{x^3 + 2x^2} \div \frac{6x^2 + 13x - 5}{2x^3 + 5x^2} = \frac{3x^2 + 5x - 2}{x^3 + 2x^2} \cdot \frac{2x^3 + 5x^2}{6x^2 + 13x - 5}$$
$$= \frac{(3x - 1)(x + 2)}{x^2(x + 2)} \cdot \frac{x^2(2x + 5)}{(2x + 5)(3x - 1)}$$
$$= \frac{(3x - 1)(x + 2) \cdot x^2(2x + 5)}{x^2(x + 2)(2x + 5)(3x - 1)}$$
$$= 1$$

33.
$$\frac{x^2 + 7x + 12}{x^3 - x^2 - 6x} \cdot \frac{x^2 - 3x - 10}{x^2 + 2x - 3} \cdot \frac{x^3 - 4x^2 + 3x}{x^2 - x - 20}$$
$$= \frac{(x + 4)(x + 3)(x - 5)(x + 2) \cdot x(x^2 - 4x + 3)}{x(x^2 - x - 6)(x + 3)(x - 1)(x - 5)(x + 4)}$$
$$= \frac{(x + 4)(x + 3)(x - 5)(x + 2) \cdot x(x - 3)(x - 1)}{x(x - 3)(x + 2)(x + 3)(x - 1)(x - 5)(x + 4)}$$
$$= 1$$

35.
$$\frac{x^3 + 27}{x^2 - 4} \div \left(\frac{x^2 + 4x + 3}{x^2 + 2x} \div \frac{x^2 + x - 6}{x^2 - 3x + 9} \right)$$
$$= \frac{x^3 + 27}{x^2 - 4} \div \left(\frac{x^2 + 4x + 3}{x^2 + 2x} \cdot \frac{x^2 - 3x + 9}{x^2 + x - 6} \right)$$
$$= \frac{(x + 3)(x^2 - 3x + 9)}{(x + 2)(x - 2)} \div \left(\frac{(x + 3)(x + 1)(x^2 - 3x + 9)}{x(x + 2)(x + 3)(x - 2)} \right)$$
$$= \frac{(x + 3)(x^2 - 3x + 9) \cdot x(x + 2)(x + 3)(x - 2)}{(x + 2)(x - 2) \cdot (x + 3)(x + 1)(x^2 - 3x + 9)}$$
$$= \frac{x(x + 3)}{x + 1}$$

37. $\dfrac{3}{x + 3} + \dfrac{x + 2}{x + 3} = \dfrac{x + 5}{x + 3}$

39. $\dfrac{4x}{x - 1} - \dfrac{4}{x - 1} = \dfrac{4x - 4}{x - 1} = \dfrac{4(x - 1)}{x - 1} = 4$

41. $\dfrac{2}{5 - x} + \dfrac{1}{x - 5} = \dfrac{2}{-(x - 5)} + \dfrac{1}{x - 5} = \dfrac{-2}{x - 5} + \dfrac{1}{x - 5} = \dfrac{-1}{x - 5} = \dfrac{1}{-(x - 5)} = \dfrac{1}{5 - x}$

43. $\dfrac{3}{x+1} + \dfrac{2}{x-1} = \dfrac{3}{(x+1)} \cdot \dfrac{(x-1)}{(x-1)} + \dfrac{2}{(x-1)} \cdot \dfrac{(x+1)}{(x+1)} = \dfrac{3x-3+2x+2}{(x+1)(x-1)}$

$$= \dfrac{5x-1}{(x+1)(x-1)}$$

45. $\dfrac{a+3}{a^2+7a+12} + \dfrac{a}{a^2-16} = \dfrac{a+3}{(a+3)(a+4)} + \dfrac{a}{(a+4)(a-4)}$

$$= \dfrac{1}{(a+4)} \cdot \dfrac{(a-4)}{(a-4)} + \dfrac{a}{(a+4)(a-4)}$$

$$= \dfrac{a-4+a}{(a+4)(a-4)}$$

$$= \dfrac{2a-4}{(a+4)(a-4)}$$

47. $\dfrac{x}{x^2-4} - \dfrac{1}{x+2} = \dfrac{x}{(x+2)(x-2)} - \dfrac{1}{(x+2)} \cdot \dfrac{(x-2)}{(x-2)}$

$$= \dfrac{x}{(x+2)(x-2)} - \dfrac{(x-2)}{(x+2)(x-2)}$$

$$= \dfrac{x-x+2}{(x+2)(x-2)}$$

$$= \dfrac{2}{(x+2)(x-2)}$$

49. $\dfrac{3x-2}{x^2+2x+1} - \dfrac{x}{x^2-1} = \dfrac{3x-2}{(x+1)(x+1)} - \dfrac{x}{(x+1)(x-1)}$

$$= \dfrac{(3x-2)}{(x+1)(x+1)} \cdot \dfrac{(x-1)}{(x-1)} - \dfrac{x}{(x+1)(x-1)} \cdot \dfrac{(x+1)}{(x+1)}$$

$$= \dfrac{3x^2-5x+2}{(x+1)^2(x-1)} - \dfrac{x^2+x}{(x+1)^2(x-1)}$$

$$= \dfrac{3x^2-5x+2-x^2-x}{(x+1)^2(x-1)}$$

$$= \dfrac{2x^2-6x+2}{(x+1)^2(x-1)}$$

$$= \dfrac{2(x^2-3x+1)}{(x+1)^2(x-1)}$$

24

51.

$$\frac{2}{y^2-1} + 3 + \frac{1}{y+1} = \frac{2}{(y+1)(y-1)} + \frac{3}{1} + \frac{1}{(y+1)}$$

$$= \frac{2}{(y+1)(y-1)} + \frac{3}{1} \cdot \frac{(y+1)(y-1)}{(y+1)(y-1)} + \frac{1}{(y+1)} \cdot \frac{(y-1)}{(y-1)}$$

$$= \frac{2}{(y+1)(y-1)} + \frac{3(y^2-1)}{(y+1)(y-1)} + \frac{y-1}{(y+1)(y-1)}$$

$$= \frac{2 + 3y^2 - 3 + y - 1}{(y+1)(y-1)}$$

$$= \frac{3y^2 + y - 2}{(y+1)(y-1)}$$

$$= \frac{(3y-2)\cancel{(y+1)}}{\cancel{(y+1)}(y-1)}$$

$$= \frac{3y-2}{y-1}$$

53.

$$\frac{1}{x-2} + \frac{3}{x+2} - \frac{3x-2}{x^2-4} = \frac{1}{x-2} + \frac{3}{x+2} - \frac{3x-2}{(x+2)(x-2)}$$

$$= \frac{1(x+2)}{(x-2)(x+2)} + \frac{3(x-2)}{(x+2)(x-2)} - \frac{3x-2}{(x+2)(x-2)}$$

$$= \frac{x+2 + 3(x-2) - (3x-2)}{(x+2)(x-2)}$$

$$= \frac{x+2 + 3x - 6 - 3x + 2}{(x+2)(x-2)}$$

$$= \frac{\cancel{x-2}}{(x+2)\cancel{(x-2)}}$$

$$= \frac{1}{x+2}$$

55.

$$\left(\frac{1}{x-2} + \frac{1}{x-3}\right) \cdot \frac{x-3}{2x} = \left(\frac{1(x-3)}{(x-2)(x-3)} + \frac{1(x-2)}{(x-2)(x-3)}\right) \cdot \frac{x-3}{2x}$$

$$= \frac{x-3 + x-2}{(x-2)(x-3)} \cdot \frac{x-3}{2x}$$

$$= \frac{(2x-5)\cancel{(x-3)}}{(x-2)\cancel{(x-3)} \cdot 2x}$$

$$= \frac{2x-5}{2x(x-2)}$$

57.

$$\frac{3x}{x-4} - \frac{x}{x+4} - \frac{3x+1}{16-x^2} = \frac{3x}{x-4} - \frac{x}{x+4} - \frac{3x+1}{-(x^2-16)}$$

$$= \frac{3x}{x-4} - \frac{x}{x+4} + \frac{3x+1}{x^2-16}$$

$$= \frac{3x(x+4)}{(x-4)(x+4)} - \frac{x(x-4)}{(x+4)(x-4)} + \frac{3x+1}{(x+4)(x-4)}$$

$$= \frac{3x(x+4) - x(x-4) + 3x+1}{(x+4)(x-4)}$$

$$= \frac{3x^2 + 12x - x^2 + 4x + 3x + 1}{(x+4)(x-4)}$$

$$= \frac{2x^2 + 19x + 1}{(x+4)(x-4)}$$

59. $\dfrac{1}{x^2+3x+2} - \dfrac{2}{x^2+4x+3} + \dfrac{1}{x^2+5x+6}$

$$= \frac{1}{(x+1)(x+2)} - \frac{2}{(x+1)(x+3)} + \frac{1}{(x+2)(x+3)}$$

$$= \frac{(x+3)}{(x+1)(x+2)(x+3)} - \frac{2(x+2)}{(x+1)(x+2)(x+3)} + \frac{(x+1)}{(x+1)(x+2)(x+3)}$$

$$= \frac{x+3 - 2(x+2) + x + 1}{(x+1)(x+2)(x+3)}$$

$$= \frac{x+3 - 2x - 4 + x + 1}{(x+1)(x+2)(x+3)}$$

$$= \frac{0}{(x+1)(x+2)(x+3)}$$

$$= 0$$

61. $\dfrac{3x-2}{x^2+x-20}-\dfrac{4x^2+2}{x^2-25}+\dfrac{3x^2-25}{x^2-16}$

$$=\frac{3x-2}{(x+5)(x-4)}-\frac{4x^2+2}{(x+5)(x-5)}+\frac{3x^2-25}{(x+4)(x-4)}$$

$$=\frac{(3x-2)(x+4)(x-5)}{(x+5)(x-4)(x+4)(x+5)}-\frac{(4x^2+2)(x+4)(x-4)}{(x+5)(x-5)(x+4)(x-4)}$$

$$+\frac{(3x^2-25)(x+5)(x-5)}{(x+4)(x-4)(x+5)(x-5)}$$

$$=\frac{3x^3-5x^2-58x+40}{(x+5)(x-5)(x+4)(x-4)}-\frac{4x^4-62x^2-32}{(x+5)(x-5)(x+4)(x-4)}$$

$$+\frac{3x^4-100x^2+625}{(x+5)(x-5)(x+4)(x-4)}$$

$$=\frac{3x^3-5x^2-58x+40-4x^4+62x^2+32+3x^4-100x^2+625}{(x+5)(x-5)(x+4)(x-4)}$$

$$=\frac{-x^4+3x^3-43x^2-58x+697}{(x+5)(x-5)(x+4)(x-4)}$$

63. $\dfrac{\frac{3a}{b}}{\frac{6ac}{b^2}}=\dfrac{3a}{b}\div\dfrac{6ac}{b^2}=\dfrac{3a}{b}\cdot\dfrac{b^2}{6ac}=\dfrac{3\not a\not b b}{6\not a\not b c}=\dfrac{b}{2c}$

65. $\dfrac{3a^2b}{\frac{ab}{27}}=\dfrac{3a^2b}{1}\div\dfrac{ab}{27}=\dfrac{3a^2b}{1}\cdot\dfrac{27}{ab}=\dfrac{81a\not a\not b}{\not a\not b}=81a$

67. $\dfrac{\frac{x-y}{ab}}{\frac{y-x}{ab}}=\dfrac{\frac{x-y}{ab}}{\frac{-(x-y)}{ab}}=\dfrac{\frac{x-y}{ab}}{-\left(\frac{x-y}{ab}\right)}=\dfrac{1}{-1}=-1$

69. $\dfrac{\frac{1}{x}+\frac{1}{y}}{xy}=\dfrac{\left(\frac{1}{x}+\frac{1}{y}\right)\cdot xy}{xy\cdot xy}=\dfrac{\frac{1}{x}\cdot xy+\frac{1}{y}\cdot xy}{x^2y^2}=\dfrac{y+x}{x^2y^2}$

71. $\dfrac{\frac{1}{x}+\frac{1}{y}}{\frac{1}{x}-\frac{1}{y}}=\dfrac{\left(\frac{1}{x}+\frac{1}{y}\right)\cdot xy}{\left(\frac{1}{x}-\frac{1}{y}\right)\cdot xy}=\dfrac{\frac{1}{x}\cdot xy+\frac{1}{y}\cdot xy}{\frac{1}{x}\cdot xy-\frac{1}{y}\cdot xy}=\dfrac{y+x}{y-x}$

73. $\dfrac{\frac{3a}{b}-\frac{4a^2}{x}}{\frac{1}{b}+\frac{1}{ax}}=\dfrac{\left(\frac{3a}{b}-\frac{4a^2}{x}\right)\cdot abx}{\left(\frac{1}{b}+\frac{1}{ax}\right)\cdot abx}=\dfrac{3a^2x-4a^3b}{ax+b}=\dfrac{a^2(3x-4ab)}{ax+b}$

75. $\dfrac{x+1-\frac{6}{x}}{x+5+\frac{6}{x}} = \dfrac{\left(x+1-\frac{6}{x}\right)\cdot x}{\left(x+5+\frac{6}{x}\right)\cdot x} = \dfrac{x^2+x-6}{x^2+5x+6} = \dfrac{\cancel{(x+3)}(x-2)}{\cancel{(x+3)}(x+2)} = \dfrac{x-2}{x+2}$

77. $\dfrac{3xy}{1-\frac{1}{xy}} = \dfrac{3xy\cdot xy}{\left(1-\frac{1}{xy}\right)\cdot xy} = \dfrac{3x^2y^2}{xy-1}$

79. $\dfrac{3x}{x+\frac{1}{x}} = \dfrac{3x\cdot x}{\left(x+\frac{1}{x}\right)\cdot x} = \dfrac{3x^2}{x^2+1}$

81. $\dfrac{\frac{x}{x+2}-\frac{2}{x-1}}{\frac{3}{x+2}+\frac{x}{x-1}} = \dfrac{\left(\frac{x}{x+2}-\frac{2}{x-1}\right)\cdot(x+2)(x-1)}{\left(\frac{3}{x+2}+\frac{x}{x-1}\right)\cdot(x+2)(x-1)} = \dfrac{x(x-1)-2(x+2)}{3(x-1)+x(x+2)}$

$$= \dfrac{x^2-x-2x-4}{3x-3+x^2+2x}$$

$$= \dfrac{x^2-3x-4}{x^2+5x-3}$$

83. $\dfrac{1}{1+x^{-1}} = \dfrac{1}{1+\frac{1}{x}} = \dfrac{1\cdot x}{\left(1+\frac{1}{x}\right)\cdot x} = \dfrac{x}{x+1}$

85. $\dfrac{3(x+2)^{-1}+2(x-1)^{-1}}{(x+2)^{-1}} = \dfrac{\frac{3}{x+2}+\frac{2}{x-1}}{\frac{1}{x+2}} = \dfrac{\left(\frac{3}{x+2}+\frac{2}{x-1}\right)\cdot(x+2)(x-1)}{\frac{1}{x+2}\cdot(x+2)(x-1)}$

$$= \dfrac{3(x-1)+2(x+2)}{x-1}$$

$$= \dfrac{3x-3+2x+4}{x-1}$$

$$= \dfrac{5x+1}{x-1}$$

87. $\dfrac{x}{1+\frac{1}{3x^{-1}}} = \dfrac{x}{1+\frac{1}{\frac{3}{x}}} = \dfrac{x\cdot\frac{3}{x}}{\left(1+\frac{1}{\frac{3}{x}}\right)\cdot\frac{3}{x}} = \dfrac{3}{\frac{3}{x}+1} = \dfrac{3\cdot x}{\left(\frac{3}{x}+1\right)\cdot x} = \dfrac{3x}{3+x}$

89. $\dfrac{1}{1+\frac{1}{1+\frac{1}{x}}} = \dfrac{1\cdot\left(1+\frac{1}{x}\right)}{\left(1+\frac{1}{1+\frac{1}{x}}\right)\cdot\left(1+\frac{1}{x}\right)} = \dfrac{1+\frac{1}{x}}{1+\frac{1}{x}+1} = \dfrac{\left(1+\frac{1}{x}\right)\cdot x}{\left(2+\frac{1}{x}\right)\cdot x} = \dfrac{x+1}{2x+1}$

28

91.
$$\dfrac{\sqrt{x^2+4}-\dfrac{5}{\sqrt{x^2+4}}}{x+1}=\dfrac{\left(\sqrt{x^2+4}-\dfrac{5}{\sqrt{x^2+4}}\right)\cdot\sqrt{x^2+4}}{(x+1)\cdot\sqrt{x^2+4}}=\dfrac{x^2+4-5}{(x+1)\sqrt{x^2+4}}$$

$$=\dfrac{x^2-1}{(x+1)\sqrt{x^2+4}}$$

$$=\dfrac{(x+1)(x-1)}{(x+1)\sqrt{x^2+4}}$$

$$=\dfrac{x-1}{\sqrt{x^2+4}}$$

$$=\dfrac{(x-1)\sqrt{x^2+4}}{\sqrt{x^2+4}\,\sqrt{x^2+4}}$$

$$=\dfrac{(x-1)\sqrt{x^2+4}}{x^2+4}$$

93.
$$\dfrac{\dfrac{3}{\sqrt{z^2-7}}+\sqrt{z^2-7}}{z+2}=\dfrac{\left(\dfrac{3}{\sqrt{z^2-7}}+\sqrt{z^2-7}\right)\cdot\sqrt{z^2-7}}{(z+2)\cdot\sqrt{z^2-7}}=\dfrac{3+z^2-7}{(z+2)\sqrt{z^2-7}}$$

$$=\dfrac{z^2-4}{(z+2)\sqrt{z^2-7}}$$

$$=\dfrac{(z+2)(z-2)}{(z+2)\sqrt{z^2-7}}$$

$$=\dfrac{z-2}{\sqrt{z^2-7}}$$

$$=\dfrac{(z-2)\sqrt{z^2-7}}{\sqrt{z^2-7}\,\sqrt{z^2-7}}$$

$$=\dfrac{(z-2)\sqrt{z^2-7}}{z^2-7}$$

95. $\dfrac{a}{b}+\dfrac{c}{d}=\dfrac{a}{b}\cdot\dfrac{d}{d}+\dfrac{c}{d}\cdot\dfrac{b}{b}=\dfrac{ad}{bd}+\dfrac{bc}{bd}=\dfrac{ad+bc}{bd}$

Chapter 1 Review Exercises (page 62)

1. $A\cap B=\{2, 3, 5, 7, 11\}\cap\{2, 4, 6, 8, 10\}=\{2\}$
The intersection is the set of elements which are in **both** sets.

3. $A\cap B\cap C=\{2, 3, 5, 7, 11\}\cap\{2, 4, 6, 8, 10\}\cap\{1, 2, 3, 4, 5\}$
$=\{2\}\cap\{1, 2, 3, 4, 5\}=\{2\}$

5.
11 13   17 19

7. $\{x\mid -3<x\le 5\}$
-3 5

9. $(-2, 4]$
$$\begin{array}{c} \longleftarrow\!\!(\!\!-\!\!-\!\!-\!\!-\!\!]\!\!\longrightarrow \\ \;\; -2 \quad\; 4 \end{array}$$

11. $|6| = 6$

13. $|1 - \sqrt{2}| = -(1 - \sqrt{2}) = -1 + \sqrt{2}$ (NOTE: $\sqrt{2} > 1$, so $1 - \sqrt{2} < 0$)

15. Associative property of addition   **17.** Associative property of multiplication

19. Transitive property of equality   **21.** $(x^3 y^2)^4 = (x^3)^4 (y^2)^4 = x^{12} y^8$

23. $(m^{-3} n^4)^2 = (m^{-3})^2 (n^4)^2 = m^{-6} n^8 = \dfrac{n^8}{m^6}$

25. $\left(\dfrac{3x^2 y^{-2}}{x^2 y^2}\right)^{-2} = (3x^0 y^{-4})^{-2} = \left(\dfrac{3}{y^4}\right)^{-2} = \left(\dfrac{y^4}{3}\right)^2 = \dfrac{(y^4)^2}{3^2} = \dfrac{y^8}{9}$

27. $\left(\dfrac{-3x^3 y}{xy^3}\right)^{-2} = (-3x^2 y^{-2})^{-2} = \left(-\dfrac{3x^2}{y^2}\right)^{-2} = \left(-\dfrac{y^2}{3x^2}\right)^2 = \dfrac{(y^2)^2}{(3x^2)^2} = \dfrac{y^4}{9x^4}$

29. $4^{1/2} = 2$ **31.** $-8^{2/3} = -(8^{1/3})^2 = -(2)^2 = -4$

33. $\left(\dfrac{16}{81}\right)^{3/4} = \dfrac{16^{3/4}}{81^{3/4}} = \dfrac{(16^{1/4})^3}{(81^{1/4})^3} = \dfrac{2^3}{3^3} = \dfrac{8}{27}$

35. $\left(\dfrac{8}{27}\right)^{-2/3} = \left(\dfrac{27}{8}\right)^{2/3} = \dfrac{(27^{1/3})^2}{(8^{1/3})^2} = \dfrac{3^2}{2^2} = \dfrac{9}{4}$

37. $\sqrt{36} = 6$ **39.** $\sqrt{\dfrac{9}{25}} = \dfrac{\sqrt{9}}{\sqrt{25}} = \dfrac{3}{5}$

41. $(x^{12} y^2)^{1/2} = (x^{12})^{1/2} (y^2)^{1/2} = x^6 y^1 = x^6 y$

43. $\left(\dfrac{-c^{2/3} c^{5/3}}{c^{-2/3}}\right)^{1/3} = \left(-\dfrac{c^{7/3}}{c^{-2/3}}\right)^{1/3} = (-c^{9/3})^{1/3} = -c$

45. $\sqrt{x^2 y^4} = \sqrt{x^2}\sqrt{y^4} = xy^2$ **47.** $\sqrt[4]{\dfrac{m^8 n^4}{p^{12}}} = \dfrac{\sqrt[4]{m^8 n^4}}{\sqrt[4]{p^{12}}} = \dfrac{m^2 n}{p^3}$

49. $(x^{16} y^4 c^2)^{1/2} = (x^{16})^{1/2} (y^4)^{1/2} (c^2)^{1/2} = |x^8|\,|y^2|\,|c| = x^8 y^2 |c|$

51. $\sqrt{x^4 y^8} = |x^2|\,|y^4| = x^2 y^4$ **53.** $\dfrac{2}{\sqrt{5}} = \dfrac{2}{\sqrt{5}} \cdot \dfrac{\sqrt{5}}{\sqrt{5}} = \dfrac{2\sqrt{5}}{\sqrt{25}} = \dfrac{2\sqrt{5}}{5}$

55. $\dfrac{1}{\sqrt[3]{2}} = \dfrac{1}{\sqrt[3]{2}} \cdot \dfrac{\sqrt[3]{4}}{\sqrt[3]{4}} = \dfrac{\sqrt[3]{4}}{\sqrt[3]{8}} = \dfrac{\sqrt[3]{4}}{2}$

57. $\dfrac{2}{\sqrt{3}-1} = \dfrac{2}{(\sqrt{3}-1)} \cdot \dfrac{(\sqrt{3}+1)}{(\sqrt{3}+1)} = \dfrac{2(\sqrt{3}+1)}{3-1} = \dfrac{2(\sqrt{3}+1)}{2} = \sqrt{3}+1$

59. $\dfrac{2x}{\sqrt{x}-2} = \dfrac{2x}{(\sqrt{x}-2)} \cdot \dfrac{(\sqrt{x}+2)}{(\sqrt{x}+2)} = \dfrac{2x(\sqrt{x}+2)}{x-4}$

61. $\dfrac{\sqrt{2}}{5} = \dfrac{\sqrt{2}}{5} \cdot \dfrac{\sqrt{2}}{\sqrt{2}} = \dfrac{2}{5\sqrt{2}}$

63. $\dfrac{\sqrt{x}+2}{5} = \dfrac{(\sqrt{x}+2)}{5} \cdot \dfrac{(\sqrt{x}-2)}{(\sqrt{x}-2)} = \dfrac{x-4}{5(\sqrt{x}-2)}$

65. $\sqrt{50} + \sqrt{8} = \sqrt{25 \cdot 2} + \sqrt{4 \cdot 2} = 5\sqrt{2} + 2\sqrt{2} = 7\sqrt{2}$

67. $(\sqrt{2}+\sqrt{3})^2 = (\sqrt{2}+\sqrt{3})(\sqrt{2}+\sqrt{3}) = \sqrt{4}+\sqrt{6}+\sqrt{6}+\sqrt{9} = 2+2\sqrt{6}+3$
$$= 5 + 2\sqrt{6}$$

69. $(\sqrt{2}+1)(\sqrt{3}+1) = \sqrt{6}+\sqrt{2}+\sqrt{3}+1$

71. 3rd degree binomial   **73.** 2nd degree monomial

75. $3x^2(x-1) - 2x(x+3) - x^2(x+2) = 3x^3 - 3x^2 - 2x^2 - 6x - x^3 - 2x^2$
$$= 2x^3 - 7x^2 - 6x$$

77. $(4a+2b)(2a-3b) = 8a^2 - 12ab + 4ab - 6b^2 = 8a^2 - 8ab - 6b^2$

79.

$$
\require{enclose}
\begin{array}{r}
x^2 \;+\; 2x \;+\; 1 \\[2pt]
2x+3 \enclose{longdiv}{2x^3 \;+\; 7x^2 \;+\; 8x \;+\; 3} \\
\underline{-(2x^3 \;+\; 3x^2)} \\
4x^2 \;+\; 8x \\
\underline{-(4x^2 \;+\; 6x)} \\
2x \;+\; 3 \\
\underline{-(2x \;+\; 3)}
\end{array}
$$

81. $3t^3 - 3t = 3t(t^2-1) = 3t(t+1)(t-1)$

83. $6x^2 + 7x - 24 = (2x-3)(3x+8)$

85. $8x^3 - 125 = (2x)^3 - 5^3 = (2x-5)[(2x)^2 + (2x)5 + 5^2] = (2x-5)(4x^2 + 10x + 25)$

87. $x^2 + 6x + 9 - t^2 = (x^2 + 6x + 9) - t^2 = (x+3)(x+3) - t^2$
$$= (x+3)^2 - t^2$$
$$= (x+3+t)(x+3-t)$$

89. $8z^3 + 343 = (2z)^3 + 7^3 = (2z + 7)[(2z)^2 - (2z)7 + 7^2] = (2z + 7)(4z^2 - 14z + 49)$

91. $121z^2 + 4 - 44z = 121z^2 - 44z + 4z = (11z - 2)(11z - 2) = (11z - 2)^2$

93. $2xy - 4zx - wy + 2zw = 2x(y - 2z) - w(y - 2z) = (2x - w)(y - 2z)$

95.
$$\frac{x^2 - 4x + 4}{x + 2} \cdot \frac{x^2 + 5x + 6}{x - 2} = \frac{(x - 2)(x - 2)}{x + 2} \cdot \frac{(x + 2)(x + 3)}{x - 2}$$
$$= \frac{(x - 2)(x - 2)(x + 2)(x + 3)}{(x + 2)(x - 2)}$$
$$= (x - 2)(x + 3)$$

97.
$$\frac{2t^2 + t - 3}{3t^2 - 7t + 4} \div \frac{10t + 15}{3t^2 - t - 4} = \frac{(2t + 3)(t - 1)}{(3t - 4)(t - 1)} \div \frac{5(2t + 3)}{(3t - 4)(t + 1)}$$
$$= \frac{2t + 3}{3t - 4} \cdot \frac{(3t - 4)(t + 1)}{5(2t + 3)}$$
$$= \frac{(2t + 3)(3t - 4)(t + 1)}{(3t - 4) \cdot 5(2t + 3)}$$
$$= \frac{t + 1}{5}$$

99.
$$\frac{x^2 + x - 6}{x^2 - x - 6} \cdot \frac{x^2 - x - 6}{x^2 + x - 2} \div \frac{x^2 - 4}{x^2 - 5x + 6} = \frac{(x^2 + x - 6)(x^2 - x - 6)}{(x^2 - x - 6)(x^2 + x - 2)} \cdot \frac{x^2 - 5x + 6}{x^2 - 4}$$
$$= \frac{x^2 + x - 6}{x^2 + x - 2} \cdot \frac{x^2 - 5x + 6}{x^2 - 4}$$
$$= \frac{(x + 3)(x - 2)(x - 2)(x - 3)}{(x + 2)(x - 1)(x + 2)(x - 2)}$$
$$= \frac{(x + 3)(x - 2)(x - 3)}{(x + 2)^2(x - 1)}$$

101.
$$\frac{2}{x - 4} + \frac{3x}{x + 5} = \frac{2}{(x - 4)} \cdot \frac{(x + 5)}{(x + 5)} + \frac{3x}{(x + 5)} \cdot \frac{(x - 4)}{(x - 4)}$$
$$= \frac{2x + 10}{(x - 4)(x + 5)} + \frac{3x^2 - 12x}{(x - 4)(x + 5)}$$
$$= \frac{3x^2 - 10x + 10}{(x - 4)(x + 5)}$$

103. $\dfrac{x}{x-1}+\dfrac{x}{x-2}+\dfrac{x}{x-3}$

$$= \frac{x}{(x-1)}\cdot\frac{(x-2)(x-3)}{(x-2)(x-3)}+\frac{x}{(x-2)}\cdot\frac{(x-1)(x-3)}{(x-1)(x-3)}+\frac{x}{(x-3)}\cdot\frac{(x-1)(x-2)}{(x-1)(x-2)}$$

$$= \frac{x(x-2)(x-3)+x(x-1)(x-3)+x(x-1)(x-2)}{(x-1)(x-2)(x-3)}$$

$$= \frac{x(x^2-5x+6)+x(x^2-4x+3)+x(x^2-3x+2)}{(x-1)(x-2)(x-3)}$$

$$= \frac{x^3-5x^2+6x+x^3-4x^2+3x+x^3-3x^2+2x}{(x-1)(x-2)(x-3)}$$

$$= \frac{3x^3-12x^2+11x}{(x-1)(x-2)(x-3)}$$

$$= \frac{x(3x^2-12x+11)}{(x-1)(x-3)}$$

105. $\dfrac{3(x+1)}{x}-\dfrac{5(x^2+3)}{x^2}+\dfrac{x}{x+1}$

$$= \frac{3(x+1)}{x}\cdot\frac{x(x+1)}{x(x+1)}-\frac{5(x^2+3)}{x^2}\cdot\frac{(x+1)}{(x+1)}+\frac{x}{(x+1)}\cdot\frac{x^2}{x^2}$$

$$= \frac{3x(x+1)(x+1)-5(x^2+3)(x+1)+x\cdot x^2}{x^2(x+1)}$$

$$= \frac{3x(x^2+2x+1)-5(x^3+x^2+3x+3)+x^3}{x^2(x+1)}$$

$$= \frac{3x^3+6x^2+3x-5x^3-5x^2-15x-15+x^3}{x^2(x+1)}$$

$$= \frac{-x^3+x^2-12x-15}{x^2(x+1)}$$

107. $\dfrac{\dfrac{5x}{2}}{\dfrac{3x^2}{8}}=\dfrac{5x}{2}\div\dfrac{3x^2}{8}=\dfrac{5x}{2}\cdot\dfrac{8}{3x^2}=\dfrac{40x}{6x^2}=\dfrac{20}{3x}$

109. $\dfrac{\dfrac{1}{x}+\dfrac{1}{y}}{x-y}=\dfrac{\dfrac{1}{x}+\dfrac{1}{y}}{\dfrac{x-y}{1}}\cdot\dfrac{\dfrac{xy}{1}}{\dfrac{xy}{1}}=\dfrac{y+x}{(x-y)xy}=\dfrac{x+y}{x^2y-xy^2}=\dfrac{x+y}{xy(x-y)}$

Chapter 1 Test   (page 64)

1. True. a is a member of the set $\{a,b,c\}$.

3. $(A \cap B) \cup C = (\{1, 2, 3\} \cap \{2, 3, 4\}) \cup \{4, 5, 6\}$
$$= \{2, 3\} \cup \{4, 5, 6\} = \{2, 3, 4, 5, 6\}$$

5. $\longleftarrow\!(\!\underset{-4}{\rule{0pt}{1pt}}\!\rule[0.5ex]{8em}{0.4pt}\!\underset{2}{]}\!\longrightarrow$

7. Commutative property of addition

9. $x^4 x^5 x^2 = x^{4+5+2} = x^{11}$

11. $\dfrac{(a^{-1}a^2)^{-2}}{a^{-3}} = \dfrac{a^2 a^{-4}}{a^{-3}} = \dfrac{a^{-2}}{a^{-3}} = a^1 = a$

13. $450{,}000 = 4.5 \times 10^5$

15. $3.7 \times 10^3 = 3700$

17. $(25a^4)^{1/2} = 25^{1/2}(a^4)^{1/2} = 5a^2$

19. $\left(\dfrac{8t^6}{27s^9}\right)^{-2/3} = \left(\dfrac{27s^9}{8t^6}\right)^{2/3} = \left[\left(\dfrac{27s^9}{8t^6}\right)^{1/3}\right]^2 = \left(\dfrac{3s^3}{2t^2}\right)^2 = \dfrac{9s^6}{4t^4}$

21. $\sqrt{12} + \sqrt{27} = \sqrt{4 \cdot 3} + \sqrt{9 \cdot 3} = 2\sqrt{3} + 3\sqrt{3} = 5\sqrt{3}$

23. $\dfrac{x}{\sqrt{x} - 2} = \dfrac{x}{(\sqrt{x} - 2)} \cdot \dfrac{(\sqrt{x} + 2)}{(\sqrt{x} + 2)} = \dfrac{x\sqrt{x} + 2x}{x - 4}$

25. $(a^2 + 3) - (2a^2 - 4) = a^2 + 3 - 2a^2 + 4 = -a^2 + 7$

27. $(3x - 4)(2x + 7) = 6x^2 + 21x - 8x - 28 = 6x^2 + 13x - 28$

29. $(x^2 + 4)(x^2 - 4) = x^4 - 4x^2 + 4x^2 - 16 = x^4 - 16$

31.
$$
\begin{array}{r}
6x \quad + \quad 19 \quad + \quad \dfrac{34}{x-3} \\[2pt]
\hline
x - 3 \,\overline{)\, 6x^2 \;\; + \quad x \;\; - \;\; 23 } \\
\underline{-(\,6x^2 \;\; - \;\; 18x\,)} \\
19x \;\; - \;\; 23 \\
\underline{-(\quad 19x \;\; - \;\; 57)} \\
34
\end{array}
$$

33. $3x + 6y = 3(x + 2y)$

35. $10t^2 - 19tw + 6w^2 = (5t - 2w)(2t - 3w)$

37. $x^4 - x^2 - 12 = (x^2 - 4)(x^2 + 3) = (x + 2)(x - 2)(x^2 + 3)$

39. $\dfrac{x}{x + 2} + \dfrac{2}{x + 2} = \dfrac{\cancel{x + 2}}{\cancel{x + 2}} = 1$

34

41. $\dfrac{x^2+x-20}{x^2-16} \cdot \dfrac{x^2-25}{x-5} = \dfrac{(x+5)(\cancel{x-4})}{(x+4)(\cancel{x-4})} \cdot \dfrac{(x+5)(\cancel{x-5})}{\cancel{x-5}} = \dfrac{(x+5)(x+5)}{x+4} = \dfrac{(x+5)^2}{x+4}$

43. $\dfrac{\frac{1}{a}+\frac{1}{b}}{\frac{1}{b}} = \dfrac{\frac{1}{a}+\frac{1}{b}}{\frac{1}{b}} \cdot \dfrac{\frac{ab}{1}}{\frac{ab}{1}} = \dfrac{b+a}{a}$

Exercise 2.1   (page 74)

1. The domain of x is the set of all real numbers.

3. The domain of x is the set of all real numbers except 0.

5. The domain of x is the set of all non-negative real numbers.

7. The domain of x is the set of all real numbers except 3 and -2.

9.
$$2x+5 = 15$$
$$2x = 10$$
$$x = 5$$
This is a conditional equation.

11.
$$2(x+2) = 2x+5$$
$$2x+4 = 2x+5$$
$$4 = 5$$
This equation has no solutions.

13.
$$\frac{x+7}{2} = 7$$
$$2\left(\frac{x+7}{2}\right) = 2(7)$$
$$x+7 = 14$$
$$x = 7$$
This is a conditional equation.

15.
$$2(a+1) = 3(a-2)-a$$
$$2a+2 = 3a-6-a$$
$$2a+2 = 2a-6$$
$$2 = -6$$
This equation has no solutions.

17.
$$3(x-3) = \frac{6x-18}{2}$$
$$2(3x-9) = 2\left(\frac{6x-18}{2}\right)$$
$$6x-18 = 6x-18$$
$$-18 = -18$$
All real numbers are solutions. The equation is an identity.

19.
$$\frac{3}{b-3} = 1$$
$$(b-3)\left(\frac{3}{b-3}\right) = (b-3)1$$
$$3 = b-3$$
$$6 = b$$
The solution checks. The equation is a conditional equation.

21.
$$2x^2+5x-3 = (2x-1)(x+3)$$
$$2x^2+5x-3 = 2x^2+5x-3$$
$$-3 = -3$$
All real numbers are solutions. The equation is an identity.

23.
$$2x+7 = 10-x$$
$$3x+7 = 10$$
$$3x = 3$$
$$x = 1$$

25.
$$\frac{5}{3}z - 8 = 7$$
$$\frac{5}{3}z = 15$$
$$5z = 45$$
$$z = 9$$

27.
$$\frac{z}{5} + 2 = 4$$
$$\frac{z}{5} = 2$$
$$z = 10$$

29.
$$\frac{3x-2}{3} = 2x + \frac{7}{3}$$
$$3\left(\frac{3x-2}{3}\right) = 3\left(2x + \frac{7}{3}\right)$$
$$3x - 2 = 6x + 7$$
$$-3x = 9$$
$$x = -3$$

31.
$$5(x-2) = 2x + 8$$
$$5x - 10 = 2x + 8$$
$$3x - 10 = 8$$
$$3x = 18$$
$$x = 6$$

33.
$$2(2x+1) - \frac{3x}{2} = \frac{-3(4+x)}{2}$$
$$2\left[2(2x+1) - \frac{3x}{2}\right] = 2\left[\frac{-3(4+x)}{2}\right]$$
$$2\left(4x + 2 - \frac{3x}{2}\right) = -3(4+x)$$
$$8x + 4 - 3x = -12 - 3x$$
$$5x + 4 = -3x - 12$$
$$8x = -16$$
$$x = -2$$

35.
$$7(2x+5) - 6(x+8) = 7$$
$$14x + 35 - 6x - 48 = 7$$
$$8x - 13 = 7$$
$$8x = 20$$
$$x = \frac{20}{8} = \frac{5}{2}$$

37.
$$(x-2)(x+5) = (x-3)(x+2)$$
$$x^2 + 3x - 10 = x^2 - x - 6$$
$$3x - 10 = -x - 6$$
$$4x - 10 = -6$$
$$4x = 4$$
$$x = 1$$

39.
$$\frac{3}{2}(3x-2) - 10x - 4 = 0$$
$$2\left[\frac{3}{2}(3x-2) - 10x - 4\right] = 2(0)$$
$$3(3x-2) - 20x - 8 = 0$$
$$9x - 6 - 20x - 8 = 0$$
$$-11x - 14 = 0$$
$$-11x = 14$$
$$x = -\frac{14}{11}$$

41.
$$x(x+2) = (x+1)^2 - 1$$
$$x^2 + 2x = x^2 + 2x + 1 - 1$$
$$x^2 + 2x = x^2 + 2x$$
$$0 = 0$$
x can be any real number.
identity

43.

$$\frac{(y+2)^2}{3} = y + 2 + \frac{y^2}{3}$$

$$(y+2)^2 = 3y + 6 + y^2$$
$$y^2 + 4y + 4 = y^2 + 3y + 6$$
$$4y + 4 = 3y + 6$$
$$y = 2$$

45.

$$2(s+2) + (s+3)^2 = s(s+5) + 2\left(\frac{17}{2} + s\right)$$
$$2s + 4 + s^2 + 6s + 9 = s^2 + 5s + 17 + 2s$$
$$s^2 + 8s + 13 = s^2 + 7s + 17$$
$$8s + 13 = 7s + 17$$
$$s = 4$$

47.

$$\frac{2}{x+1} + \frac{1}{3} = \frac{1}{x+1}$$
$$3(x+1)\left(\frac{2}{x+1} + \frac{1}{3}\right) = 3(x+1)\left(\frac{1}{x+1}\right)$$
$$3(2) + (x+1)1 = 3(1)$$
$$6 + x + 1 = 3$$
$$x = -4 \qquad \text{The solution checks.}$$

49.

$$\frac{9t+6}{t(t+3)} = \frac{7}{t+3}$$
$$t(t+3)\left(\frac{9t+6}{t(t+3)}\right) = t(t+3)\left(\frac{7}{t+3}\right)$$
$$9t + 6 = 7t$$
$$2t + 6 = 0$$
$$2t = -6 \qquad \text{The solution does not check.}$$
$$t = -3 \qquad \text{The equation has no solutions.}$$

51.

$$\frac{2}{(a-7)(a+2)} = \frac{4}{(a+3)(a+2)}$$
$$(a-7)(a+2)(a+3)\left[\frac{2}{(a-7)(a+2)}\right] = (a-7)(a+2)(a+3)\left[\frac{4}{(a+3)(a+2)}\right]$$
$$2(a+3) = 4(a-7)$$
$$2a + 6 = 4a - 28$$
$$-2a = -34$$
$$a = 17 \qquad \text{The solution checks.}$$

53.

$$\frac{2x+3}{x^2+5x+6}+\frac{3x-2}{x^2+x-6}=\frac{5x-2}{x^2-4}$$

$$\frac{2x+3}{(x+2)(x+3)}+\frac{3x-2}{(x+3)(x-2)}=\frac{5x-2}{(x+2)(x-2)}$$

$$\left(\frac{2x+3}{(x+2)(x+3)}+\frac{3x-2}{(x+3)(x-2)}=\frac{5x-2}{(x+2)(x-2)}\right)\cdot(x+2)(x+3)(x-2)$$

$$(x-2)(2x+3)+(x+2)(3x-2)=(x+3)(5x-2)$$
$$2x^2-x-6+3x^2+4x-4=5x^2+13x-6$$
$$5x^2+3x-10=5x^2+13x-6$$
$$3x-10=13x-6$$
$$-10x=4$$
$$x=-\frac{4}{10}=-\frac{2}{5}$$

55.

$$\frac{3x+5}{x^3+8}+\frac{3}{x^2-4}=\frac{2(3x-2)}{(x-2)(x^2-2x+4)}$$

$$\frac{3x+5}{(x+2)(x^2-2x+4)}+\frac{3}{(x+2)(x-2)}=\frac{6x-4}{(x-2)(x^2-2x+4)}$$ {multiply {through by {the least {common {denom.
$$(x-2)(3x+5)+(x^2-2x+4)(3)=(x+2)(6x-4)$$
$$3x^2-x-10+3x^2-6x+12=6x^2+8x-8$$
$$6x^2-7x+2=6x^2+8x-8$$
$$-7x+2=8x-8$$
$$-15x+2=-8$$
$$-15x=-10$$
$$x=\frac{-10}{-15}=\frac{2}{3}$$

57.

$$\frac{1}{11-n}-\frac{2(3n-1)}{-7n^2+74n+33}=\frac{1}{7n+3}$$

$$\frac{1}{-(n-11)}-\frac{6n-2}{-(7n^2-74n-33)}=\frac{1}{7n+3}$$

$$-\frac{1}{n-11}+\frac{6n-2}{7n^2-74n-33}=\frac{1}{7n+3}$$

$$\left(-\frac{1}{n-11}+\frac{6n-2}{(7n+3)(n-11)}=\frac{1}{7n+3}\right)\cdot(n-11)(7n+3)$$

$$-(7n+3)+6n-2=1(n-11)$$
$$-7n-3+6n-2=n-11$$
$$-2n-5=-11$$
$$-2n=-6$$
$$n=3$$

38

59.
$$\frac{5}{y+4} + \frac{2}{y+2} = \frac{6}{y+2} - \frac{1}{y^2+6y+8}$$
$$\frac{5}{y+4} = \frac{6}{y+2} - \frac{2}{y+2} - \frac{1}{y^2+6y+8}$$
$$\frac{5}{y+4} = \frac{4}{y+2} - \frac{1}{(y+2)(y+4)}$$
$$\left(\frac{5}{y+4} = \frac{4}{y+2} - \frac{1}{(y+2)(y+4)} \right) \cdot (y+4)(y+2)$$
$$5(y+2) = 4(y+4) - 1$$
$$5y + 10 = 4y + 16 - 1$$
$$y = 5$$

61.
$$\frac{3y}{6-3y} + \frac{2y}{2y+4} = \frac{8}{4-y^2}$$
$$\frac{3y}{-(3y-6)} + \frac{2y}{2(y+2)} = \frac{8}{-(y^2-4)}$$
$$-\frac{3y}{3(y-2)} + \frac{2y}{2(y+2)} = -\frac{8}{(y+2)(y-2)}$$
$$\left(-\frac{y}{y-2} + \frac{y}{y+2} = -\frac{8}{(y+2)(y-2)} \right) \cdot (y-2)(y+2)$$
$$-y(y+2) + y(y-2) = -8$$
$$-y^2 - 2y + y^2 - 2y = -8$$
$$-4y = -8$$
$$y = 2$$

This solution does not check.
The equation has no solutions.

63.
$$\frac{a}{a+2} - 1 = -\frac{3a+2}{a^2+4a+4}$$
$$\frac{a}{a+2} - 1 = -\frac{3a+2}{(a+2)(a+2)}$$
$$\left(\frac{a}{a+2} - 1 = -\frac{3a+2}{(a+2)(a+2)} \right) \cdot (a+2)(a+2)$$
$$a(a+2) - (a+2)(a+2) = -(3a+2)$$
$$a^2 + 2a - (a^2 + 4a + 4) = -3a - 2$$
$$a^2 + 2a - a^2 - 4a - 4 = -3a - 2$$
$$-2a - 4 = -3a - 2$$
$$a - 4 = -2$$
$$a = 2$$

65.

$$k = 2.2p$$

$$\frac{k}{2.2} = \frac{2.2p}{2.2}$$

$$\frac{k}{2.2} = p, \text{ or } p = \frac{k}{2.2}$$

67.

$$A = \tfrac{1}{2}h(b_1 + b_2)$$

$$2A = 2\left[\tfrac{1}{2}h(b_1 + b_2)\right]$$

$$2A = h(b_1 + b_2)$$

$$\frac{2A}{h} = \frac{h(b_1 + b_2)}{h}$$

$$\frac{2A}{h} = b_1 + b_2$$

$$\frac{2A}{h} - b_1 = b_2, \text{ or } b_2 = \frac{2A}{h} - b_1$$

69.

$$V = \tfrac{1}{3}\pi r^2 h$$

$$3V = \pi r^2 h$$

$$\frac{3V}{\pi h} = \frac{\pi r^2 h}{\pi h}$$

$$\frac{3V}{\pi h} = r^2, \text{ or } r^2 = \frac{3V}{\pi h}$$

71.

$$P_n = L + \frac{si}{f}$$

$$fP_n = fL + si$$

$$fP_n - fL = si$$

$$f(P_n - L) = si$$

$$\frac{f(P_n - L)}{i} = s, \text{ or } s = \frac{f(P_n - L)}{i}$$

73.

$$A = 2(lw + hl + hw)$$

$$A = 2lw + 2hl + 2hw$$

$$A - 2hl = 2lw + 2hw$$

$$A - 2hl = w(2l + 2h)$$

$$\frac{A - 2hl}{2l + 2h} = w, \text{ or } w = \frac{A - 2hl}{2(l + h)}$$

75.

$$F = \frac{mMg}{r^2}$$

$$Fr^2 = mMg$$

$$\frac{Fr^2}{Mg} = m, \text{ or } m = \frac{Fr^2}{Mg}$$

77.

$$\frac{x}{a} + \frac{y}{b} = 1$$

$$ab\left(\frac{x}{a} + \frac{y}{b}\right) = ab(1)$$

$$bx + ay = ab$$

$$ay = ab - bx$$

$$y = \frac{b(a - x)}{a}$$

79.

$$\frac{x^2}{a^2} - \frac{y^2}{b^2} = 1$$

$$a^2 b^2\left(\frac{x^2}{a^2} - \frac{y^2}{b^2}\right) = a^2 b^2(1)$$

$$b^2 x^2 - a^2 y^2 = a^2 b^2$$

$$b^2 x^2 = a^2 b^2 + a^2 y^2$$

$$x^2 = \frac{a^2(b^2 + y^2)}{b^2}$$

81.

$$\frac{1}{r} = \frac{1}{r_1} + \frac{1}{r_2}$$

$$rr_1r_2\left(\frac{1}{r}\right) = rr_1r_2\left(\frac{1}{r_1} + \frac{1}{r_2}\right)$$

$$r_1r_2 = rr_2 + rr_1$$
$$r_1r_2 = r(r_2 + r_1)$$

$$\frac{r_1r_2}{r_1 + r_2} = r, \text{ or } r = \frac{r_1r_2}{r_1 + r_2}$$

83.

$$l = a + (n-1)d$$
$$l = a + nd - d$$
$$l - a + d = nd$$
$$\frac{l - a + d}{d} = n, \text{ or } n = \frac{l - a + d}{d}$$

85.

$$a = (n-2)\frac{180}{n}$$

$$an = (n-2)180$$
$$an = 180n - 360$$
$$an - 180n = -360$$
$$n(a - 180) = -360$$

$$n = -\frac{360}{a - 180}$$

$$n = \frac{360}{180 - a}$$

87.

$$S = \frac{a - lr}{1 - r}$$

$$S(1 - r) = a - lr$$
$$S - Sr + lr = a, \text{ or } a = S - Sr + lr$$

89.

$$R = \frac{1}{\frac{1}{r_1} + \frac{1}{r_2} + \frac{1}{r_3}}$$

$$\left(\frac{1}{r_1} + \frac{1}{r_2} + \frac{1}{r_3}\right)R = 1$$

$$r_1r_2r_3\left(\frac{1}{r_1} + \frac{1}{r_2} + \frac{1}{r_3}\right)R = r_1r_2r_3$$

$$Rr_2r_3 + Rr_1r_3 + Rr_1r_2 = r_1r_2r_3$$
$$Rr_1r_3 + Rr_1r_2 - r_1r_2r_3 = -Rr_2r_3$$
$$r_1(Rr_3 + Rr_2 + r_2r_3) = -Rr_2r_3$$

$$r_1 = \frac{-Rr_2r_3}{Rr_3 + Rr_2 + r_2r_3}$$

Exercise 2.2 (page 81)

1. Let x = his score on the first exam.
   $x + 5$ = his score on the midterm.
   $x + 13$ = his score on the final.

$$\boxed{\frac{\text{Sum of scores}}{3}} = 90$$

$$\frac{x + x + 5 + x + 13}{3} = 90$$

$$\frac{3x + 18}{3} = 90$$

$$3x + 18 = 270$$

$$x = 84$$

He scored 84 on the first exam.

3. Let x = the first number.
   $2x + 3$ = the other number.

$$\boxed{\text{Sum of numbers}} = 54$$

$$x + 2x + 3 = 54$$

$$3x + 3 = 54$$

$$3x = 51$$

$$x = 17$$

The numbers are 17 and 37.

5. Let x = the number of locks.
   Then $40 + 28x$ = the total charge.

$$\boxed{\begin{array}{c}\text{Total} \\ \text{charge}\end{array}} = 236$$

$$40 + 28x = 236$$

$$28x = 196$$

$$x = 7$$

7 locks can be replaced.

7. The number of feet of framing material equals the perimeter of the frame.

$$\boxed{\text{Perimeter}} = 14$$

$$x + x + 2 + x + x + 2 = 14$$

$$4x + 4 = 14$$

$$x = \frac{10}{4} = \frac{5}{2} = 2\frac{1}{2}$$

The width is $2\frac{1}{2}$ feet.

9. Area of triangle = $\frac{1}{2}(16)(20) = 160$. Area of rectangle = $20x$.

$$2 \cdot \boxed{\begin{array}{c}\text{Area of} \\ \text{triangle}\end{array}} = \boxed{\begin{array}{c}\text{Area of} \\ \text{triangle}\end{array}} + \boxed{\begin{array}{c}\text{Area of} \\ \text{rectangle}\end{array}}$$

$$2(160) = 160 + 20x$$

$$320 = 160 + 20x$$

$$160 = 20x$$

$$8 = x \qquad \text{The dimensions of the wading pool are 20 feet by 8 feet.}$$

11. Old living area = $12(x + 8) = 12x + 96$. Porch area = $12(x - 2) = 12x - 24$.
    New living area = $12x + 96 + 12x - 24 = 24x + 72$

$$\boxed{\begin{array}{c}\text{New living} \\ \text{area}\end{array}} = 1.5 \cdot \boxed{\begin{array}{c}\text{Old living} \\ \text{area}\end{array}}$$

$$24x + 72 = 1.5(12x + 96)$$

$$24x + 72 = 18x + 144$$

$$6x = 72$$

$$x = 12 \qquad \text{The length of the room is } x + 8 = 20 \text{ feet.}$$

13. Let $x =$ amount invested at 7%, and $22,000 - x =$ amount invested at 6%.

$$\boxed{\begin{array}{c}\text{Income from}\\ \text{7\% investment}\end{array}} + \boxed{\begin{array}{c}\text{Income from}\\ \text{6\% investment}\end{array}} = \boxed{\begin{array}{c}\text{Total}\\ \text{income}\end{array}}$$

$$0.07x + 0.06(22,000 - x) = 1420$$
$$7x + 6(22,000 - x) = 142,000$$
$$7x + 132,000 - 6x = 142,000$$
$$x + 132,000 = 142,000$$
$$x = 10,000$$

$10,000 is invested at 7%, while $12,000 is invested at 6%.

15. Let $x =$ number of student tickets and $585 - x =$ adult tickets.

$$\boxed{\begin{array}{c}\text{Student}\\ \text{sales}\end{array}} + \boxed{\begin{array}{c}\text{Adult}\\ \text{sales}\end{array}} = \boxed{\begin{array}{c}\text{Total}\\ \text{sales}\end{array}}$$

$$1.75x + 2.50(585 - x) = 1217.25$$
$$1.75x + 1462.50 - 2.5x = 1217.25$$
$$-0.75x + 1462.50 = 1217.25$$
$$-0.75x = -245.25$$
$$x = \frac{-245.25}{-0.75} = 327 \quad \text{There were 327 student tickets sold.}$$

17. Let $x =$ amount invested at 8% and $37,000 - x =$ amount invested at $9\frac{1}{2}\%$.

$$\boxed{\begin{array}{c}\text{Income at}\\ \text{9.5\%}\end{array}} = \boxed{\begin{array}{c}\text{Income at}\\ \text{8\%}\end{array}} + 452.50$$

$$0.095(37,000 - x) = 0.08x + 452.50$$
$$3515 - 0.095x = 0.08x + 452.50$$
$$-0.095x - 0.08x + 3515 = 452.50$$
$$-0.175x = 452.50 - 3515$$
$$-0.175x = -3062.50$$
$$x = \frac{-3062.50}{-0.175} = 17,500$$

$17,500 is invested at 8%, and $19,500 is invested at $9\frac{1}{2}\%$.

19. Let $x =$ the original price.

$$\boxed{\begin{array}{c}\text{Selling}\\ \text{price}\end{array}} = \boxed{\begin{array}{c}\text{Original}\\ \text{price}\end{array}} - \boxed{\begin{array}{c}\text{20\%}\\ \text{discount}\end{array}}$$

$$63.96 = x - 0.20x$$
$$63.96 = 0.80x$$
$$\frac{63.96}{0.80} = x$$
$$79.95 = x \qquad \text{The original price was \$79.95.}$$

21. The break point is the number of plates which would have the same manufacturing cost using either process.

Let x = number of plates made at the break point.

Cost of first machine = $600 + 3x$.   Cost of bigger machine = $800 + 2x$.

$$\boxed{\begin{array}{c}\text{Cost of}\\\text{first machine}\end{array}} = \boxed{\begin{array}{c}\text{Cost of}\\\text{bigger machine}\end{array}}$$

$$600 + 3x = 800 + 2x$$
$$600 + x = 800$$
$$x = 200 \quad \text{The break point is 200 plates.}$$

23. Let x = number of computers sold at the break-even point.

Costs = $8925 + 850x$.   Income = $1275x$.

$$\boxed{\text{Costs}} = \boxed{\text{Income}}$$

$$8925 + 850x = 1275x$$
$$8925 = 425x$$
$$\frac{8925}{425} = x$$
$$21 = x \quad \text{The store must sell 21 computers to break even.}$$

25. Let x = the number of hours needed for both to mow the lawn. Then $\frac{1}{x}$ is the amount of the lawn they can mow in 1 hour.

	Hours to mow lawn	Amount done in 1 hour
Woman	2	$\frac{1}{2}$
Man	4	$\frac{1}{4}$
Both		$\frac{1}{x}$

The amount they both can do in 1 hour is simply the sum of what they can do separately in 1 hour:

$$\boxed{\begin{array}{c}\text{Amount woman}\\\text{does in 1 hour}\end{array}} + \boxed{\begin{array}{c}\text{Amount man}\\\text{does in 1 hour}\end{array}} = \boxed{\begin{array}{c}\text{Amount both can}\\\text{do in 1 hour}\end{array}}$$

$$\frac{1}{2} + \frac{1}{4} = \frac{1}{x}$$
$$8x\left(\frac{1}{2} + \frac{1}{4}\right) = 8x\left(\frac{1}{x}\right)$$
$$4x + 2x = 8$$
$$6x = 8$$
$$x = \frac{8}{6} = \frac{4}{3} = 1\frac{1}{3}$$

It will take them $1\frac{1}{3}$ hours to mow the lawn together.

27. Let x = the number of hours needed to fill the pool while the drain is open.
Then $\frac{1}{x}$ is the amount of the pool which is filled in 1 hour with the drain open.
The pool fills in 10 hours, so $\frac{1}{10}$ is filled in 1 hour.

The pool drains in 19 hours, so $\frac{1}{19}$ is drained in 1 hour.

$$\boxed{\begin{array}{c}\text{Amount filled in 1 hour}\\\text{with the drain open}\end{array}} = \boxed{\begin{array}{c}\text{Amount filled}\\\text{in 1 hour}\end{array}} - \boxed{\begin{array}{c}\text{Amount drained}\\\text{in 1 hour}\end{array}}$$

$$\frac{1}{x} = \frac{1}{10} - \frac{1}{19}$$
$$190x\left(\frac{1}{x}\right) = 190x\left(\frac{1}{10} - \frac{1}{19}\right)$$
$$190 = 19x - 10x$$
$$190 = 9x$$
$$\frac{190}{9} = x, \text{ or } x = \frac{190}{9} = 21\frac{1}{9}$$

It will take $21\frac{1}{9}$ hours to fill the pool if the drain is open.

29. Note the steps which would have to occur to bring the percentage up to 50%:
At the start, 40% of the 6 liters is antifreeze, so there are $0.40(6) = 2.4$ liters
of antifreeze at the start, and the rest other. Remove x liters of the 40%
solution. You are then removing $0.40x$ liters of antifreeze and $0.60x$ liters of
other. Add x liters of pure antifreeze. You are adding 0 liters of other. Now,
take Start − Remove + Add to get the ending amount:

	Start	Remove	Add	End
Antifreeze (liters)	2.4	0.40x	x	2.4 − 0.40x + x
Other (liters)	3.6	0.60x	0	3.6 − 0.60x
Total liquid (liters)	6.0	x	x	6.0 − x + x

To get an equation, note that 50% of the End total liquid should be antifreeze:

$$\boxed{\text{50\% of ending total}} = \boxed{\text{Ending antifreeze}}$$
$$0.50(6) = 2.4 + 0.60x$$
$$3 = 2.4 + 0.60x$$
$$0.6 = 0.6x$$
$$1 = x$$

1 liter of the 40% solution should be replaced with pure antifreeze.

31. Keep track of the amounts of alcohol and other. At the start, 20% of the 1 liter is alcohol, so there is $0.20(1) = 0.20$ liters of alcohol. Add x liters of pure alcohol (100% solution). Add across to get the ending amounts.

	Start	Add	End
Alcohol (liters)	0.20	x	$0.20 + x$
Other (liters)	0.80	0	0.80
Total (liters)	1.00	x	$1 + x$

To get an equation, note that 25% of the ending amount should be alcohol.

$$\boxed{\text{25\% of ending amount}} = \boxed{\text{Ending amount of alcohol}}$$

$$0.25(1 + x) = 0.20 + x$$
$$0.25 + 0.25x = 0.20 + x$$
$$0.25 = 0.20 + 0.75x$$
$$0.05 = 0.75x$$
$$5 = 75x$$

$$\frac{5}{75} = x, \text{ or } x = \frac{5}{75} = \frac{1}{15}$$

The nurse should add $\frac{1}{15}$ of a liter of pure alcohol.

33. At the start, there are 15,000 gallons of water and 0 gallons of chlorine. Then x gallons of chlorine are added (and 0 gallons of water are added). Add across each row to get the ending amounts:

	Start	Add	End
Chlorine (gallons)	0	x	x
Water (gallons)	15,000	0	15,000
Total (gallons)	15,000	x	$15,000 + x$

Note that $\frac{3}{100}\%$ (0.0003) of the ending total amount should be chlorine:

$$\boxed{\text{3/100\% of ending total amount}} = \boxed{\text{Ending amount of chlorine}}$$

$$0.0003(15000 + x) = x$$
$$4.5 + 0.0003x = x$$
$$4.5 = 0.9997x$$

$$\frac{4.5}{0.9997} = x$$

$$4.5 \approx x$$

About 4.5 gallons of chlorine should be added to the pool.

35. At the start, there is 24% salt, or $0.24(12) = 2.88$ liters of salt and 9.12 liters of water. Then x liters of water evaporates (but all the salt remains).

	Start	Evaporate	End
Salt (liters)	2.88	0	2.88
Water (liters)	9.12	x	$9.12 - x$
Total (liters)	12.00	x	$12 - x$

Note that 36% of the ending total amount should be salt:

$$\boxed{\text{36\% of ending total amount}} = \boxed{\text{Ending amount of salt}}$$
$$0.36(12 - x) = 2.88$$
$$4.32 - 0.36x = 2.88$$
$$-0.36x = -1.44$$
$$x = \frac{-1.44}{-0.36} = \frac{144}{36} = 4$$

4 liters of water should be evaporated.

37. Let $r =$ his first rate of speed, so $r + 26 =$ his second rate.

	r	t	d
First	r	5	$5r$
Second	$r + 26$	3	$3(r + 26)$

The distance traveled each way is the same:

$$\boxed{\text{First distance}} = \boxed{\text{Second distance}}$$
$$5r = 3(r + 26)$$
$$5r = 3r + 78$$
$$r = 39$$

He drove 39 miles per hour to go, and 65 miles per hour to return.

39. Let $t =$ the number of hours each drives.

	r	t	d
First	60	t	$60t$
Second	64	t	$64t$

Since they are traveling in opposite directions, their distance apart is the sum of their separate distances.

$$\boxed{\text{Total distance}} = \boxed{\text{First distance}} + \boxed{\text{Second distance}}$$
$$310 = 60t + 64t$$
$$310 = 124t$$
$$\frac{310}{124} = t, \text{ or } t = \frac{310}{124} = \frac{155}{62} = \frac{5}{2} = 2\frac{1}{2}$$

The cars will be 310 miles apart after $2\frac{1}{2}$ hours.

41. Let t = the number of hours each jogs.

	r	t	d
First	8	t	$8t$
Second	10	t	$10t$

$$440 \text{ yds} = \frac{440 \text{ yds}}{1} \cdot \frac{3 \text{ ft}}{1 \text{ yd}} \cdot \frac{1 \text{ mi}}{5280 \text{ ft}}$$

$$= \frac{1320}{5280} \text{ mi} = \frac{1}{4} \text{ mi}$$

Since they are jogging towards each other, the distance they start apart will equal the sum of the distances they jog separately.

$$\boxed{\text{Starting distance}} = \boxed{\text{First distance}} + \boxed{\text{Second distance}}$$

$$\frac{1}{4} = 8t + 10t$$

$$\frac{1}{4} = 18t$$

$$\frac{1}{18} \cdot \frac{1}{4} = \frac{1}{18} \cdot 18t$$

$$\frac{1}{72} = t, \text{ or } t = \frac{1}{72} \text{ hour}$$

$\frac{1}{72}$ hr $= \frac{1 \text{ hr}}{72} \cdot \frac{60 \text{ min}}{1 \text{ hr}} \cdot \frac{60 \text{ sec}}{1 \text{ min}} = \frac{3600}{72}$ sec $= 50$ sec   They will meet in 50 seconds.

43. Let x = the speed of the boat in still water.

	r	t	d
Up	$x - 2$	$\dfrac{5}{x-2}$	5
Down	$x + 2$	$\dfrac{7}{x+2}$	7

$$\boxed{\text{Time upstream}} = \boxed{\text{Time downstream}}$$

$$\frac{5}{x-2} = \frac{7}{x+2}$$

$$5(x+2) = 7(x-2)$$

$$5x + 10 = 7x - 14$$

$$x = 12 \quad \text{The boat can travel 12 miles per hour in still water.}$$

45. Let x = the number of nickels, the number of dimes, and the number of quarters.

$$0.05x + 0.10x + 0.25x = 3.20$$

$$0.40x = 3.20$$

$$x = \frac{3.20}{0.40} = 8 \quad \text{There are 8 of each type of coin.}$$

47. If 25% of the mixture (2400 pounds total) is barley, then there are 0.25(2400) or 600 pounds of barley, and then 1800 pounds of oats and soybean meal. 14% protein corresponds to 0.14(2400) or 336 pounds of protein. Let x = pounds of oats and $1800 - x$ = pounds of soybean meal.

$$\boxed{\text{Total pounds protein}} = \boxed{\text{Barley protein}} + \boxed{\text{Oats protein}} + \boxed{\text{Soybean protein}}$$

$$336 = 0.117(600) + 0.118x + 0.445(1800 - x)$$
$$336 = 70.2 + 0.118x + 801 - 0.445x$$
$$336 = 871.2 - 0.327x$$
$$-535.2 = -0.327x$$
$$\frac{-535.2}{-0.327} = x, \text{ or } x = \frac{-535.2}{-0.327} \approx 1636.7$$

The farmer should use about 600 pounds of barley, 1636.7 pounds of oats, and 163.3 pounds of soybean meal.

49.

$$V = \pi r^2 h$$
$$712.51 = \pi(4.5)^2 d$$
$$\frac{712.51}{\pi(4.5)^2} = d$$

Use the following keystrokes on the TI-81 calculator:

[7] [1] [2] [.] [5] [1] [÷] [( ] [2nd] [π] [×] [4] [.] [5] [x^2] [ )] [ENTER]

{11.19994948} The depth is about 11.2 mm.

Exercise 2.3 (page 93)

1.
$$x^2 - x - 6 = 0$$
$$(x - 3)(x + 2) = 0$$
$x - 3 = 0 \quad$ or $\quad x + 2 = 0$
$x = 3 \qquad\qquad x = -2$

3.
$$x^2 - 144 = 0$$
$$(x + 12)(x - 12) = 0$$
$x + 12 = 0 \quad$ or $\quad x - 12 = 0$
$x = -12 \qquad\qquad x = 12$

5.
$$2x^2 + x - 10 = 0$$
$$(2x + 5)(x - 2) = 0$$
$2x + 5 = 0 \quad$ or $\quad x - 2 = 0$
$2x = -5 \qquad\qquad x = 2$
$x = -\frac{5}{2}$

7.
$$5x^2 - 13x + 6 = 0$$
$$(5x - 3)(x - 2) = 0$$
$5x - 3 = 0 \quad$ or $\quad x - 2 = 0$
$5x = 3 \qquad\qquad x = 2$
$x = \frac{3}{5}$

9.
$$15x^2 + 16x = 15$$
$$15x^2 + 16x - 15 = 0$$
$$(3x + 5)(5x - 3) = 0$$
$3x + 5 = 0 \quad$ or $\quad 5x - 3 = 0$
$3x = -5 \qquad\qquad 5x = 3$
$x = -\frac{5}{3} \qquad\qquad x = \frac{3}{5}$

11.
$$12x^2 + 9 = 24x$$
$$12x^2 - 24x + 9 = 0$$
$$3(2x - 1)(2x - 3) = 0$$
$2x - 1 = 0 \quad$ or $\quad 2x - 3 = 0$
$2x = 1 \qquad\qquad 2x = 3$
$x = \frac{1}{2} \qquad\qquad x = \frac{3}{2}$

13.
$$x^2 = 9$$
$$x = \sqrt{9} \quad \text{or} \quad x = -\sqrt{9}$$
$$x = 3 \quad \bigm| \quad x = -3$$

15.
$$y^2 - 50 = 0$$
$$y^2 = 50$$
$$y = \sqrt{50} \quad \text{or} \quad y = -\sqrt{50}$$
$$y = 5\sqrt{2} \quad \bigm| \quad y = -5\sqrt{2}$$

17.
$$(x-1)^2 = 4$$
$$x - 1 = \sqrt{4} \quad \text{or} \quad x - 1 = -\sqrt{4}$$
$$x - 1 = 2 \quad \bigm| \quad x - 1 = -2$$
$$x = 3 \quad \bigm| \quad x = -1$$

19.
$$a^2 + 2a + 1 = 9$$
$$(a+1)^2 = 9$$
$$a + 1 = \sqrt{9} \quad \text{or} \quad a + 1 = -\sqrt{9}$$
$$a + 1 = 3 \quad \bigm| \quad a + 1 = -3$$
$$a = 2 \quad \bigm| \quad a = -4$$

21. $x^2 + 6x + \left(\frac{1}{2} \cdot 6\right)^2 = x^2 + 6x + (3)^2 = x^2 + 6x + 9$

23. $x^2 - 4x + \left(\frac{1}{2} \cdot -4\right)^2 = x^2 - 4x + (-2)^2 = x^2 - 4x + 4$

25. $a^2 + 5a + \left(\frac{1}{2} \cdot 5\right)^2 = a^2 + 5a + \left(\frac{5}{2}\right)^2 = a^2 + 5a + \frac{25}{4}$

27. $r^2 - 11r + \left(\frac{1}{2} \cdot -11\right)^2 = r^2 - 11r + \left(-\frac{11}{2}\right)^2 = r^2 - 11r + \frac{121}{4}$

29. $y^2 + \frac{3}{4}y + \left(\frac{1}{2} \cdot \frac{3}{4}\right)^2 = y^2 + \frac{3}{4}y + \left(\frac{3}{8}\right)^2 = y^2 + \frac{3}{4}y + \frac{9}{64}$

31. $q^2 - \frac{1}{5}q + \left(\frac{1}{2} \cdot -\frac{1}{5}\right)^2 = q^2 - \frac{1}{5}q + \left(-\frac{1}{10}\right)^2 = q^2 - \frac{1}{5}q + \frac{1}{100}$

33.
$$x^2 - 8x + 15 = 0$$
$$x^2 - 8x = -15$$
$$x^2 - 8x + 16 = -15 + 16$$
$$(x-4)^2 = 1$$
$$x - 4 = \sqrt{1} \quad \text{or} \quad x - 4 = -\sqrt{1}$$
$$x - 4 = 1 \quad \bigm| \quad x - 4 = -1$$
$$x = 5 \quad \bigm| \quad x = 3$$

35.
$$x^2 + x - 6 = 0$$
$$x^2 + x = 6$$
$$x^2 + x + \frac{1}{4} = \frac{24}{4} + \frac{1}{4}$$
$$\left(x + \frac{1}{2}\right)^2 = \frac{25}{4}$$
$$x + \frac{1}{2} = \sqrt{\frac{25}{4}} \quad \text{or} \quad x + \frac{1}{2} = -\sqrt{\frac{25}{4}}$$
$$x + \frac{1}{2} = \frac{5}{2} \quad \bigm| \quad x + \frac{1}{2} = -\frac{5}{2}$$
$$x = \frac{4}{2} \quad \bigm| \quad x = -\frac{6}{2}$$
$$x = 2 \quad \bigm| \quad x = -3$$

37.
$$x^2 - 25x = 0$$
$$x^2 - 25x + \frac{625}{4} = 0 + \frac{625}{4}$$
$$\left(x - \frac{25}{2}\right)^2 = \frac{625}{4}$$
$$x - \frac{25}{2} = \sqrt{\frac{625}{4}} \quad \text{or} \quad x - \frac{25}{2} = -\sqrt{\frac{625}{4}}$$
$$x - \frac{25}{2} = \frac{25}{2} \qquad\qquad x - \frac{25}{2} = -\frac{25}{2}$$
$$x = \frac{50}{2} \qquad\qquad\qquad x = 0$$
$$x = 25$$

39.
$$3x^2 + 4x = 4$$
$$x^2 + \frac{4}{3}x = \frac{4}{3}$$
$$x^2 + \frac{4}{3}x + \frac{4}{9} = \frac{12}{9} + \frac{4}{9}$$
$$\left(x + \frac{2}{3}\right)^2 = \frac{16}{9}$$
$$x + \frac{2}{3} = \sqrt{\frac{16}{9}} \quad \text{or} \quad x + \frac{2}{3} = -\sqrt{\frac{16}{9}}$$
$$x + \frac{2}{3} = \frac{4}{3} \qquad\qquad x + \frac{2}{3} = -\frac{4}{3}$$
$$x = \frac{2}{3} \qquad\qquad\qquad x = -\frac{6}{3}$$
$$x = -2$$

41.
$$x^2 + 5 = -5x$$
$$x^2 + 5x = -5$$
$$x^2 + 5x + \frac{25}{4} = -\frac{20}{4} + \frac{25}{4}$$
$$\left(x + \frac{5}{2}\right)^2 = \frac{5}{4}$$
$$x + \frac{5}{2} = \sqrt{\frac{5}{4}} \quad \text{or} \quad x + \frac{5}{2} = -\sqrt{\frac{5}{4}}$$
$$x + \frac{5}{2} = \frac{\sqrt{5}}{2} \qquad\qquad x + \frac{5}{2} = -\frac{\sqrt{5}}{2}$$
$$x = -\frac{5}{2} + \frac{\sqrt{5}}{2} \qquad\qquad x = -\frac{5}{2} - \frac{\sqrt{5}}{2}$$
$$x = \frac{-5 + \sqrt{5}}{2} \qquad\qquad x = \frac{-5 - \sqrt{5}}{2}$$

43.

$$3x^2 = 1 - 4x$$

$$3x^2 + 4x = 1$$

$$x^2 + \frac{4}{3}x = \frac{1}{3}$$

$$x^2 + \frac{4}{3}x + \frac{4}{9} = \frac{3}{9} + \frac{4}{9}$$

$$\left(x + \frac{2}{3}\right)^2 = \frac{7}{9}$$

$$x + \frac{2}{3} = \sqrt{\frac{7}{9}} \qquad \text{or} \qquad x + \frac{2}{3} = -\sqrt{\frac{7}{9}}$$

$$x + \frac{2}{3} = \frac{\sqrt{7}}{3} \qquad\qquad\qquad x + \frac{2}{3} = -\frac{\sqrt{7}}{3}$$

$$x = -\frac{2}{3} + \frac{\sqrt{7}}{3} \qquad\qquad x = -\frac{2}{3} - \frac{\sqrt{7}}{3}$$

$$x = \frac{-2 + \sqrt{7}}{3} \qquad\qquad x = \frac{-2 - \sqrt{7}}{3}$$

45.

$$x^2 - 12 = 0$$

$$a = 1, \quad b = 0, \quad c = -12$$

$$x = \frac{-(0) \pm \sqrt{(0)^2 - 4(1)(-12)}}{2(1)}$$

$$x = \frac{0 \pm \sqrt{0 + 48}}{2}$$

$$x = \frac{\pm\sqrt{48}}{2}$$

$$x = \frac{\pm 4\sqrt{3}}{2}$$

$$x = \frac{4\sqrt{3}}{2} \quad \text{or} \quad x = \frac{-4\sqrt{3}}{2}$$

$$x = 2\sqrt{3} \qquad\qquad x = -2\sqrt{3}$$

47.

$$2x^2 - x - 15 = 0$$

$$a = 2, \quad b = -1, \quad c = -15$$

$$x = \frac{-(-1) \pm \sqrt{(-1)^2 - 4(2)(-15)}}{2(2)}$$

$$x = \frac{1 \pm \sqrt{1 + 120}}{4}$$

$$x = \frac{1 \pm \sqrt{121}}{4}$$

$$x = \frac{1 \pm 11}{4}$$

$$x = \frac{1 + 11}{4} \quad \text{or} \quad x = \frac{1 - 11}{4}$$

$$x = \frac{12}{4} \qquad\qquad x = \frac{-10}{4}$$

$$x = 3 \qquad\qquad x = -\frac{5}{2}$$

49.

$$5x^2 - 9x - 2 = 0$$
$$a = 5, \quad b = -9, \quad c = -2$$

$$x = \frac{-(-9) \pm \sqrt{(-9)^2 - 4(5)(-2)}}{2(5)}$$

$$x = \frac{9 \pm \sqrt{121}}{10}$$

$$x = \frac{9 \pm 11}{10}$$

$$x = \frac{9 + 11}{10} \qquad \text{or} \qquad x = \frac{9 - 11}{10}$$

$$x = \frac{20}{10} \qquad\qquad\qquad x = \frac{-2}{10}$$

$$x = 2 \qquad\qquad\qquad x = -\frac{1}{5}$$

51.

$$2x^2 + 2x - 4 = 0$$
$$a = 2, \quad b = 2, \quad c = -4$$

$$x = \frac{-(2) \pm \sqrt{(2)^2 - 4(2)(-4)}}{2(2)}$$

$$x = \frac{-2 \pm \sqrt{36}}{4}$$

$$x = \frac{-2 \pm 6}{4}$$

$$x = \frac{-2 + 6}{4} \qquad \text{or} \qquad x = \frac{-2 - 6}{4}$$

$$x = \frac{4}{4} \qquad\qquad\qquad x = \frac{-8}{4}$$

$$x = 1 \qquad\qquad\qquad x = -2$$

53.

$$-3x^2 = 5x + 1$$
$$0 = 3x^2 + 5x + 1$$
$$3x^2 + 5x + 1 = 0$$
$$a = 3, \quad b = 5, \quad c = 1$$

$$x = \frac{-(5) \pm \sqrt{(5)^2 - 4(3)(1)}}{2(3)}$$

$$x = \frac{-5 \pm \sqrt{25 - 12}}{6}$$

$$x = \frac{-5 \pm \sqrt{13}}{6}$$

$$x = \frac{-5 + \sqrt{13}}{6} \qquad \text{or} \qquad x = \frac{-5 - \sqrt{13}}{6}$$

55.
$$5x\left(x + \tfrac{1}{5}\right) = 3$$
$$5x^2 + x = 3$$
$$5x^2 + x - 3 = 0$$
$$a = 5, \quad b = 1, \quad c = -3$$
$$x = \frac{-(1) \pm \sqrt{(1)^2 - 4(5)(-3)}}{2(5)}$$
$$x = \frac{-1 \pm \sqrt{1 + 60}}{10}$$
$$x = \frac{-1 + \sqrt{61}}{10} \quad \text{or} \quad x = \frac{-1 - \sqrt{61}}{10}$$

57.
$$x^2 + 6x + 9 = 0$$
$$a = 1, \quad b = 6, \quad c = 9$$
$$b^2 - 4ac = (6)^2 - 4(1)(9)$$
$$= 0$$
Since a, b and c are rational and the discriminant is 0, the solutions are equal rational numbers.

59.
$$3x^2 - 2x + 5 = 0$$
$$a = 3, \quad b = -2, \quad c = 5$$
$$b^2 - 4ac = (-2)^2 - 4(3)(5)$$
$$= -56$$
Since the discriminant is negative, the solutions are not real numbers.

61.
$$10x^2 + 29x = 21$$
$$10x^2 + 29x - 21 = 0$$
$$a = 10, \quad b = 29, \quad c = -21$$
$$b^2 - 4ac = (29)^2 - 4(10)(-21)$$
$$= 1681$$
Since a, b and c are rational and the discriminant is a nonzero perfect square ($41^2 = 1681$), the solutions are unequal rational numbers.

63.
$$-3x^2 + 2x = 21$$
$$-3x^2 + 2x - 21 = 0$$
$$a = -3, \quad b = 2, \quad c = -21$$
$$b^2 - 4ac = (2)^2 - 4(-3)(-21)$$
$$= -248$$
Since the discriminant is negative, the solutions are not real numbers.

65.
$$x^2 + kx + 3k - 5 = 0$$
$$x^2 + kx + (3k - 5) = 0$$
$$a = 1, \quad b = k, \quad c = 3k - 5$$

Find the discriminant:
$$b^2 - 4ac = k^2 - 4(1)(3k - 5)$$
$$= k^2 - 12k + 20$$

If the roots are equal, then the discriminant equals 0:
$$k^2 - 12k + 20 = 0$$
$$(k - 2)(k - 10) = 0$$
So when $k = 2$ or $k = 10$ the equation will have equal solutions.

67.
$$1492x^2 + 1984x - 1776 = 0$$
$$a = 1492, \quad b = 1984, \quad c = -1776$$
$$b^2 - 4ac = (1984)^2 - 4(1492)(-1776)$$
$$= 3{,}936{,}256 + 10{,}599{,}168 > 0, \text{ so the equation has real solutions.}$$

69.

$$x + 1 = \frac{12}{x}$$

$$x(x+1) = x\left(\frac{12}{x}\right)$$

$$x^2 + x = 12$$

$$x^2 + x - 12 = 0$$

$$(x+4)(x-3) = 0$$

$$x + 4 = 0 \quad \text{or} \quad x - 3 = 0$$

$$x = -4 \quad \Big| \quad x = 3$$

Both solutions check.

71.

$$8x - \frac{3}{x} = 10$$

$$x\left(8x - \frac{3}{x}\right) = x(10)$$

$$8x^2 - 3 = 10x$$

$$8x^2 - 10x - 3 = 0$$

$$(4x+1)(2x-3) = 0$$

$$4x + 1 = 0 \quad \text{or} \quad 2x - 3 = 0$$

$$4x = -1 \quad \Big| \quad 2x = 3$$

$$x = -\frac{1}{4} \quad \Big| \quad x = \frac{3}{2}$$

Both solutions check.

73.

$$\frac{5}{x} = \frac{4}{x^2} - 6$$

$$x^2\left(\frac{5}{x}\right) = x^2\left(\frac{4}{x^2} - 6\right)$$

$$5x = 4 - 6x^2$$

$$6x^2 + 5x - 4 = 0$$

$$(2x-1)(3x+4) = 0$$

$$2x - 1 = 0 \quad \text{or} \quad 3x + 4 = 0$$

$$2x = 1 \quad \Big| \quad 3x = -4$$

$$x = \frac{1}{2} \quad \Big| \quad x = -\frac{4}{3}$$

Both solutions check.

75.

$$x\left(30 - \frac{13}{x}\right) = \frac{10}{x}$$

$$30x - 13 = \frac{10}{x}$$

$$x(30x - 13) = x\left(\frac{10}{x}\right)$$

$$30x^2 - 13x = 10$$

$$30x^2 - 13x - 10 = 0$$

$$(6x-5)(5x+2) = 0$$

$$6x - 5 = 0 \quad \text{or} \quad 5x + 2 = 0$$

$$6x = 5 \quad \Big| \quad 5x = -2$$

$$x = \frac{5}{6} \quad \Big| \quad x = -\frac{2}{5}$$

Both solutions check.

77.

$$(a-2)(a+4) = 2a(a-3)$$

$$a^2 + 2a - 8 = 2a^2 - 6a$$

$$-a^2 + 8a - 8 = 0$$

$$a^2 - 8a + 8 = 0$$

$$a^2 - 8a = -8$$

$$a^2 - 8a + 16 = -8 + 16$$

$$(a-4)^2 = 8$$

$$a - 4 = \pm\sqrt{8}$$

$$a - 4 = \pm 2\sqrt{2}$$

$$a - 4 = 2\sqrt{2} \quad \text{or} \quad a - 4 = -2\sqrt{2}$$

$$a = 4 + 2\sqrt{2} \quad \Big| \quad a = 4 - 2\sqrt{2}$$

79.

$$\frac{1}{x} + \frac{3}{x+2} = 2$$

$$x(x+2)\left(\frac{1}{x} + \frac{3}{x+2}\right) = 2x(x+2)$$

$$x + 2 + 3x = 2x^2 + 4x$$

$$4x + 2 = 2x^2 + 4x$$

$$2 = 2x^2$$

$$1 = x^2$$

$$x = \pm\sqrt{1}$$

$$x = \sqrt{1} \quad \text{or} \quad x = -\sqrt{1}$$

$$x = 1 \quad \Big| \quad x = -1$$

Both solutions check.

81.

$$\frac{1}{x+1} + \frac{5}{2x-4} = 1$$

$$(x+1)(2x-4)\left(\frac{1}{x+1} + \frac{5}{2x-4}\right) = (x+1)(2x-4)(1)$$

$$(2x-4)(1) + (x+1)(5) = (x+1)(2x-4)$$
$$2x - 4 + 5x + 5 = 2x^2 - 2x - 4$$
$$7x + 1 = 2x^2 - 2x - 4$$
$$0 = 2x^2 - 9x - 5$$
$$0 = (2x+1)(x-5)$$

$$2x + 1 = 0 \qquad \text{or} \qquad x - 5 = 0$$
$$2x = -1 \qquad\qquad\qquad x = 5$$
$$x = -\frac{1}{2}$$

Both solutions check.

83.

$$x + 1 + \frac{x+2}{x-1} = \frac{3}{x-1}$$

$$(x-1)\left(x + 1 + \frac{x+2}{x-1}\right) = (x-1)\left(\frac{3}{x-1}\right)$$

$$(x-1)(x+1) + x + 2 = 3$$
$$x^2 - 1 + x + 2 - 3 = 0$$
$$x^2 + x - 2 = 0$$
$$(x+2)(x-1) = 0$$
$$x = -2 \text{ or } x = 1$$

Since $x = 1$ makes a denominator equal $= 0$, the only solution is $x = -2$.

85.

$$\frac{24}{a} - 11 = \frac{-12}{a+1}$$

$$a(a+1)\left(\frac{24}{a} - 11\right) = a(a+1)\left(\frac{-12}{a+1}\right)$$

$$24(a+1) - 11a(a+1) = -12a$$
$$24a + 24 - 11a^2 - 11a + 12a = 0$$
$$-11a^2 + 25a + 24 = 0$$
$$11a^2 - 25a - 24 = 0$$
$$(11a + 8)(a - 3) = 0$$

$$11a + 8 = 0 \qquad \text{or} \qquad a - 3 = 0$$
$$11a = -8 \qquad\qquad\qquad a = 3$$
$$a = -\frac{8}{11}$$

87.

$$h = \tfrac{1}{2}gt^2$$

$$\tfrac{2}{g} \cdot h = \tfrac{2}{g} \cdot \left(\tfrac{1}{2}gt^2\right)$$

$$\frac{2h}{g} = t^2$$

$$\pm\sqrt{\frac{2h}{g}} = t, \text{ or } t = \pm\sqrt{\frac{2h}{g}}$$

89.

$$h = 64t - 16t^2$$
$$16t^2 - 64t + h = 0$$
$$a = 16, \ b = -64, \ c = h$$

$$t = \frac{-(-64) \pm \sqrt{(-64)^2 - 4(16)(h)}}{2(16)}$$

$$t = \frac{64 \pm \sqrt{4096 - 64h}}{32}$$

$$t = \frac{64 \pm \sqrt{64(64 - h)}}{32}$$

$$t = \frac{64 \pm 8\sqrt{64 - h}}{32}$$

$$t = \frac{8(8 \pm \sqrt{64 - h})}{32}$$

$$t = \frac{8 \pm \sqrt{64 - h}}{4}$$

91.

$$\frac{x^2}{a^2} + \frac{y^2}{b^2} = 1$$

$$a^2b^2\left(\frac{x^2}{a^2} + \frac{y^2}{b^2}\right) = a^2b^2$$

$$b^2x^2 + a^2y^2 = a^2b^2$$
$$a^2y^2 = -b^2x^2 + a^2b^2$$

$$y^2 = \frac{b^2(-x^2 + a^2)}{a^2}$$

$$y = \pm\sqrt{\frac{b^2(a^2 - x^2)}{a^2}}$$

$$y = \pm\frac{b}{a}\sqrt{(a^2 - x^2)}$$

93.

$$\frac{x^2}{a^2} - \frac{y^2}{b^2} = 1$$

$$a^2b^2\left(\frac{x^2}{a^2} - \frac{y^2}{b^2}\right) = a^2b^2$$

$$b^2x^2 - a^2y^2 = a^2b^2$$
$$b^2x^2 = a^2y^2 + a^2b^2$$
$$b^2x^2 = a^2(y^2 + b^2)$$

$$\frac{b^2x^2}{y^2 + b^2} = a^2$$

$$\pm\sqrt{\frac{b^2x^2}{y^2 + b^2}} = a$$

$$a = \pm bx\sqrt{\frac{1}{y^2 + b^2}}$$

$$a = \pm bx \cdot \frac{1}{\sqrt{y^2 + b^2}} \cdot \frac{\sqrt{y^2 + b^2}}{\sqrt{y^2 + b^2}}$$

$$a = \pm\frac{bx\sqrt{y^2 + b^2}}{y^2 + b^2}$$

95. $x^2 + xy - y^2 = 0$

$a = 1, \quad b = y, \quad c = -y^2$

$$x = \frac{-y \pm \sqrt{y^2 - 4(1)(-y^2)}}{2(1)}$$

$$x = \frac{-y \pm \sqrt{y^2 + 4y^2}}{2}$$

$$x = \frac{-y \pm \sqrt{5y^2}}{2}$$

$$x = \frac{-y \pm y\sqrt{5}}{2}$$

97. $ax^2 + bx + c = 0$

$$x = \frac{-b \pm \sqrt{b^2 - 4ac}}{2a}$$

$$r_1 = \frac{-b + \sqrt{b^2 - 4ac}}{2a} \quad \text{or} \quad r_2 = \frac{-b - \sqrt{b^2 - 4ac}}{2a}$$

Add the solutions together:

$$r_1 + r_2 = \frac{-b + \sqrt{b^2 - 4ac}}{2a} + \frac{-b - \sqrt{b^2 - 4ac}}{2a}$$

$$= \frac{-b + \sqrt{b^2 - 4ac} + -b - \sqrt{b^2 - 4ac}}{2a}$$

$$= \frac{-2b}{2a} = -\frac{b}{a}$$

99. Rewrite the equation and use the quadratic formula to solve for t:

$$h = v_0 t - 16t^2$$
$$16t^2 - v_0 t + h = 0 \quad \{a = 16, \quad b = -v_0, \quad c = h\}$$

$$t = \frac{-(-v_0) \pm \sqrt{(-v_0)^2 - 4(16)(h)}}{2(16)}$$

$$t = \frac{v_0 \pm \sqrt{v_0^2 - 64h}}{32}$$

The solutions to this equation are t_1 and t_2, since the height is h.

$$t_1 = \frac{v_0 + \sqrt{v_0^2 - 64h}}{32} \quad \text{and} \quad t_2 = \frac{v_0 - \sqrt{v_0^2 - 64h}}{32}$$

Compute $16t_1 t_2$:

$$16t_1 t_2 = 16 \cdot \frac{\left(v_0 + \sqrt{v_0^2 - 64h}\right)}{32} \cdot \frac{\left(v_0 - \sqrt{v_0^2 - 64h}\right)}{32}$$

$$= 16 \cdot \frac{v_0^2 - \left(\sqrt{v_0^2 - 64h}\right)^2}{32^2}$$

$$= 16 \cdot \frac{v_0^2 - v_0^2 + 64h}{1024}$$

$$= \frac{16 \cdot 64h}{1024}$$

$$= \frac{1024h}{1024} = h$$

Thus the height of the tree, h, is equal to $16t_1 t_2$.

Exercise 2.4   (page 98)

1. Let $x =$ the first consecutive even natural number.
$x + 2 =$ the next consecutive even natural number.

$$x(x + 2) = 48$$
$$x^2 + 2x = 48$$
$$x^2 + 2x - 48 = 0$$
$$(x + 8)(x - 6) = 0$$

$$x + 8 = 0 \qquad \text{or} \qquad x - 6 = 0$$
$$x = -8 \qquad\qquad\qquad\quad x = 6$$

Since the numbers must be natural numbers, $x = 6$ is the only valid solution, and the numbers are 6 and 8.

3. Let $w =$ the width of the rectangle.
$w + 4 =$ the length of the rectangle.

$$\boxed{\text{Area}} = \boxed{\text{Length}} \cdot \boxed{\text{Width}}$$
$$32 = (w+4)w$$
$$32 = w^2 + 4w$$
$$0 = w^2 + 4w - 32$$
$$0 = (w+8)(w-4)$$

$$w + 8 = 0 \qquad \text{or} \qquad w - 4 = 0$$
$$w = -8 \qquad\qquad\qquad w = 4$$

Since the width must be positive, $w = 4$ is the only valid solution, and the dimensions are 4 feet by 8 feet.

5. Let $s =$ the side of the second square, and its area $= s^2$.
$s - 4 =$ the side of the first square, and its area $= (s-4)^2$.

$$\boxed{\begin{array}{c}\text{Area of} \\ \text{1st square}\end{array}} + \boxed{\begin{array}{c}\text{Area of} \\ \text{2nd square}\end{array}} = 106$$

$$(s-4)^2 + s^2 = 106$$
$$s^2 - 8s + 16 + s^2 = 106$$
$$2s^2 - 8s - 90 = 0$$
$$2(s^2 - 4s - 45) = 0$$
$$2(s-9)(s+5) = 0$$

$$s - 9 = 0 \qquad \text{or} \qquad s + 5 = 0$$
$$s = 9 \qquad\qquad\qquad s = -5$$

Since the length of a side must be positive, the only valid solution is $s = 9$, and so 9 centimeters is the length of one side of the larger square.

7. The floor area is a square with each side having a length of $12 - 2x$. So then the floor area is $(12 - 2x)^2$.

$$\boxed{\text{Floor area}} = 64$$
$$(12 - 2x)^2 = 64$$
$$12 - 2x = \pm\sqrt{64}$$
$$12 - 2x = \pm 8$$

$$12 - 2x = 8 \qquad \text{or} \qquad 12 - 2x = -8$$
$$-2x = -4 \qquad\qquad\qquad -2x = -20$$
$$x = 2 \qquad\qquad\qquad\qquad x = 10$$

The side of the floor area must be positive, so $12 - 2x$ must be positive. Then $x = 10$ is not a valid solution, so x must equal 2. The depth is 2 inches.

9. Let $r =$ his first rate, and $r - 10 =$ his return rate.

	r	t	d
First	r	$\frac{40}{r}$	40
Return	$r - 10$	$\frac{40}{r - 10}$	40

Use $\quad time = \dfrac{distance}{rate}$

to complete the table.

$$\boxed{\text{Return time}} = \boxed{\text{First time}} + 2$$

$$\frac{40}{r - 10} = \frac{40}{r} + 2$$

$$r(r - 10)\left(\frac{40}{r - 10}\right) = r(r - 10)\left(\frac{40}{r} + 2\right)$$

$$40r = 40(r - 10) + 2r(r - 10)$$

$$40r = 40r - 400 + 2r^2 - 20r$$

$$0 = 2r^2 - 20r - 400$$

$$0 = 2(r^2 - 10r - 200)$$

$$0 = 2(r - 20)(r + 10)$$

$$r - 20 = 0 \qquad \text{or} \qquad r + 10 = 0$$

$$r = 20 \qquad \qquad \qquad r = -10$$

Since the rate must be positive, the only valid solution is $r = 20$. His rates are then 20 miles per hour and 10 miles per hour.

11. Let $r =$ the actual rate and $r + 10 =$ the faster rate.

	r	t	d
Actual	r	$\frac{420}{r}$	420
Faster	$r + 10$	$\frac{420}{r + 10}$	420

Use $\quad time = \dfrac{distance}{rate}$

to complete the table.

$$\boxed{\text{Faster time}} = \boxed{\text{Actual time}} - 1$$

$$\frac{420}{r + 10} = \frac{420}{r} - 1$$

$$r(r + 10)\left(\frac{420}{r + 10}\right) = r(r + 10)\left(\frac{420}{r} - 1\right)$$

$$420r = 420(r + 10) - r(r + 10)$$

$$420r = 420r + 4200 - r^2 - 10r$$

$$r^2 + 10r - 4200 = 0$$

$$(r + 70)(r - 60) = 0$$

$$r + 70 = 0 \qquad \text{or} \qquad r - 60 = 0$$

$$r = -70 \qquad \qquad \qquad r = 60$$

Since the rates must be positive, the only valid solution is $r = 60$. The trip took $420/60 = 7$ hours at the slower rate.

13. When the projectile returns to earth, $h = 0$. Set $h = 0$ and solve for t:

$$h = -16t^2 + 400t$$
$$0 = -16t^2 + 400t$$
$$0 = -16t(t - 25)$$

$$-16t = 0 \qquad \text{or} \qquad t - 25 = 0$$
$$t = 0 \qquad \qquad \qquad t = 25$$

$t = 0$ is when the projectile is launched (and has a height of 0), so $t = 25$ is when the projectile returns to earth (and has a height of 0 again). It takes 25 seconds.

15. If the coin falls 1454 feet, then set $s = 1454$ and solve for t:

$$s = 16t^2$$
$$1454 = 16t^2$$
$$\frac{1454}{16} = t^2$$
$$t^2 = \pm\sqrt{\frac{1454}{16}} = \pm\frac{\sqrt{1454}}{4} \approx \pm 9.53$$

Since the time must be positive, it will take about 9.53 seconds.

17. Let $x =$ the number of nickel increases made in the fare.

$25 + 5x =$ the new bus fare in cents.

$3000 - 80x =$ the new number of passengers.

$$\boxed{\text{Revenue (in cents)}} = \boxed{\text{Number of passengers}} \cdot \boxed{\text{Bus fare (in cents)}}$$

$$99{,}400 = (3000 - 80x)(25 + 5x)$$
$$99{,}400 = 75{,}000 + 15{,}000x - 2000x - 400x^2$$
$$0 = -400x^2 + 13{,}000x - 24{,}400$$
$$0 = -200(2x^2 - 65x + 122)$$
$$0 = -200(2x - 61)(x - 2)$$

$$2x - 61 = 0 \qquad \text{or} \qquad x - 2 = 0$$
$$2x = 61 \qquad \qquad \qquad x = 2$$
$$x = \frac{61}{2}$$

30.5 nickels is not a valid increase, so the company should increase the fare by 2 nickels, or by 10 cents.

19. Let $x =$ the number of 50 cent decreases in the ticket price.

$15 - 0.50x =$ the new ticket price.

$1200 + 40x =$ the new attendance at the concert.

$$\boxed{\text{Revenue}} = \boxed{\text{Attendance}} \cdot \boxed{\text{Ticket price}}$$

$$17{,}280 = (1200 + 40x)(15 - 0.50x)$$
$$17{,}280 = 18{,}000 - 600x + 600x - 20x^2$$
$$-720 = -20x^2$$
$$36 = x^2$$
$$\pm\sqrt{36} = x$$
$$\pm 6 = x$$

$x = 6$ is the only valid answer, so the attendance was 1440.

21. The first pipe fills the tank in 4 hours, so it fills $\frac{1}{4}$ of the tank in 1 hour.

Let $x =$ the number of hours needed for the second pipe to fill the pool. Together, then, the pipes fill the tank in $x - 2$ hours.

$$\boxed{\begin{array}{c}\text{Amount filled by}\\\text{first in 1 hour}\end{array}} + \boxed{\begin{array}{c}\text{Amount filled by}\\\text{second in 1 hour}\end{array}} = \boxed{\begin{array}{c}\text{Amount filled by}\\\text{both in 1 hour}\end{array}}$$

$$\frac{1}{4} + \frac{1}{x} = \frac{1}{x-2}$$

$$4x(x-2)\left(\frac{1}{4} + \frac{1}{x}\right) = 4x(x-2)\left(\frac{1}{x-2}\right)$$

$$x(x-2) + 4(x-2) = 4x$$

$$x^2 - 2x + 4x - 8 - 4x = 0$$

$$x^2 - 2x - 8 = 0$$

$$(x-4)(x+2) = 0$$

$$x - 4 = 0 \qquad \text{or} \qquad x + 2 = 0$$

$$x = 4 \qquad\qquad\qquad x = -2$$

The only valid solution is $x = 4$, so it would take the second hose 4 hours alone.

23. Let $x =$ the number of hours needed for Steven to mow the lawn.
Then $x - 1 =$ the number of hours needed for Kristy to mow the lawn

$$\boxed{\begin{array}{c}\text{Amount mowed by}\\\text{Steven in 1 hour}\end{array}} + \boxed{\begin{array}{c}\text{Amount mowed by}\\\text{Kristy in 1 hour}\end{array}} = \boxed{\begin{array}{c}\text{Amount mowed by}\\\text{both in 1 hour}\end{array}}$$

$$\frac{1}{x} + \frac{1}{x-1} = \frac{1}{5}$$

$$5x(x-1)\left(\frac{1}{x} + \frac{1}{x-1}\right) = 5x(x-1)\left(\frac{1}{5}\right)$$

$$5(x-1) + 5x = x(x-1)$$

$$5x - 5 + 5x = x^2 - x$$

$$10x - 5 = x^2 - x$$

$$0 = x^2 - 11x + 5$$

$$\{a = 1, b = -11, c = 5\}$$

$$x = \frac{-(-11) \pm \sqrt{(-11)^2 - 4(1)(5)}}{2(1)}$$

$$x = \frac{11 \pm \sqrt{101}}{2} \approx \frac{11 \pm 10.05}{2}$$

$$x \approx \frac{11 + 10.05}{2} \qquad \text{or} \qquad x \approx \frac{11 - 10.05}{2}$$

$$x \approx 10.525 \qquad\qquad\qquad x \approx 0.475$$

Since Kristy's time must be positive, $x - 1$ must be positive, and the only valid solution is $x \approx 10.525$. It would take her about $10.5 - 1 = 9.5$ hours alone.

25. Let $x =$ the length of the diagonal. Then $x - 3 =$ the width and $x + 4 =$ the length. The diagonal is the hypotenuse of a right triangle with the width and length as legs, so the Pythagorean Theorem can be used:

$$\text{width}^2 + \text{length}^2 = \text{diagonal}^2$$
$$(x - 3)^2 + (x + 4)^2 = x^2$$
$$x^2 - 6x + 9 + x^2 + 8x + 16 = x^2$$
$$2x^2 + 2x + 25 - x^2 = 0$$
$$x^2 + 2x + 25 = 0 \quad \{a = 1,\ b = 2,\ c = 25\}$$

$b^2 - 4ac = (2)^2 - 4(1)(25) = 4 - 100 = -96$. Since the discriminant is negative, there are no real solutions to the equation, and no such rectangle exists.

27. Let $r =$ Maude's interest rate and then $r + 1 =$ Matilda's rate.

	I	p	r
Maude	280	$\frac{280}{r}$	r
Matilda	240	$\frac{240}{r + 0.01}$	$r + 0.01$

Use $\dfrac{Interest}{rate} = principal$ to complete the table.

$$\boxed{\text{Maude's principal}} = \boxed{\text{Matilda's principal}} + 1000$$

$$\frac{280}{r} = \frac{240}{r + 0.01} + 1000$$

$$r(r + 0.01)\left(\frac{280}{r}\right) = r(r + 0.01)\left(\frac{240}{r + 0.01} + 1000\right)$$

$$280(r + 0.01) = 240r + 1000r(r + 0.01)$$
$$280r + 2.8 = 240r + 1000r^2 + 10r$$
$$0 = 1000r^2 + 240r + 10r - 280r - 2.8$$
$$0 = 1000r^2 - 30r - 2.8$$
$$10(0) = 10(1000r^2 - 30r - 2.8)$$
$$0 = 10{,}000r^2 - 300r - 28$$
$$0 = 4(2500r^2 - 75r - 7)$$
$$0 = 4(100r - 7)(25r + 1)$$

$$100r - 7 = 0 \qquad \text{or} \qquad 25r + 1 = 0$$
$$100r = 7 \qquad\qquad\qquad 25r = -1$$
$$r = \frac{7}{100} = 0.07 \qquad\qquad r = -\frac{1}{25} = -0.04$$

Since r must be positive, $r = 0.07$ is the only valid solution. Maude's rate was 7% while Matilda's rate was 8%.

29. Let $x =$ the total number of professors. Then $x - 4$ actually contribute.

$$\boxed{\begin{array}{c}\text{Actual}\\\text{share}\end{array}} = \boxed{\begin{array}{c}\text{Planned}\\\text{share}\end{array}} + 10$$

$$\frac{150}{x-4} = \frac{150}{x} + 10$$

$$x(x-4)\left(\frac{150}{x-4}\right) = x(x-4)\left(\frac{150}{x} + 10\right)$$

$$150x = 150(x-4) + 10x(x-4)$$
$$0 = 150x - 600 + 10x^2 - 40x - 150x$$
$$0 = 10x^2 - 40x - 600$$
$$0 = 10(x^2 - 4x - 60)$$
$$0 = 10(x-10)(x+6)$$

$$x - 10 = 0 \qquad \text{or} \qquad x + 6 = 0$$
$$x = 10 \qquad\qquad\qquad x = -6$$

The only valid solution is $x = 10$, so 10 professors are in the department.

31. Let $x =$ the number of spokes.

$$\boxed{\begin{array}{c}\text{Degrees between spokes}\\\text{with } x \text{ spokes}\end{array}} = \boxed{\begin{array}{c}\text{Degrees between spokes}\\\text{with } x + 10 \text{ spokes}\end{array}} + 6$$

$$\frac{360}{x} = \frac{360}{x+10} + 6$$

$$x(x+10)\left(\frac{360}{x}\right) = x(x+10)\left(\frac{360}{x+10} + 6\right)$$

$$360(x+10) = 360x + 6x(x+10)$$
$$360x + 3600 = 360x + 6x^2 + 60x$$
$$0 = 6x^2 + 60x - 3600$$
$$0 = 6(x^2 + 10x - 600)$$
$$0 = 6(x+30)(x-20)$$

$$x + 30 = 0 \qquad \text{or} \qquad x - 20 = 0$$
$$x = -30 \qquad\qquad\qquad x = 20$$

Since $x = 20$ is the only valid solution, there are 20 spokes.

33. Let $x =$ the length of one leg and $x - 14 =$ the length of the other.

$$\text{leg}^2 + \text{leg}^2 = \text{hypotenuse}^2$$
$$x^2 + (x-14)^2 = 26^2$$
$$x^2 + x^2 - 28x + 196 = 676$$
$$2x^2 - 28x - 480 = 0$$
$$2(x^2 - 14x - 240) = 0$$
$$2(x-24)(x+10) = 0$$

$$x - 24 = 0 \qquad \text{or} \qquad x + 10 = 0$$
$$x = 24 \qquad\qquad\qquad x = -10$$

Since $x = 24$ is the only valid solution, the legs are 24 meters and 10 meters.

Exercise 2.5 (page 107)

1. $i^9 = (i^4)^2 \cdot i = 1^2 \cdot i = i$

3. $i^{38} = (i^4)^9 i^2 = 1^9 \cdot i^2 = i^2 = -1$

5. $i^{-6} = \dfrac{1}{i^6} = \dfrac{1}{i^4 \cdot i^2} = \dfrac{1}{1 \cdot (-1)} = -1$

7. $i^{-10} = \dfrac{1}{i^{10}} = \dfrac{1}{i^4 \cdot i^4 \cdot i^2} = \dfrac{1}{(1)(1)(-1)} = -1$

9. $x + (x+y)i = 3 + 8i$

   $x = 3$ $x + y = 8$

                  $3 + y = 8$

                     $y = 5$

   $x = 3$ and $y = 5$

11. $3x - 2yi = 2 + (x+y)i$

   $3x = 2$ $-2y = x + y$

                  $-3y = x$

   $x = \dfrac{2}{3}$

                  $-3y = \dfrac{2}{3}$

           $-\dfrac{1}{3}(-3y) = -\dfrac{1}{3} \cdot \dfrac{2}{3}$

                    $y = -\dfrac{2}{9}$

   $x = \dfrac{2}{3}$ and $y = -\dfrac{2}{9}$

13. $(2 - 7i) + (3 + i) = 2 - 7i + 3 + i = 5 - 6i$

15. $(5 - 6i) - (7 + 4i) = 5 - 6i - 7 - 4i = -2 - 10i$

17. $\quad (14i + 2) + (2 - \sqrt{-16}) = 14i + 2 + 2 - \sqrt{-1 \cdot 16} = 14i + 4 - \sqrt{i^2 \cdot 16}$

$$= 14i + 4 - 4i$$
$$= 4 + 10i$$

19. $\quad (3 + \sqrt{-4}) - (2 + \sqrt{-9}) = (3 + \sqrt{i^2 \cdot 4}) - (2 + \sqrt{i^2 \cdot 9}) = (3 + 2i) - (2 + 3i)$

$$= 3 + 2i - 2 - 3i$$
$$= 1 - i$$

21. $(2 + 3i)(3 + 5i) = 6 + 10i + 9i + 15i^2 = 6 + 19i + 15(-1) = -9 + 19i$

23. $(2 + 3i)^2 = (2 + 3i)(2 + 3i) = 4 + 12i + 9i^2 = 4 + 12i + 9(-1) = -5 + 12i$

25. $\quad (11 + \sqrt{-25})(2 - \sqrt{-36}) = (11 + 5i)(2 - 6i) = 22 - 66i + 10i - 30i^2$

$$= 22 - 56i - 30(-1)$$
$$= 52 - 56i$$

27. $\quad (\sqrt{-16} + 3)(2 + \sqrt{-9}) = (4i + 3)(2 + 3i) = 8i + 12i^2 + 6 + 9i$

$$= 17i + 12(-1) + 6$$
$$= -6 + 17i$$

29. $\dfrac{1}{i^3} = \dfrac{1}{i^3}\cdot\dfrac{i}{i} = \dfrac{i}{i^4} = \dfrac{i}{1} = i = 0 + i$

31. $\dfrac{-4}{i^{10}} = \dfrac{-4}{i^{10}}\cdot\dfrac{i^2}{i^2} = \dfrac{-4i^2}{i^{12}} = \dfrac{-4i^2}{(i^4)^3} = \dfrac{-4i^2}{1^3} = -4i^2 = 4 = 4 + 0i$

33. $\dfrac{1}{2+i} = \dfrac{1}{(2+i)}\cdot\dfrac{(2-i)}{(2-i)} = \dfrac{2-i}{4-i^2} = \dfrac{2-i}{4-(-1)} = \dfrac{2-i}{5} = \dfrac{2}{5} - \dfrac{1}{5}i$

35. $\dfrac{2i}{7+i} = \dfrac{2i}{(7+i)}\cdot\dfrac{(7-i)}{(7-i)} = \dfrac{14i - 2i^2}{49 - i^2} = \dfrac{14i - 2(-1)}{49 - (-1)} = \dfrac{2+14i}{50} = \dfrac{1}{25} + \dfrac{7}{25}i$

37. $\dfrac{2+i}{3-i} = \dfrac{(2+i)}{(3-i)}\cdot\dfrac{(3+i)}{(3+i)} = \dfrac{6 + 5i + i^2}{9 - i^2} = \dfrac{6 + 5i + (-1)}{9 - (-1)} = \dfrac{5+5i}{10} = \dfrac{1}{2} + \dfrac{1}{2}i$

39. $\dfrac{4-5i}{2+3i} = \dfrac{(4-5i)}{(2+3i)}\cdot\dfrac{(2-3i)}{(2-3i)} = \dfrac{8 - 22i + 15i^2}{4 - 9i^2} = \dfrac{-7 - 22i}{13} = -\dfrac{7}{13} - \dfrac{22}{13}i$

41.
$$\dfrac{5 - \sqrt{-16}}{-8 + \sqrt{-4}} = \dfrac{5 - 4i}{-8 + 2i} = \dfrac{(5 - 4i)}{(-8 + 2i)}\cdot\dfrac{(-8 - 2i)}{(-8 - 2i)} = \dfrac{-40 + 22i + 8i^2}{64 - 4i^2}$$
$$= \dfrac{-48 + 22i}{68}$$
$$= -\dfrac{48}{68} + \dfrac{22}{68}i$$
$$= -\dfrac{12}{17} + \dfrac{11}{34}i$$

43.
$$\dfrac{2 + i\sqrt{3}}{3 + i} = \dfrac{(2 + i\sqrt{3})}{(3 + i)}\cdot\dfrac{(3 - i)}{(3 - i)} = \dfrac{6 - 2i + (3\sqrt{3})i - i^2\sqrt{3}}{9 - i^2}$$
$$= \dfrac{(6 + \sqrt{3}) + (3\sqrt{3} - 2)i}{10}$$
$$= \dfrac{6 + \sqrt{3}}{10} + \dfrac{3\sqrt{3} - 2}{10}\, i$$

45. $|\,3 + 4i\,| = \sqrt{3^2 + 4^2} = \sqrt{9 + 16} = \sqrt{25} = 5$

47. $|\,2 + 3i\,| = \sqrt{2^2 + 3^2} = \sqrt{4 + 9} = \sqrt{13}$

49. $\left|\,-7 + \sqrt{-49}\,\right| = |\,-7 + 7i\,| = \sqrt{(-7)^2 + 7^2} = \sqrt{49 + 49} = \sqrt{98} = 7\sqrt{2}$

51. $\left|\,\dfrac{1}{2} + \dfrac{1}{2}i\,\right| = \sqrt{\left(\dfrac{1}{2}\right)^2 + \left(\dfrac{1}{2}\right)^2} = \sqrt{\dfrac{1}{4} + \dfrac{1}{4}} = \sqrt{\dfrac{2}{4}} = \dfrac{\sqrt{2}}{\sqrt{4}} = \dfrac{\sqrt{2}}{2}$

53. $|-6i| = |0-6i| = \sqrt{0^2 + (-6)^2} = \sqrt{0+36} = \sqrt{36} = 6$

55. $\left|\dfrac{2}{1+i}\right| = \left|\dfrac{2}{(1+i)} \cdot \dfrac{(1-i)}{(1-i)}\right| = \left|\dfrac{2-2i}{1-i^2}\right| = \left|\dfrac{2-2i}{2}\right| = |1-i| = \sqrt{1^2 + (-1)^2} = \sqrt{2}$

57.
$$\left|\frac{-3i}{2+i}\right| = \left|\frac{-3i}{(2+i)} \cdot \frac{(2-i)}{(2-i)}\right|$$

$$= \left|\frac{-6i + 3i^2}{4 - i^2}\right|$$

$$= \left|\frac{-3 - 6i}{5}\right|$$

$$= \left|-\frac{3}{5} - \frac{6}{5}i\right|$$

$$= \sqrt{\left(-\frac{3}{5}\right)^2 + \left(-\frac{6}{5}\right)^2}$$

$$= \sqrt{\frac{9}{25} + \frac{36}{25}}$$

$$= \sqrt{\frac{45}{25}}$$

$$= \frac{\sqrt{45}}{\sqrt{25}}$$

$$= \frac{3\sqrt{5}}{5}$$

59.
$$\left|\frac{i+2}{i-2}\right| = \left|\frac{(i+2)}{(i-2)} \cdot \frac{(i+2)}{(i+2)}\right|$$

$$= \left|\frac{i^2 + 4i + 4}{i^2 - 4}\right|$$

$$= \left|\frac{3 + 4i}{-5}\right|$$

$$= \left|-\frac{3}{5} - \frac{4}{5}i\right|$$

$$= \sqrt{\left(-\frac{3}{5}\right)^2 + \left(-\frac{4}{5}\right)^2}$$

$$= \sqrt{\frac{9}{25} + \frac{16}{25}}$$

$$= \sqrt{\frac{25}{25}}$$

$$= \sqrt{1}$$

$$= 1$$

61. $x^2 + 2x + 2 = 0$
$a = 1,\ b = 2,\ c = 2$

$$x = \frac{-(2) \pm \sqrt{(2)^2 - 4(1)(2)}}{2(1)}$$

$$x = \frac{-2 \pm \sqrt{4 - 8}}{2}$$

$$x = \frac{-2 \pm \sqrt{-4}}{2}$$

$$x = \frac{-2 \pm 2i}{2}$$

$$x = -1 \pm i$$

63. $y^2 + 4y + 5 = 0$
$a = 1,\ b = 4,\ c = 5$

$$y = \frac{-(4) \pm \sqrt{(4)^2 - 4(1)(5)}}{2(1)}$$

$$y = \frac{-4 \pm \sqrt{16 - 20}}{2}$$

$$y = \frac{-4 \pm \sqrt{-4}}{2}$$

$$y = \frac{-4 \pm 2i}{2}$$

$$y = -2 \pm i$$

65.
$$x^2 - 2x = -5$$
$$x^2 - 2x + 5 = 0$$
$$a = 1, \ b = -2, \ c = 5$$

$$x = \frac{-(-2) \pm \sqrt{(-2)^2 - 4(1)(5)}}{2(1)}$$

$$x = \frac{2 \pm \sqrt{4 - 20}}{2}$$

$$x = \frac{2 \pm \sqrt{-16}}{2}$$

$$x = \frac{2 \pm 4i}{2}$$

$$x = 1 \pm 2i$$

67.
$$x^2 - \frac{2}{3}x = -\frac{2}{9}$$
$$9\left(x^2 - \frac{2}{3}x\right) = 9\left(-\frac{2}{9}\right)$$

$$9x^2 - 6x = -2$$
$$9x^2 - 6x + 2 = 0$$
$$a = 9, \ b = -6, \ c = 2$$

$$x = \frac{-(-6) \pm \sqrt{(-6)^2 - 4(9)(2)}}{2(9)}$$

$$x = \frac{6 \pm \sqrt{36 - 72}}{18}$$

$$x = \frac{6 \pm \sqrt{-36}}{18}$$

$$x = \frac{6 \pm 6i}{18}$$

$$x = \frac{1}{3} \pm \frac{1}{3}i$$

69. $x^2 + 4 = x^2 - (-4) = x^2 - (i^2 \cdot 4) = x^2 - (2i)^2 = (x + 2i)(x - 2i)$

71. $\quad 25p^2 + 36q^2 = 25p^2 - (-36q^2) = 25p^2 - (i^2 \cdot 36q^2) = (5p)^2 - (6qi)^2$
$$= (5p + 6qi)(5p - 6qi)$$

73. $\quad 2y^2 + 8z^2 = 2(y^2 + 4z^2) = 2[y^2 - (-4z^2)] = 2[y^2 - (i^2 \cdot 4z^2)]$
$$= 2[y^2 - (2zi)^2]$$
$$= 2(y + 2zi)(y - 2zi)$$

75. $\quad 50m^2 + 2n^2 = 2(25m^2 + n^2) = 2[25m^2 - (-n^2)] = 2[25m^2 - (i^2 \cdot n^2)]$
$$= 2[(5m)^2 - (ni)^2]$$
$$= 2(5m + ni)(5m - ni)$$

77. $\quad (a + bi) + (c + di) = a + bi + c + di = a + c + bi + di = \boxed{(a + c) + (b + d)i}$

$$(c + di) + (a + bi) = c + di + a + bi = c + a + di + bi = a + c + bi + di$$
$$= \boxed{(a + c) + (b + d)i}$$

79.

$$[(a+bi)+(c+di)]+(x+yi) = [a+bi+c+di]+x+yi$$
$$= (a+c)+(b+d)i+x+yi$$
$$= \boxed{(a+c+x)+(b+d+y)i}$$

$$(a+bi)+[(c+di)+(x+yi)] = a+bi+[c+di+x+yi]$$
$$= a+bi+(c+x)+(d+y)i$$
$$= \boxed{(a+c+x)+(b+d+y)i}$$

Exercise 2.6   (page 113)

1.
$$x^3 + 9x^2 + 20x = 0$$
$$x(x^2 + 9x + 20) = 0$$
$$x(x+5)(x+4) = 0$$

$x = 0$	or	$x+5 = 0$	or	$x+4 = 0$
		$x = -5$		$x = -4$

3.
$$6a^3 - 5a^2 - 4a = 0$$
$$a(6a^2 - 5a - 4) = 0$$
$$a(2a+1)(3a-4) = 0$$

$a = 0$	or	$2a+1 = 0$	or	$3a-4 = 0$
		$2a = -1$		$3a = 4$
		$a = -\frac{1}{2}$		$a = \frac{4}{3}$

5.
$$y^4 - 26y^2 + 25 = 0$$
$$(y^2 - 25)(y^2 - 1) = 0$$
$$(y+5)(y-5)(y+1)(y-1) = 0$$

$y+5 = 0$	or	$y-5 = 0$	or	$y+1 = 0$	or	$y-1 = 0$
$y = -5$		$y = 5$		$y = -1$		$y = 1$

7.
$$x^4 - 37x^2 + 36 = 0$$
$$(x^2 - 36)(x^2 - 1) = 0$$
$$(x+6)(x-6)(x+1)(x-1) = 0$$

$x+6 = 0$	or	$x-6 = 0$	or	$x+1 = 0$	or	$x-1 = 0$
$x = -6$		$x = 6$		$x = -1$		$x = 1$

9.
$$2y^4 - 46y^2 = -180$$
$$2y^4 - 46y^2 + 180 = 0$$
$$2(y^4 - 23y^2 + 90) = 0$$
$$2(y^2 - 5)(y^2 - 18) = 0$$

$y^2 - 5 = 0$	or	$y^2 - 18 = 0$
$y^2 = 5$		$y^2 = 18$
$y = \pm\sqrt{5}$		$y = \pm\sqrt{18} = \pm3\sqrt{2}$

11.
$$z^{3/2} - z^{1/2} = 0$$
$$z^{1/2}(z^{2/2} - 1) = 0$$
$$z^{1/2}(z - 1) = 0$$

$z^{1/2} = 0$	or	$z - 1 = 0$
$(z^{1/2})^2 = 0^2$		$z = 1$
$z = 0$		NOTE: Both solutions check.

13.
$$2m^{2/3} + 3m^{1/3} - 2 = 0$$
$$2(2m^{1/3} - 1)(m^{1/3} + 2) = 0$$

$2m^{1/3} - 1 = 0$	or	$m^{1/3} + 2 = 0$
$2m^{1/3} = 1$		$m^{1/3} = -2$
$m^{1/3} = \frac{1}{2}$		$(m^{1/3})^3 = (-2)^3$
$(m^{1/3})^3 = (\frac{1}{2})^3$		$m = -8$
$m = \frac{1}{8}$		NOTE: Both solutions check.

15.
$$x - 13x^{1/2} + 12 = 0$$
$$(x^{1/2} - 12)(x^{1/2} - 1) = 0$$

$x^{1/2} - 12 = 0$	or	$x^{1/2} - 1 = 0$
$x^{1/2} = 12$		$x^{1/2} = 1$
$(x^{1/2})^2 = 12^2$		$(x^{1/2})^2 = 1^2$
$x = 144$		$x = 1$

NOTE: Both solutions check.

17.
$$2t^{1/3} + 3t^{1/6} - 2 = 0$$
$$(2t^{1/6} - 1)(t^{1/6} + 2) = 0$$

$2t^{1/6} - 1 = 0$	or	$t^{1/6} + 2 = 0$
$2t^{1/6} = 1$		$t^{1/6} = -2$
$t^{1/6} = \frac{1}{2}$		$(t^{1/6})^6 = (-2)^6$
$(t^{1/6})^6 = (\frac{1}{2})^6$		$t = 64$
$t = \frac{1}{64}$		NOTE: $t = 64$ DOES NOT check and is an extraneous solution. $t = \frac{1}{64}$ is a solution which checks.

19.
$$6p + p^{1/2} - 1 = 0$$
$$(2p^{1/2} + 1)(3p^{1/2} - 1) = 0$$

$$2p^{1/2} + 1 = 0 \quad \text{or} \quad 3p^{1/2} - 1 = 0$$

$$p^{1/2} = -\tfrac{1}{2} \qquad\qquad p^{1/2} = \tfrac{1}{3}$$

$$\left(p^{1/2}\right)^2 = \left(-\tfrac{1}{2}\right)^2 \qquad \left(p^{1/2}\right)^2 = \left(\tfrac{1}{3}\right)^2$$

$$p = \tfrac{1}{4} \qquad\qquad\qquad p = \tfrac{1}{9}$$

NOTE: $p = \tfrac{1}{4}$ DOES NOT check and is an extraneous solution. $p = \tfrac{1}{9}$ is a solution which checks.

21.
$$\sqrt{x-2} = 5$$
$$\left(\sqrt{x-2}\right)^2 = 5^2$$
$$x - 2 = 25$$
$$x = 27$$
The solution checks.

23.
$$\sqrt{x+3} = 2\sqrt{x}$$
$$\left(\sqrt{x+3}\right)^2 = \left(2\sqrt{x}\right)^2$$
$$x + 3 = 4x$$
$$-3x = -3$$
$$x = 1$$
The solution checks.

25.
$$2\sqrt{x^2+3} = \sqrt{-16x-3}$$
$$\left(2\sqrt{x^2+3}\right)^2 = \left(\sqrt{-16x-3}\right)^2$$
$$4(x^2+3) = -16x - 3$$
$$4x^2 + 12 = -16x - 3$$
$$4x^2 + 16x + 15 = 0$$
$$(2x+5)(2x+3) = 0$$
$$2x + 5 = 0 \quad \text{or} \quad 2x + 3 = 0$$
$$2x = -5 \qquad\qquad 2x = -3$$
$$x = -\tfrac{5}{2} \qquad\qquad x = -\tfrac{3}{2}$$
The solutions both check.

27.
$$\sqrt[3]{7x+1} = 4$$
$$\left(\sqrt[3]{7x+1}\right)^3 = 4^3$$
$$7x + 1 = 64$$
$$7x = 63$$
$$x = 9$$
The solution checks.

29.
$$\sqrt[4]{30t + 25} = 5$$
$$\left(\sqrt[4]{30t+25}\right)^4 = 5^4$$
$$30t + 25 = 625$$
$$30t = 600$$
$$t = 20$$
The solution checks.

31.
$$\sqrt{x^2+21} = x + 3$$
$$\left(\sqrt{x^2+21}\right)^2 = (x+3)^2$$
$$x^2 + 21 = x^2 + 6x + 9$$
$$21 = 6x + 9$$
$$12 = 6x$$
$$2 = x$$
The solution checks.

33.
$$\sqrt{y+2} = 4 - y$$
$$\left(\sqrt{y+2}\right)^2 = (4-y)^2$$
$$y + 2 = 16 - 8y + y^2$$
$$0 = y^2 - 9y + 14$$
$$0 = (y-2)(y-7)$$
$$y - 2 = 0 \quad \text{or} \quad y - 7 = 0$$
$$y = 2 \quad | \quad y = 7$$
$y = 7$ is an extraneous solution.

35.
$$x - \sqrt{7x - 12} = 0$$
$$x = \sqrt{7x - 12}$$
$$x^2 = \left(\sqrt{7x - 12}\right)^2$$
$$x^2 = 7x - 12$$
$$x^2 - 7x + 12 = 0$$
$$(x-3)(x-4) = 0$$
$$x - 3 = 0 \quad \text{or} \quad x - 4 = 0$$
$$x = 3 \quad | \quad x = 4$$
Both solutions check.

37.
$$x + 4 = \sqrt{\frac{6x+6}{5}} + 3$$
$$x + 1 = \sqrt{\frac{6x+6}{5}}$$
$$(x+1)^2 = \left(\sqrt{\frac{6x+6}{5}}\right)^2$$
$$x^2 + 2x + 1 = \frac{6x+6}{5}$$
$$5(x^2 + 2x + 1) = 5\left(\frac{6x+6}{5}\right)$$
$$5x^2 + 10x + 5 = 6x + 6$$
$$5x^2 + 4x - 1 = 0$$
$$(5x-1)(x+1) = 0$$
$$5x - 1 = 0 \quad \text{or} \quad x + 1 = 0$$
$$5x = 1 \quad | \quad x = -1$$
$$x = \frac{1}{5} \quad | \quad \text{Both solutions check.}$$

39.
$$\sqrt{\frac{x^2-1}{x-2}} = 2\sqrt{2}$$
$$\left(\sqrt{\frac{x^2-1}{x-2}}\right)^2 = (2\sqrt{2})^2$$
$$\frac{x^2-1}{x-2} = 8$$
$$(x-2)\left(\frac{x^2-1}{x-2}\right) = (x-2)8$$
$$x^2 - 1 = 8x - 16$$
$$x^2 - 8x + 15 = 0$$
$$(x-3)(x-5) = 0$$
$$x - 3 = 0 \quad \text{or} \quad x - 5 = 0$$
$$x = 3 \quad | \quad x = 5$$
Both solutions check.

41.
$$\sqrt[3]{x^3 + 7} = x + 1$$
$$\left(\sqrt[3]{x^3+7}\right)^3 = (x+1)^3$$
$$x^3 + 7 = x^3 + 3x^2 + 3x + 1$$
$$0 = 3x^2 + 3x - 6$$
$$0 = 3(x^2 + x - 2)$$
$$0 = 3(x+2)(x-1)$$
$$x + 2 = 0 \quad \text{or} \quad x - 1 = 0$$
$$x = -2 \quad | \quad x = 1$$
Both solutions check.

43.

$$\sqrt[3]{8x^3 - 61} = 2x + 1$$
$$\left(\sqrt[3]{8x^3 + 61}\right)^3 = (2x + 1)^3$$
$$8x^3 + 61 = (2x + 1)(2x + 1)(2x + 1)$$
$$8x^3 + 61 = (4x^2 + 4x + 1)(2x + 1)$$
$$8x^3 + 61 = 8x^3 + 4x^2 + 8x^2 + 4x + 2x + 1$$
$$8x^3 + 61 = 8x^3 + 12x^2 + 6x + 1$$
$$0 = 12x^2 + 6x - 60$$
$$0 = 6(2x + 5)(x - 2)$$
$$2x + 5 = 0 \qquad \text{or} \qquad x - 2 = 0$$
$$x = -\frac{5}{2} \qquad\qquad x = 2 \qquad \text{Both solutions check.}$$

45.

$$\sqrt{x + 3} = \sqrt{2x + 8} - 1$$
$$\left(\sqrt{x + 3}\right)^2 = \left(\sqrt{2x + 8} - 1\right)^2$$
$$x + 3 = (\sqrt{2x + 8} - 1)(\sqrt{2x + 8} - 1)$$
$$x + 3 = 2x + 8 - 2\sqrt{2x + 8} + 1$$
$$2\sqrt{2x + 8} = x + 6$$
$$\left(2\sqrt{2x + 8}\right)^2 = (x + 6)^2$$
$$4(2x + 8) = x^2 + 12x + 36$$
$$8x + 32 = x^2 + 12x + 36$$
$$0 = x^2 + 4x + 4$$
$$0 = (x + 2)(x + 2)$$
$$x + 2 = 0 \qquad \text{or} \qquad x + 2 = 0$$
$$x = -2 \qquad\qquad x = -2$$
The solution checks.

47.

$$\sqrt{y + 8} - \sqrt{y - 4} = -2$$
$$\sqrt{y + 8} = \sqrt{y - 4} - 2$$
$$\left(\sqrt{y + 8}\right)^2 = \left(\sqrt{y - 4} - 2\right)^2$$
$$y + 8 = (\sqrt{y - 4} - 2)(\sqrt{y - 4} - 2)$$
$$y + 8 = y - 4 - 4\sqrt{y - 4} + 4$$
$$y + 8 = -4\sqrt{y - 4} + y$$
$$8 = -4\sqrt{y - 4}$$
$$8^2 = \left(-4\sqrt{y - 4}\right)^2$$
$$64 = 16(y - 4)$$
$$128 = 16y$$
$$8 = y \qquad \text{This solution is extraneous, so there is no solution.}$$

49.
$$\sqrt{2b+3} - \sqrt{b+1} = \sqrt{b-2}$$
$$\left(\sqrt{2b+3} - \sqrt{b+1}\right)^2 = \left(\sqrt{b-2}\right)^2$$
$$\left(\sqrt{2b+3} - \sqrt{b+1}\right)\left(\sqrt{2b+3} - \sqrt{b+1}\right) = b-2$$
$$2b+3 - 2\sqrt{2b+3}\sqrt{b+1} + b+1 = b-2$$
$$3b+4 - 2\sqrt{(2b+3)(b+1)} = b-2$$
$$-2\sqrt{2b^2 + 5b + 3} = -2b-6$$
$$\left(-2\sqrt{2b^2 + 5b + 3}\right)^2 = (-2b-6)^2$$
$$4(2b^2 + 5b + 3) = 4b^2 + 24b + 36$$
$$4b^2 - 4b - 24 = 0$$
$$4(b-3)(b+2) = 0$$

$b - 3 = 0$ or $b + 2 = 0$

$b = 3$ | $b = -2$

$b = -2$ is an extraneous solution.

51.
$$\sqrt{\sqrt{b} + \sqrt{b+8}} = 2$$
$$\left(\sqrt{\sqrt{b} + \sqrt{b+8}}\right)^2 = 2^2$$
$$\sqrt{b} + \sqrt{b+8} = 4$$
$$\sqrt{b} = 4 - \sqrt{b+8}$$
$$\left(\sqrt{b}\right)^2 = (4 - \sqrt{b+8})^2$$
$$b = (4 - \sqrt{b+8})(4 - \sqrt{b+8})$$
$$b = 16 - 8\sqrt{b+8} + b + 8$$
$$8\sqrt{b+8} = 24$$
$$\sqrt{b+8} = 3$$
$$\left(\sqrt{b+8}\right)^2 = 3^2$$
$$b + 8 = 9$$
$$b = 1 \quad \text{The solution checks.}$$

Exercise 2.7   (page 123)

1. $3x + 2 < 5$
$3x < 3$
$x < 1$

$\boxed{x \in (-\infty, 1)}$

3. $3x + 2 \geq 5$
$3x \geq 3$
$x \geq 1$

$\boxed{x \in [1, \infty)}$

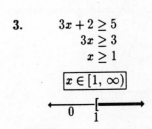

5. $\quad -5x + 3 > -2$
$$-5x > -5$$
$$x < 1$$

$$\boxed{x \in (-\infty,\, 1)}$$

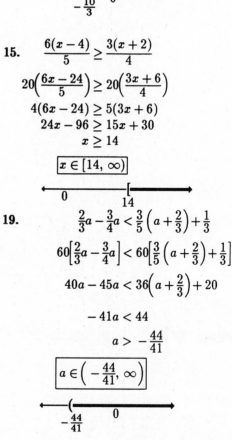

7. $\quad -5x + 3 \leq -2$
$$-5x \leq -5$$
$$x \geq 1$$

$$\boxed{x \in [1,\, \infty)}$$

9. $\quad 2(x-3) \leq -2(x-3)$
$$2x - 6 \leq -2x + 6$$
$$4x \leq 12$$
$$x \leq 3$$

$$\boxed{x \in (-\infty,\, 3]}$$

11. $\quad \dfrac{3}{5}x + 4 > 2$
$$5\left(\dfrac{3}{5}x + 4\right) > 5(2)$$
$$3x + 20 > 10$$
$$3x > -10$$

$$\boxed{x \in \left(-\dfrac{10}{3},\, \infty\right)}$$

13. $\quad \dfrac{x+3}{4} < \dfrac{2x-4}{3}$
$$12\left(\dfrac{x+3}{4}\right) < 12\left(\dfrac{2x-4}{3}\right)$$
$$3(x+3) < 4(2x-4)$$
$$3x + 9 < 8x - 16$$
$$x > 5$$

$$\boxed{x \in (5,\, \infty)}$$

15. $\quad \dfrac{6(x-4)}{5} \geq \dfrac{3(x+2)}{4}$
$$20\left(\dfrac{6x-24}{5}\right) \geq 20\left(\dfrac{3x+6}{4}\right)$$
$$4(6x - 24) \geq 5(3x + 6)$$
$$24x - 96 \geq 15x + 30$$
$$x \geq 14$$

$$\boxed{x \in [14,\, \infty)}$$

17. $\quad \dfrac{5}{9}(a+3) - a \geq \dfrac{4}{3}(a-3) - 1$
$$9\left[\dfrac{5}{9}(a+3) - a\right] \geq 9\left[\dfrac{4}{3}(a-3) - 1\right]$$
$$5(a+3) - 9a \geq 12(a-3) - 9$$
$$5a + 15 - 9a \geq 12a - 36 - 9$$
$$-16a \geq -60$$
$$a \leq \dfrac{60}{16}$$
$$a \leq \dfrac{15}{4}$$

$$\boxed{a \in \left(-\infty,\, \dfrac{15}{4}\right]}$$

19. $\quad \dfrac{2}{3}a - \dfrac{3}{4}a < \dfrac{3}{5}\left(a + \dfrac{2}{3}\right) + \dfrac{1}{3}$
$$60\left[\dfrac{2}{3}a - \dfrac{3}{4}a\right] < 60\left[\dfrac{3}{5}\left(a + \dfrac{2}{3}\right) + \dfrac{1}{3}\right]$$
$$40a - 45a < 36\left(a + \dfrac{2}{3}\right) + 20$$
$$-41a < 44$$
$$a > -\dfrac{44}{41}$$

$$\boxed{a \in \left(-\dfrac{44}{41},\, \infty\right)}$$

21. $4 < 2x - 8 \le 10$

$$12 < 2x \le 18$$
$$6 < x \le 9$$
$$\boxed{x \in (6, 9]}$$

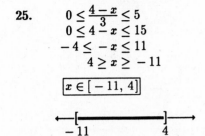

23. $9 \ge \dfrac{x-4}{2} > 2$

$$18 \ge x - 4 > 4$$
$$22 \ge x > 8, \text{ or } 8 < x \le 22$$
$$\boxed{x \in (8, 22]}$$

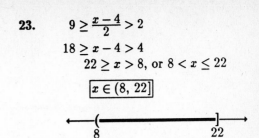

25. $0 \le \dfrac{4-x}{3} \le 5$

$$0 \le 4 - x \le 15$$
$$-4 \le -x \le 11$$
$$4 \ge x \ge -11$$
$$\boxed{x \in [-11, 4]}$$

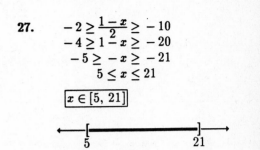

27. $-2 \ge \dfrac{1-x}{2} \ge -10$

$$-4 \ge 1 - x \ge -20$$
$$-5 \ge -x \ge -21$$
$$5 \le x \le 21$$
$$\boxed{x \in [5, 21]}$$

29. $-3x > -2x > -x$

$-3x > -2x$	and	$-2x > -x$
$0 > x$	$\mid$	$0 > x$

$x < 0$

$x < 0$

$x < 0$ and
$x < 0$

$x \in (-\infty, 0)$

31. $x < 2x < 3x$

$x < 2x$	and	$2x < 3x$
$0 < x$	$\mid$	$0 < x$

$x > 0$

$x > 0$

$x > 0$ and
$x > 0$

$x \in (0, \infty)$

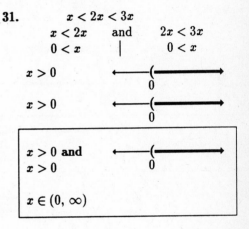

33.

$$2x + 1 < 3x - 2 < 12$$

$2x + 1 < 3x - 2$	and	$3x - 2 < 12$
$3 < x$		$3x < 14$
		$x < \dfrac{14}{3}$

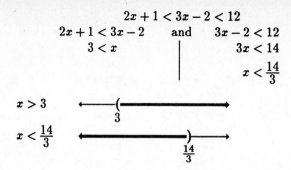

$x > 3$

$x < \dfrac{14}{3}$

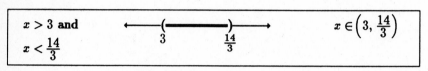

$x > 3$ and $x < \dfrac{14}{3}$ $x \in \left(3, \dfrac{14}{3}\right)$

35.

$$2 + x < 3x - 2 < 5x + 2$$

$2 + x < 3x - 2$	and	$3x - 2 < 5x + 2$
$4 < 2x$		$-4 < 2x$
$2 < x$		$-2 < x$

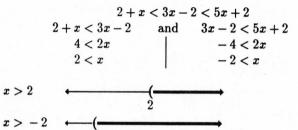

$x > 2$

$x > -2$

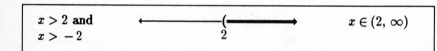

$x > 2$ and $x > -2$ $x \in (2, \infty)$

37.

$$3 + x > 7x - 2 > 5x - 10$$

$3 + x > 7x - 2$	and	$7x - 2 > 5x - 10$
$5 > 6x$		$2x > -8$
$\dfrac{5}{6} > x$		$x > -4$

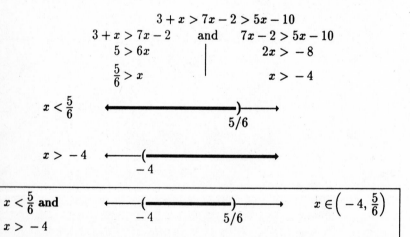

$x < \dfrac{5}{6}$

$x > -4$

$x < \dfrac{5}{6}$ and $x > -4$ $x \in \left(-4, \dfrac{5}{6}\right)$

39.

$$x \le x + 1 \le 2x + 3$$

$$
\begin{array}{ccc}
x \le x + 1 & \text{and} & x + 1 \le 2x + 3 \\
0 \le 1 & & -2 \le x
\end{array}
$$

$0 \le 1$ ⟵———————————⟶ {true for all x}

$x \ge -2$ ⟵——[———————⟶
$\qquad\qquad\; -2$

$\boxed{\begin{array}{l} 0 \le 1 \text{ and} \\ x \ge -2 \end{array} \qquad \longleftarrow [\!\!-\!\!-\!\!-\!\!-\!\!-\!\!\longrightarrow \qquad x \in [-2, \infty)}$
$\qquad\qquad\qquad\qquad\quad -2$

41.
$$x^2 + 7x + 12 < 0$$
$$(x + 3)(x + 4) < 0$$

$x + 3 \quad$ -----------0++++
$x + 4 \quad$ ----0+++++++++++

⟵——(———)——⟶
$\quad\; -4 \quad\;\; -3$

$x \in (-4, -3)$

43.
$$x^2 - 5x + 6 \ge 0$$
$$(x - 2)(x - 3) \ge 0$$

$x - 2 \quad$ -----0+++++++++++++
$x - 3 \quad$ -----------0++++++

⟵——]————[——⟶
$\qquad 2 \qquad\; 3$

$x \in (-\infty, 2] \cup [3, \infty)$

45.
$$x^2 + 5x + 6 < 0$$
$$(x + 2)(x + 3) < 0$$

$x + 2 \quad$ -----------0++++
$x + 3 \quad$ ----0+++++++++++

⟵——(———)——⟶
$\quad\; -3 \quad\;\; -2$

$x \in (-3, -2)$

47.
$$6x^2 + 5x + 1 \ge 0$$
$$(2x + 1)(3x + 1) \ge 0$$

$2x + 1 \quad$ -----0+++++++++++++
$3x + 1 \quad$ ------------0+++++

⟵——]————[——⟶
$\quad -1/2 \quad -1/3$

$x \in \left(-\infty, -\tfrac{1}{2}\right] \cup \left[-\tfrac{1}{3}, \infty\right)$

49.
$$6x^2 - 5x < -1$$
$$6x^2 - 5x + 1 < 0$$
$$(2x - 1)(3x - 1) < 0$$

$2x - 1 \quad$ -----------0++++
$3x - 1 \quad$ ----0+++++++++++

⟵——(———)——⟶
$\quad\; 1/3 \quad\;\; 1/2$

$x \in \left(\tfrac{1}{3}, \tfrac{1}{2}\right)$

51.
$$2x^2 \ge 3 - x$$
$$2x^2 + x - 3 \ge 0$$
$$(2x + 3)(x - 1) \ge 0$$

$2x + 3 \quad$ ----0+++++++++++++
$x - 1 \quad$ ------------0++++

⟵——]————[——⟶
$\quad -3/2 \qquad 1$

$x \in \left(-\infty, -\tfrac{3}{2}\right] \cup [1, \infty)$

53. $\dfrac{x+3}{x-2} < 0$

$x+3 \quad \text{----}0\text{+++++++++++}$
$x-2 \quad \text{-----------}0\text{++++}$

$\longleftarrow\!\!-\!(\underline{\quad\quad})\!-\!\longrightarrow$
-32

$x \in (-3,\, 2)$

55. $\dfrac{x^2+x}{x^2-1} > 0$

$\dfrac{x(x+1)}{(x+1)(x-1)} > 0$

$x \quad\quad \text{---------}0\text{++++++++++}$
$x+1 \quad \text{----}0\text{+++++++++++++++}$
$x+1 \quad \text{----}0\text{+++++++++++++++}$
$x-1 \quad \text{--------------}0\text{++++}$

$\longleftarrow\!\!-\!)(\underline{\quad})\!-\!-(\underline{\quad}\longrightarrow$
-101

$x \in (-\infty,\, -1) \cup (-1,\, 0) \cup (1,\, \infty)$

57. $\dfrac{x^2+5x+6}{x^2+x-6} \geq 0$

$\dfrac{(x+2)(x+3)}{(x+3)(x-2)} \geq 0$

$x+2 \quad \text{------}0\text{++++++++++}$
$x+3 \quad \text{--}0\text{++++++++++++++++}$
$x+3 \quad \text{--}0\text{++++++++++++++++}$
$x-2 \quad \text{------------}0\text{++++}$

$\longleftarrow\!\!-\!)(\underline{\quad}]\!-\!-\!-(\underline{\quad}\longrightarrow$
$-3-22$

$x \in (-\infty,\, -3) \cup (-3,\, -2] \cup$
$(2, \infty)$

59. $\dfrac{6x^2-x-1}{x^2+4x+4} > 0$

$\dfrac{(2x-1)(3x+1)}{(x+2)(x+2)} > 0$

$2x-1 \quad \text{----------------}0\text{+++}$
$3x+1 \quad \text{---------}0\text{++++++++++}$
$x+2 \quad \text{---}0\text{++++++++++++++++}$
$x+2 \quad \text{---}0\text{++++++++++++++++}$

$\longleftarrow\!\!-\!)(\underline{\quad\quad})\!-\!-(\underline{\quad}\longrightarrow$
$-2-1/31/2$

$x \in (-\infty,\, -2) \cup (-2,\, -\tfrac{1}{3}) \cup (\tfrac{1}{2}, \infty)$

61. $\dfrac{3}{x} > 2$

$\dfrac{3}{x} - 2 > 0$

$\dfrac{3}{x} - \dfrac{2x}{x} > 0$

$\dfrac{3-2x}{x} > 0$

$\dfrac{2x-3}{x} < 0$

$2x-3 \quad \text{-----------}0\text{++++}$

$x \quad\quad \text{----}0\text{+++++++++++}$

$\longleftarrow\!\!-\!(\underline{\quad\quad})\!-\!\longrightarrow$
$03/2$

$x \in \left(0, \tfrac{3}{2}\right)$

63. $\dfrac{6}{x} < 4$

$\dfrac{6}{x} - 4 < 0$

$\dfrac{6}{x} - \dfrac{4x}{x} < 0$

$\dfrac{6-4x}{x} < 0$

$\dfrac{4x-6}{x} > 0$

$4x-6 \quad \text{--------------}0\text{++++}$

$x \quad\quad \text{-------}0\text{++++++++++}$

$\longleftarrow\!\!-\!(\underline{\quad})\!-\!-\!-(\underline{\quad}\longrightarrow$
$03/2$

$x \in (-\infty,\, 0) \cup \left(\tfrac{3}{2},\, \infty\right)$

65.
$$\frac{3}{x-2} \le 5$$

$$\frac{3}{x-2} - 5 \le 0$$

$$\frac{3}{x-2} - \frac{5(x-2)}{x-2} \le 0$$

$$\frac{3 - 5x + 10}{x-2} \le 0$$

$$\frac{13 - 5x}{x-2} \le 0$$

$$\frac{5x - 13}{x-2} \ge 0$$

$5x - 13$ $---------0+++++$
$x - 2$ $-----0+++++++++$

    2 $13/5$

$$x \in (-\infty, 2) \cup \left[\frac{13}{5}, \infty\right)$$

67.
$$\frac{6}{x^2 - 1} < 1$$

$$\frac{6}{x^2 - 1} - 1 < 0$$

$$\frac{6}{x^2 - 1} - \frac{x^2 - 1}{x^2 - 1} < 0$$

$$\frac{6 - x^2 + 1}{x^2 - 1} < 0$$

$$\frac{7 - x^2}{x^2 - 1} < 0$$

$$\frac{x^2 - 7}{x^2 - 1} > 0$$

$$\frac{(x + \sqrt{7})(x - \sqrt{7})}{(x + 1)(x - 1)} > 0$$

$x + \sqrt{7}$ $---0+++++++++++++++++$
$x - \sqrt{7}$ $----------------0+++$
$x + 1$ $--------0++++++++++$
$x - 1$ $-----------0++++++$

  $-\sqrt{7}$ -1 1 $\sqrt{7}$

$$x \in (-\infty, -\sqrt{7}) \cup (-1, 1) \cup (\sqrt{7}, \infty)$$

69.   27,720 ft < alts (ft) < 33,000 ft

$$\frac{27,720 \text{ ft}}{5280} < \frac{\text{alts (ft)}}{5280} < \frac{33,000 \text{ ft}}{5280}$$

$$5\tfrac{1}{4} \text{ mi} < \text{alts (mi)} < 6\tfrac{1}{4} \text{ mi}$$

71.   50 cm < perim < 60 cm

$$\frac{50 \text{ cm}}{3} < \frac{\text{perim}}{3} < \frac{60 \text{ cm}}{3}$$

$$16\tfrac{2}{3} \text{ cm} < \text{side} < 20 \text{ cm}$$

73.
$$20 < l < 30$$
$$40 < 2l < 60$$
$$40 + 2w < 2l + 2w < 60 + 2w$$
$$40 + 2w < P < 60 + 2w$$

75. Let x = number of compact disks.

$$\boxed{\begin{array}{c}\text{Total}\\\text{cost}\end{array}} = \boxed{\begin{array}{c}\text{Cost of}\\\text{player}\end{array}} + \boxed{\begin{array}{c}\text{Cost of}\\\text{disks}\end{array}}$$

$$= 150 + 9.75x$$

$$\boxed{\text{Total cost}} \le 275$$

$$150 + 9.75x \le 275$$
$$9.75x \le 125$$
$$x \le 12.8205...$$

He can by at most 12 disks.

77.
$$3x - 4 \geq 0$$
$$3x \geq 4$$
$$x \geq \frac{4}{3}$$
$$x \in \left[\frac{4}{3}, \infty\right)$$

79. $3x + 2 > 0$ (denominator)
$$3x > -2 \quad \text{(cannot equal 0)}$$
$$x > -\frac{2}{3}$$
$$x \in \left(-\frac{2}{3}, \infty\right)$$

81.
$$x^2 - 9 \geq 0$$
$$(x + 3)(x - 3) \geq 0$$

$x + 3$ ----0+++++++++++

$x - 3$ -----------0++++

$$\xleftarrow{\hspace{1cm}}]\underset{-3}{\hspace{1cm}}\underset{3}{[}\xrightarrow{\hspace{1cm}}$$

$$x \in (-\infty, -3] \cup [3, \infty)$$

83. $x^2 + 4x + 3 \geq 0$ **and** $x \neq 0$
$$(x + 3)(x + 1) \geq 0 \quad \text{and} \quad x \neq 0$$

$x + 3$ ----0+++++++++++++

$x + 1$ -----------0+++++++

$$\xleftarrow{\hspace{1cm}}]\underset{-3}{\hspace{1cm}}\underset{-1}{[}\underset{0}{)(}\xrightarrow{\hspace{1cm}}$$

$$x \in (-\infty, -3] \cup [-1, 0) \cup (0, \infty)$$

Exercise 2.8   (page 130)

1. $|7| = 7$

3. $|0| = 0$

5. $|5| - |-3| = 5 - 3 = 2$

7. $|\pi - 2| = \pi - 2$ {NOTE: $\pi > 2$}

9. If $x > 5$, then $x - 5 > 0$, so
$|x - 5| = x - 5$.

11. $|x^3| = x^3$ if $x \geq 0$.
$|x^3| = -x^3$ if $x < 0$.

13.
$$|x + 2| = 2$$

$x + 2 = 2$	or	$x + 2 = -2$
$x = 0$		$x = -4$

15.
$$|3x - 1| = 5$$

$3x - 1 = 5$	or	$3x - 1 = -5$
$3x = 6$		$3x = -4$
$x = 2$		$x = -\frac{4}{3}$

17.
$$\left|\frac{3x - 4}{2}\right| = 5$$

$\frac{3x - 4}{2} = 5$	or	$\frac{3x - 4}{2} = -5$
$3x - 4 = 10$		$3x - 4 = -10$
$3x = 14$		$3x = -6$
$x = \frac{14}{3}$		$x = -2$

19.
$$\left|\frac{2x - 4}{5}\right| = 2$$

$\frac{2x - 4}{5} = 2$	or	$\frac{2x - 4}{5} = -2$
$2x - 4 = 10$		$2x - 4 = -10$
$2x = 14$		$2x = -6$
$x = 7$		$x = -3$

21.
$$\left|\frac{x - 3}{4}\right| = -2$$

This is impossible. An absolute

value is never negative and so cannot equal -2.

23.
$$\left|\frac{4x - 2}{x}\right| = 3$$

$\frac{4x - 2}{x} = 3$	or	$\frac{4x - 2}{x} = -3$
$4x - 2 = 3x$		$4x - 2 = -3x$
$x = 2$		$7x = 2$
		$x = \frac{2}{7}$

25.

$$|x| = x$$

$x = x$ or $x = -x$

$0 = 0$ $2x = 0$

{only if $x \geq 0$} $x = 0$

27.

$$|x+3| = |x|$$

$x + 3 = x$ or $x + 3 = -x$

$3 = 0$ $2x = -3$

impossible $x = -\dfrac{3}{2}$

29.

$$|x-3| = |2x+3|$$

$x - 3 = 2x + 3$ or $x - 3 = -(2x+3)$

$-6 = x$ $x - 3 = -2x - 3$

                     $3x = 0$

                     $x = 0$

31.

$$|x+2| = |x-2|$$

$x + 2 = x - 2$ or $x + 2 = -(x-2)$

$0 = -4$ $x + 2 = -x + 2$

impossible $2x = 0$

                     $x = 0$

33.

$$\left|\frac{x+3}{2}\right| = |2x-3|$$

$\dfrac{x+3}{2} = 2x - 3$ or $\dfrac{x+3}{2} = -(2x-3)$

$x + 3 = 2(2x-3)$ $x + 3 = -2(2x-3)$

$x + 3 = 4x - 6$ $x + 3 = -4x + 6$

$-3x = -9$ $5x = 3$

$x = 3$ $x = \dfrac{3}{5}$

35.

$$\left|\frac{3x-1}{2}\right| = \left|\frac{2x+3}{3}\right|$$

$\dfrac{3x-1}{2} = \dfrac{2x+3}{3}$ or $\dfrac{3x-1}{2} = -\dfrac{2x+3}{3}$

$6\left(\dfrac{3x-1}{2}\right) = 6\left(\dfrac{2x+3}{3}\right)$ $6\left(\dfrac{3x-1}{2}\right) = -6\left(\dfrac{2x+3}{3}\right)$

$3(3x-1) = 2(2x+3)$ $3(3x-1) = -2(2x+3)$

$9x - 3 = 4x + 6$ $9x - 3 = -4x - 6$

$5x = 9$ $13x = -3$

$x = \dfrac{9}{5}$ $x = -\dfrac{3}{13}$

37.

$$|x-3| < 6$$

$-6 < x - 3 < 6$

$-3 < x < 9$

$x \in (-3, 9)$

39.

$$|x+3| > 6$$

$x + 3 > 6$ or $x + 3 < -6$

$x > 3$ $x < -9$

$x \in (-\infty, -9) \cup (3, \infty)$

41. $|2x+4| \geq 10$

$2x+4 \geq 10$ or $2x+4 \leq -10$

$2x \geq 6$ | $2x \leq -14$

$x \geq 3$ | $x \leq -7$

$$x \in (-\infty, -7] \cup [3, \infty)$$

43. $|3x+5| + 1 \leq 9$

$|3x+5| \leq 8$

$-8 \leq 3x+5 \leq 8$

$-13 \leq 3x \leq 3$

$$-\frac{13}{3} \leq x \leq 1$$

$$x \in \left[-\frac{13}{3}, 1\right]$$

45. $|x+3| > 0$

$x+3 > 0$ or $x+3 < -0$

$x > -3$ | $x < -3$

$$x \in (-\infty, -3) \cup (-3, \infty)$$

47. $\left|\frac{5x+2}{3}\right| < 1$

$-1 < \frac{5x+2}{3} < 1$

$-3 < 5x+2 < 3$

$-5 < 5x < 1$

$$-1 < x < \frac{1}{5}$$

$$x \in \left(-1, \frac{1}{5}\right)$$

49. $3\left|\frac{3x-1}{2}\right| > 5$

$\left|\frac{3x-1}{2}\right| > \frac{5}{3}$

$\frac{3x-1}{2} > \frac{5}{3}$ or $\frac{3x-1}{2} < -\frac{5}{3}$

$3x-1 > \frac{10}{3}$ | $3x-1 < -\frac{10}{3}$

$3x > \frac{13}{3}$ | $3x < -\frac{7}{3}$

$x > \frac{13}{9}$ | $x < -\frac{7}{9}$

$$x \in \left(-\infty, -\frac{7}{9}\right) \cup \left(\frac{13}{9}, \infty\right)$$

51. $\frac{|x-1|}{-2} > 3 \implies |x-1| > -6$

This is true for all real x, since $|x-1|$ is always greater than or equal to 0.

53.

$$0 < |2x+1| < 3$$

$$0 < |2x+1| \quad \text{and} \quad |2x+1| < 3$$

(1) $\|2x+1\| > 0$	**(2)** $\|2x+1\| < 3$

$2x+1 > 0 \quad$ **or** $\quad 2x+1 < 0$

$\qquad 2x > -1 \qquad\qquad 2x < -1$

$\qquad\quad x > -\frac{1}{2} \qquad\qquad x < -\frac{1}{2}$

(2) $|2x+1| < 3$

$-3 < 2x+1 < 3$

$-4 < 2x < 2$

$-2 < x < 1$

$x > -\frac{1}{2}$

$-1/2$

$x < -\frac{1}{2}$

$-1/2$

$-2 < x < 1$

$-2 \qquad\qquad 1$
(2)

(1) (or)

$-1/2$

(1) $|2x+1| > 0$
$\qquad\qquad -1/2$

(2) $|2x+1| < 3$
$\qquad -2 \qquad\qquad 1$

(1) and (2) $0 < |2x+1| < 3$
$\qquad -2 \quad -1/2 \quad 1$

$$\boxed{x \in \left(-2,\, -\tfrac{1}{2}\right) \cup \left(-\tfrac{1}{2},\, 1\right)}$$

55.

$$8 > |3x-1| > 3$$

$$8 > |3x-1| \quad \text{and} \quad |3x-1| > 3$$

(1) $|3x-1| < 8$

$-8 < 3x-1 < 8$

$-7 < 3x < 9$

$-\frac{7}{3} < x < 3$

(2) $|3x-1| > 3$

$3x-1 > 3 \quad$ **or** $\quad 3x-1 < -3$

$\qquad 3x > 4 \qquad\qquad 3x < -2$

$\qquad\quad x > \frac{4}{3} \qquad\qquad x < -\frac{2}{3}$

$-\frac{7}{3} < x < 3$
$\qquad -7/3 \qquad 3$

(1)

$x > \frac{4}{3}$
$\qquad\qquad 4/3$

$x < -\frac{2}{3}$
$\qquad\qquad -2/3$

(2) (or)
$\qquad\qquad -2/3 \quad 4/3$

(1) $|3x-1| < 8$
$\qquad -7/3 \qquad\qquad 3$

(2) $|3x-1| > 3$
$\qquad -2/3 \quad 4/3$

(1) and (2) $3 < |3x-1| < 8$
$\qquad -7/3 \quad -2/3 \quad 4/3 \quad 3$

$$\boxed{x \in \left(-\tfrac{7}{3},\, -\tfrac{2}{3}\right) \cup \left(\tfrac{4}{3},\, 3\right)}$$

57.

$$2 < \left|\frac{x-5}{3}\right| < 4$$

$$2 < \left|\frac{x-5}{3}\right| \qquad \textbf{and} \qquad \left|\frac{x-5}{3}\right| < 4$$

(1) $\left|\frac{x-5}{3}\right| > 2$ **(2)** $\left|\frac{x-5}{3}\right| < 4$

$\frac{x-5}{3} > 2$ **or** $\frac{x-5}{3} < -2$ $-4 < \frac{x-5}{3} < 4$

$x - 5 > 6$ $x - 5 < -6$ $-12 < x - 5 < 12$

$x > 11$ $x < -1$ $-7 < x < 17$

$x > 11$ ⟵————(————⟶ $-7 < x < 17$
                 11   **(2)**

$x < -1$ ⟵————)————⟶   ⟵——(————————)——⟶
           −1   −7   17

(1) (or)   ⟵————)————(————⟶
               −1   11

(1)   ⟵————————)————(————————⟶ $\left|\frac{x-5}{3}\right| > 2$
                        −1   11

(2)   ⟵——(————————————————)——⟶ $\left|\frac{x-5}{3}\right| < 4$
            −7   17

(1) and (2)   ⟵——(————)————(————)——⟶ $\boxed{x \in (-7, 1) \cup (11, 17)}$
               −7   −1   11   17

59.

$$10 > \left|\frac{x-2}{2}\right| > 4$$

$$10 > \left|\frac{x-2}{2}\right| \qquad \textbf{and} \qquad \left|\frac{x-2}{2}\right| > 4$$

(1) $\left|\frac{x-2}{2}\right| < 10$ **(2)** $\left|\frac{x-2}{2}\right| > 4$

$-10 < \frac{x-2}{2} < 10$ $\frac{x-2}{2} > 4$ **or** $\frac{x-2}{2} < -4$

$-20 < x - 2 < 20$ $x - 2 > 8$ $x - 2 < -8$

$-18 < x < 22$ $x > 10$ $x < -6$

⟵——(————————————)——⟶ $x > 10$ ⟵————————(————⟶
  −18   22   10

(1) $x < -6$ ⟵————)————————⟶
                                           −6

                               (2) (or)   ⟵————)————(————⟶
                                         −6   10

(1)   ⟵——(————————————)——⟶
            −18   22

(2)   ⟵————————)————(————————⟶
                    −6   10

(1) and (2)   ⟵——(————)————(————)——⟶ $\boxed{x \in (-18, -6) \cup (10, 22)}$
               −18   −6   10   22

61.

$$2 \le \left|\frac{x+1}{3}\right| < 3$$

$$2 \le \left|\frac{x+1}{3}\right| \qquad \textbf{and} \qquad \left|\frac{x+1}{3}\right| < 3$$

$$(1)\ \left|\frac{x+1}{3}\right| \ge 2 \qquad\qquad\qquad (2)\ \left|\frac{x+1}{3}\right| < 3$$

$$\frac{x+1}{3} \ge 2 \quad \text{or} \quad \frac{x+1}{3} \le -2 \qquad -3 < \frac{x+2}{3} < 3$$

$$x+1 \ge 6 \qquad\qquad x+1 \le -6 \qquad\qquad -9 < x+2 < 9$$

$$x \ge 5 \qquad\qquad\qquad x \le -7 \qquad\qquad\qquad -11 < x < 7$$

$x \ge 5$

$x \le -7$

(2)

$(1)\ (\text{or})$

(1)

(2)

$(1) \text{ and } (2)$ $\boxed{x \in (-11, -7] \cup [5, 7)}$

63.

$$|x+1| \ge |x|$$

$$\sqrt{(x+1)^2} \ge \sqrt{x^2}$$

$$(x+1)^2 \ge x^2$$

$$x^2 + 2x + 1 \ge x^2$$

$$2x \ge -1$$

$$x \ge -\frac{1}{2}$$

$$\boxed{x \in \left[-\frac{1}{2}, \infty\right)}$$

65.

$$|2x+1| < |2x-1|$$

$$\sqrt{(2x+1)^2} < \sqrt{(2x-1)^2}$$

$$(2x+1)^2 < (2x-1)^2$$

$$4x^2 + 4x + 1 < 4x^2 - 4x + 1$$

$$8x < 0$$

$$x < 0$$

$$\boxed{x \in (-\infty, 0)}$$

67.

$$|x+1| < |x|$$

$$\sqrt{(x+1)^2} < \sqrt{x^2}$$

$$(x+1)^2 < x^2$$

$$x^2 + 2x + 1 < x^2$$

$$2x < -1$$

$$x < -\frac{1}{2}$$

$$\boxed{x \in \left(-\infty, -\frac{1}{2}\right)}$$

69.

$$|2x+1| \ge |2x-1|$$

$$\sqrt{(2x+1)^2} \ge \sqrt{(2x-1)^2}$$

$$(2x+1)^2 \ge (2x-1)^2$$

$$4x^2 + 4x + 1 \ge 4x^2 - 4x + 1$$

$$8x \ge 0$$

$$x \ge 0$$

$$\boxed{x \in [0, \infty)}$$

Chapter 2 Review Exercises (page 132)

1. The domain of x is the set of all real numbers.

3. The domain of x is the set of all real numbers greater than or equal to 0.

5.
$$3(9x + 4) = 28$$
$$27x + 12 = 28$$
$$27x = 16$$
$$x = \frac{16}{27}$$
conditional equation

7.
$$8(3x - 5) - 4(x + 3) = 12$$
$$24x - 40 - 4x - 12 = 12$$
$$20x - 52 = 12$$
$$20x = 64$$
$$x = \frac{64}{20} = \frac{16}{5}$$
conditional equation

9.
$$\frac{3}{x - 1} = \frac{1}{2}$$
$$2(x - 1)\left(\frac{3}{x - 1}\right) = 2(x - 1)\left(\frac{1}{2}\right)$$
$$6 = x - 1$$
$$7 = x \qquad \text{conditional equation}$$

11.
$$\frac{3x}{x - 1} - \frac{5}{x + 3} = 3$$
$$(x - 1)(x + 3)\left(\frac{3x}{x - 1} - \frac{5}{x + 3}\right) = (x - 1)(x + 3) \cdot 3$$
$$3x(x + 3) - 5(x - 1) = (x^2 + 2x - 3) \cdot 3$$
$$3x^2 + 9x - 5x + 5 = 3x^2 + 6x - 9$$
$$4x + 5 = 6x - 9$$
$$-2x = -14$$
$$x = 7 \qquad \text{conditional equation}$$

13.
$$C = \frac{5}{9}(F - 32)$$
$$\frac{9}{5}C = \frac{9}{5} \cdot \frac{5}{9}(F - 32)$$
$$\frac{9}{5}C = F - 32$$
$$\frac{9}{5}C + 32 = F, \text{ or } F = \frac{9}{5}C + 32$$

15.
$$\frac{1}{f} = \frac{1}{f_1} + \frac{1}{f_2}$$
$$f f_1 f_2\left(\frac{1}{f}\right) = f f_1 f_2\left(\frac{1}{f_1} + \frac{1}{f_2}\right)$$
$$f_1 f_2 = f f_2 + f f_1$$
$$f_1 f_2 - f f_1 = f f_2$$
$$f_1(f_2 - f) = f f_2$$
$$f_1 = \frac{f f_2}{f_2 - f}$$

17. Keep track of the amounts of alcohol and other. At the start, 50% of the 1 liter is alcohol, so there is $0.50(1) = 0.50$ liters of alcohol. Add x liters of water (0% solution). Add across to get the ending amounts.

	Start	Add	End
Alcohol (liters)	0.50	0	0.50
Other (liters)	0.50	x	$0.50 + x$
Total (liters)	1.00	x	$1 + x$

To get an equation, note that 20% of the ending amount should be alcohol.

$$\boxed{20\% \text{ of ending amount}} = \boxed{\text{Ending amount of alcohol}}$$

$$0.20(1 + x) = 0.50$$
$$0.20 + 0.20x = 0.50$$
$$0.20x = 0.30$$
$$2x = 3$$
$$x = \frac{3}{2} = 1\frac{1}{2} \qquad 1\frac{1}{2} \text{ liters of water should be added.}$$

19. Let $x =$ the number of hours both pipes work together. Then together the pipes can fill $\frac{1}{x}$ of the tank in 1 hour.

$$\boxed{\begin{array}{c}\text{Amount the pipes do} \\ \text{together in 1 hour}\end{array}} = \boxed{\begin{array}{c}\text{Amount 1st pipe} \\ \text{does in 1 hour}\end{array}} + \boxed{\begin{array}{c}\text{Amount 2nd pipe} \\ \text{does in 1 hour}\end{array}}$$

$$\frac{1}{x} = \frac{1}{9} + \frac{1}{12}$$
$$108x\left(\frac{1}{x}\right) = 108x\left(\frac{1}{9} + \frac{1}{12}\right)$$
$$108 = 12x + 9x$$
$$108 = 21x$$
$$\frac{108}{21} = x, \text{ or } x = \frac{36}{7} = 5\frac{1}{7}$$

It will take them $5\frac{1}{7}$ hours to fill the tank working together.

21. Let $x =$ the amount at 11%, then $10,000 - x =$ the amount at 14%.

$$\boxed{\begin{array}{c}\text{Total} \\ \text{income}\end{array}} = \boxed{\begin{array}{c}\text{Interest} \\ \text{at 11\%}\end{array}} + \boxed{\begin{array}{c}\text{Interest} \\ \text{at 14\%}\end{array}}$$

$$1265 = 0.11x + 0.14(10,000 - x)$$
$$1265 = 0.11x + 1400 - 0.14x$$
$$-135 = -0.03x$$
$$4500 = x \qquad \$4500 \text{ is at 11\% and } \$5500 \text{ is at 14\%.}$$

23. See the figure to the right.

$$\text{Area of rectangle} = \text{Length} \cdot \text{Width}$$
$$10{,}450 = (300 - 2x)x$$
$$10{,}450 = 300x - 2x^2$$
$$2x^2 - 300x + 10{,}450 = 0$$
$$2(x^2 - 150x + 5{,}225) = 0$$
$$2(x - 55)(x - 95) = 0$$
$$x - 55 = 0 \quad \text{or} \quad x - 95 = 0$$
$$x = 55 \quad | \quad x = 95$$

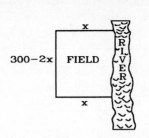

The dimensions are 55 yards by 190 yards, or 95 yards by 110 yards.

25.
$$2x^2 - x - 6 = 0$$
$$(2x + 3)(x - 2) = 0$$
$$2x + 3 = 0 \quad \text{or} \quad x - 2 = 0$$
$$2x = -3 \quad | \quad x = 2$$
$$x = -\frac{3}{2}$$

27.
$$5x^2 - 8x = 0$$
$$x(5x - 8) = 0$$
$$x = 0 \quad \text{or} \quad 5x - 8 = 0$$
$$5x = 8$$
$$x = \frac{8}{5}$$

29.
$$x^2 - 8x + 15 = 0$$
$$x^2 - 8x = -15$$
$$x^2 - 8x + 16 = -15 + 16$$
$$(x - 4)^2 = 1$$
$$x - 4 = \pm\sqrt{1}$$
$$x = 4 \pm 1$$
$$x = 4 + 1 \quad \text{or} \quad x = 4 - 1$$
$$x = 5 \quad | \quad x = 3$$

31.
$$5x^2 - x - 1 = 0$$
$$x^2 - \frac{1}{5}x - \frac{1}{5} = 0$$
$$x^2 - \frac{1}{5}x = \frac{1}{5}$$
$$x^2 - \frac{1}{5} + \frac{1}{100} = \frac{20}{100} + \frac{1}{100}$$
$$\left(x - \frac{1}{10}\right)^2 = \frac{21}{100}$$
$$x - \frac{1}{10} = \pm\sqrt{\frac{21}{100}}$$
$$x = \frac{1 \pm \sqrt{21}}{10}$$

33. $x^2 + 5x - 14 = 0$
$$a = 1, \, b = 5, \, c = -14$$

$$x = \frac{-(5) \pm \sqrt{5^2 - 4(1)(-14)}}{2(1)}$$
$$x = \frac{-5 \pm \sqrt{25 + 56}}{2}$$
$$x = \frac{-5 \pm \sqrt{81}}{2}$$
$$x = 2 \quad \text{or} \quad x = -7$$

35.
$$5x^2 = 1 - x$$
$$5x^2 + x - 1 = 0$$
$$a = 5, \, b = 1, \, c = -1$$

$$x = \frac{-(1) \pm \sqrt{1^2 - 4(5)(-1)}}{2(5)}$$
$$x = \frac{-1 \pm \sqrt{1 + 20}}{10}$$
$$x = \frac{-1 \pm \sqrt{21}}{10}$$

37. $kx^2 + 4x + 12 = 0$
$a = k,\ b = 4,\ c = 12$
$b^2 - 4ac = 4^2 - 4(k)(12)$
$\qquad\qquad = 16 - 48k$
If the roots are equal, then the
discriminant = 0.
$\qquad 16 - 48k = 0$
$\qquad\quad -48k = -16$
$$k = \frac{-16}{-48} = \frac{1}{3}$$

39. Set $h = 48$:
$$h = -16t^2 + 64t$$
$$48 = -16t^2 + 64t$$
$$16t^2 - 64t + 48 = 0$$
$$16(t^2 - 4t + 3) = 0$$
$$16(t - 3)(t - 1) = 0$$
$t - 3 = 0 \qquad$ or $\qquad t - 1 = 0$
$\quad t = 3 \qquad\qquad\qquad t = 1$

It will be 48 feet high in 1 second.

41. $i^{53} = i^{52} \cdot i = (i^4)^{13} \cdot i = 1^{13} \cdot i = i = 0 + i$

43. $(2 - 3i) + (-4 + 2i) = 2 - 3i - 4 + 2i = -2 - i$

45. $(3 - \sqrt{-36}) + (\sqrt{-16} + 2) = 3 - i\sqrt{36} + i\sqrt{16} + 2 = 5 - 6i + 4i = 5 - 2i$

47. $\dfrac{3}{i} = \dfrac{3}{i} \cdot \dfrac{i}{i} = \dfrac{3i}{i^2} = \dfrac{3i}{-1} = -3i = 0 - 3i$

49. $\dfrac{3}{1+i} = \dfrac{3}{(1+i)} \cdot \dfrac{(1-i)}{(1-i)} = \dfrac{3-3i}{1-i^2} = \dfrac{3-3i}{2} = \dfrac{3}{2} - \dfrac{3}{2}i$

51. $\dfrac{3+i}{3-i} = \dfrac{(3+i)}{(3-i)} \cdot \dfrac{(3+i)}{(3+i)} = \dfrac{9+6i+i^2}{9-i^2} = \dfrac{8+6i}{10} = \dfrac{4}{5} + \dfrac{3}{5}i$

53. $|3 - i| = \sqrt{3^2 + (-1)^2} = \sqrt{9+1} = \sqrt{10}$

55.
$$\frac{3x}{2} - \frac{2x}{x-1} = x - 3$$
$$2(x-1)\left(\frac{3x}{2} - \frac{2x}{x-1}\right) = 2(x-1)(x-3)$$
$$3x(x-1) - 2(2x) = 2(x^2 - 4x + 3)$$
$$3x^2 - 3x - 4x = 2x^2 - 8x + 6$$
$$x^2 + x - 6 = 0$$
$$(x+3)(x-2) = 0$$
$x + 3 = 0 \qquad$ or $\qquad x - 2 = 0$
$\quad x = -3 \qquad\qquad\qquad x = 2 \qquad$ Both solutions check.

57.
$$x^4 - 2x^2 + 1 = 0$$
$$(x^2 - 1)(x^2 - 1) = 0$$
$$(x+1)(x-1)(x+1)(x-1) = 0$$
$x + 1 = 0 \qquad$ or $\qquad x - 1 = 0$
$\quad x = -1 \qquad\qquad\qquad x = 1$
Both solutions check.

59.
$$a - a^{1/2} - 6 = 0$$
$$(a^{1/2} - 3)(a^{1/2} + 2) = 0$$
$a^{1/2} - 3 = 0 \quad$ or $\quad a^{1/2} + 2 = 0$
$\quad a^{1/2} = 3 \qquad\qquad\quad a^{1/2} = -2$
$\qquad a = 9 \qquad\qquad\qquad\quad a = 4$
$a = 9$ checks, but $a = 4$ is extraneous.

61.

$$\sqrt{x-1} + x = 7$$
$$\sqrt{x-1} = 7 - x$$
$$\left(\sqrt{x-1}\right)^2 = (7-x)^2$$
$$x - 1 = 49 - 14x + x^2$$
$$0 = x^2 - 15x + 50$$
$$0 = (x-5)(x-10)$$

$x - 5 = 0$ or $x - 10 = 0$

$x = 5$ | $x = 10$ $x = 5$ checks, but $x = 10$ is extraneous.

63.

$$\sqrt{5-x} + \sqrt{5+x} = 4$$
$$\sqrt{5-x} = 4 - \sqrt{5+x}$$
$$\left(\sqrt{5-x}\right)^2 = \left(4 - \sqrt{5+x}\right)^2$$
$$5 - x = 16 - 8\sqrt{5+x} + 5 + x$$
$$8\sqrt{5+x} = 2x + 16 \qquad \{\text{multiply through by } 1/2\}$$
$$\left(4\sqrt{5+x}\right)^2 = (x+8)^2$$
$$16(5+x) = x^2 + 16x + 64$$
$$80 + 16x = x^2 + 16x + 64$$
$$16 = x^2$$
$$x = \pm\sqrt{16} = \pm 4 \qquad \text{Both solutions check.}$$

65.

$$2x - 9 < 5$$
$$2x < 14$$
$$x < 7$$

$$\longleftarrow\!\overset{)}{\underset{7}{}}\!\!\longrightarrow$$

67.

$$\frac{5(x-1)}{2} < x$$
$$2\left(\frac{5x-5}{2}\right) < 2x$$
$$5x - 5 < 2x$$
$$3x < 5$$
$$x < \frac{5}{3}$$

$$\longleftarrow\!\overset{)}{\underset{5/3}{}}\!\!\longrightarrow$$

69.

$$(x+2)(x-4) > 0$$

$x + 2$ $---0+++++++++++$
$x - 4$ $----------0++++$

$$\longleftarrow\!\!\overset{)}{\underset{-2}{}}\!\!\!\!-\!\!\!\!\overset{(}{\underset{4}{}}\!\!\longrightarrow$$

$$x \in (-\infty, -2) \cup (4, \infty)$$

71.

$$x^2 - 2x - 3 < 0$$
$$(x+1)(x-3) < 0$$

$x + 1$ $---0++++++++++++++$
$x - 3$ $-------------0++++$

$$\longleftarrow\!\!\overset{(}{\underset{-1}{}}\!\!\!\!-\!\!\!\!\overset{)}{\underset{3}{}}\!\!\longrightarrow$$

$$x \in (-1, 3)$$

92

73.

$$\frac{x+2}{x-3} \geq 0$$

$x+2$ ---0+++++++++++++
$x-3$ -----------0++++

$$x \in (-\infty, -2] \cup (3, \infty)$$

75.

$$\frac{x^2+x-2}{x-3} \geq 0$$

$$\frac{(x+2)(x-1)}{x-3} \geq 0$$

$x+2$ ---0++++++++++++++
$x-1$ --------0+++++++++
$x-3$ -------------0+++

$$x \in [-2, 1] \cup (3, \infty)$$

77.

$$|x+1| = 6$$

$x+1=6$ or $x+1=-6$
$x=5$ | $x=-7$

79.

$$|x+3| < 3$$
$$-3 < x+3 < 3$$
$$-6 < x < 0$$

81.

$$\left|\frac{x+2}{3}\right| < 1$$

$$-1 < \frac{x+2}{3} < 1$$
$$-3 < x+2 < 3$$
$$-5 < x < 1$$

83.

$$1 < |2x+3| < 4$$

$1 < |2x+3|$ **and** $|2x+3| < 4$

(1) $|2x+3| > 1$ **(2)** $|2x+3| < 4$

$2x+3 > 1$ or $2x+3 < -1$ $-4 < 2x+3 < 4$
   $2x > -2$ $2x < -4$ $-7 < 2x < 1$

     $x > -1$ $x < -2$ $-\frac{7}{2} < x < \frac{1}{2}$

$x > -1$

$-\frac{7}{2} < x < \frac{1}{2}$

$x < -2$

(1) (or)

(1)

(2)

(1) and **(2)**

$$x \in \left(-\frac{7}{2}, -2\right) \cup \left(-1, \frac{1}{2}\right)$$

$|2x+3| > 1$

$|2x+3| < 4$

$1 < |2x+3| < 4$

93

Chapter 2 Test (page 134)

1. The domain of x is the set of all real numbers except 0 and 1.

3.
$$7(2a+5) - 7 = 6(a+8)$$
$$14a + 35 - 7 = 6a + 48$$
$$14a + 28 = 6a + 48$$
$$8a = 20$$
$$a = \frac{20}{8} = \frac{5}{2}$$

5.
$$\frac{3}{x^2 - 5x - 14} = \frac{4}{x^2 + 5x + 6}$$
$$\frac{3}{(x-7)(x+2)} = \frac{4}{(x+2)(x+3)}$$
$$(x-7)(x+2)(x+3)\left(\frac{3}{(x-7)(x+2)}\right) = (x-7)(x+2)(x+3)\left(\frac{4}{(x+2)(x+3)}\right)$$
$$3(x+3) = 4(x-7)$$
$$3x + 9 = 4x - 28$$
$$37 = x$$

7. Let x = the student's score on the final exam.

Average grade $= \dfrac{\text{sum of tests} + \text{final} + \text{final}}{5}$

$\qquad = \dfrac{3 \cdot \text{average test} + x + x}{5}$

$\qquad = \dfrac{3 \cdot 75 + 2x}{5}$

$\qquad = \dfrac{225 + 2x}{5}$

Solve:
$$\frac{225 + 2x}{5} = 80$$
$$225 + 2x = 400$$
$$2x = 175$$
$$x = \frac{175}{2}$$
$$= 87\frac{1}{2}$$

He must score 87.5 on the final.

9.
$$4x^2 - 8x + 3 = 0$$
$$a = 4, \ b = -8, \ c = 3$$

$$x = \frac{-(-8) \pm \sqrt{(-8)^2 - 4(4)(3)}}{2(4)}$$

$$x = \frac{8 \pm \sqrt{64 - 48}}{8}$$

$$x = \frac{8 \pm \sqrt{16}}{8}$$

$$x = \frac{8 \pm 4}{8}$$

$$x = \frac{3}{2} \text{ or } x = \frac{1}{2}$$

11. If $ax^2 + bx + c = 0$, then
$$x = \frac{-b \pm \sqrt{b^2 - 4ac}}{2a}$$

13.
$$x^2 + (k+1)x + k + 4 = 0$$
$$a = 1,\ b = k+1,\ c = k+4$$
$$\begin{aligned}
b^2 - 4ac &= (k+1)^2 - 4(1)(k+4) \\
&= k^2 + 2k + 1 - 4k - 16 \\
&= k^2 - 2k - 15
\end{aligned}$$

If the roots are equal, then the discriminant is equal to 0.
$$k^2 - 2k - 15 = 0$$
$$(k-5)(k+3) = 0$$
$$k = 5 \text{ or } k = -3$$

15. $i^{13} = i^{12}i = (i^4)^3 i = 1^3 i = i$

17. $(4 - 5i) - (-3 + 7i) = 4 - 5i + 3 - 7i = 7 - 12i$

19. $\dfrac{2}{2-i} = \dfrac{2}{(2-i)} \cdot \dfrac{(2+i)}{(2+i)} = \dfrac{4+2i}{2^2 - i^2} = \dfrac{4+2i}{5} = \dfrac{4}{5} + \dfrac{2}{5}i$

21. $|5 - 12i| = \sqrt{5^2 + (-12)^2} = \sqrt{25 + 144} = \sqrt{169} = 13$

23.
$$z^4 - 13z^2 + 36 = 0$$
$$(z^2 - 4)(z^2 - 9) = 0$$
$$(z+2)(z-2)(z+3)(z-3) = 0$$
$$z = -2,\ z = 2,\ z = -3 \text{ or } z = 3$$

25.
$$\sqrt{x+5} = 12$$
$$\left(\sqrt{x+5}\right)^2 = 12^2$$
$$x + 5 = 144$$
$$x = 139$$

27.
$$5x - 3 \le 7$$
$$5x \le 10$$
$$x \le 2$$

29.
$$1 + x < 3x - 3 < 4x - 2$$
$$1 + x < 3x - 3 \quad \text{and} \quad 3x - 3 < 4x - 2$$
$$-2x < -4 \qquad\qquad -x < 1$$
$$x > 2 \qquad\qquad\quad x > -1$$

$x > 2$

$x > -1$

$$1 + x < 3x - 3 < 4x - 2$$

31.
$$\left|\frac{3x+2}{2}\right| = 4$$

$$\frac{3x+2}{2} = 4 \quad \text{or} \quad \frac{3x+2}{2} = -4$$
$$3x + 2 = 8 \qquad\qquad 3x + 2 = -8$$
$$3x = 6 \qquad\qquad\quad 3x = -10$$
$$x = 2 \qquad\qquad\quad x = -\frac{10}{3}$$

33.
$$|2x - 5| > 2$$

$2x - 5 > 2$ or $2x - 5 < -2$

$2x > 7$ $2x < 3$

$x > \dfrac{7}{2}$ $x < \dfrac{3}{2}$

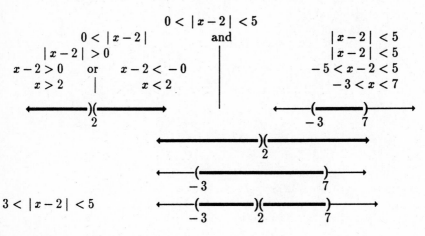

35.
$$0 < |x - 2| < 5$$

$0 < |x - 2|$ and $|x - 2| < 5$

$|x - 2| > 0$ $|x - 2| < 5$

$x - 2 > 0$ or $x - 2 < -0$ $-5 < x - 2 < 5$

$x > 2$ $x < 2$ $-3 < x < 7$

$3 < |x - 2| < 5$

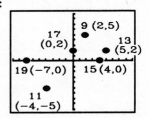

Exercise 3.1   (page 145)

1. $A\,(2, 3)$ **3.** $C\,(-2, -3)$ **5.** $E\,(0, 0)$ **7.** $G\,(-5, -5)$

9 − 19. All points are graphed on this coordinate plane:

9. Quadrant I   **11.** Quadrant III

13. Quadrant I   **15.** x-axis

17. y-axis   **19.** x-axis

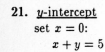

21.   <u>y-intercept</u>   <u>x-intercept</u>

    set $x = 0$:   set $y = 0$:

      $x + y = 5$ $x + y = 5$

      $0 + y = 5$ $x + 0 = 5$

        $y = 5$ $x = 5$

       $(0, 5)$ $(5, 0)$

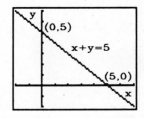

23. y-intercept   x-intercept

$$2x - y = 4 \qquad 2x - y = 4$$
$$2(0) - y = 4 \qquad 2x - 0 = 4$$
$$-y = 4 \qquad 2x = 4$$
$$y = -4 \qquad x = 2$$
$$(0, -4) \qquad (2, 0)$$

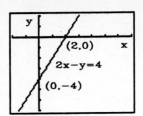

25. y-intercept   x-intercept

$$3x + 2y = 6 \qquad 3x + 2y = 6$$
$$3(0) + 2y = 6 \qquad 3x + 2(0) = 6$$
$$2y = 6 \qquad 3x = 6$$
$$y = 3 \qquad x = 2$$
$$(0, 3) \qquad (2, 0)$$

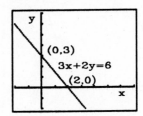

27. y-intercept   x-intercept

$$4x - 5y = 20 \qquad 4x - 5y = 20$$
$$4(0) - 5y = 20 \qquad 4x - 5(0) = 20$$
$$-5y = 20 \qquad 4x = 20$$
$$y = -4 \qquad x = 5$$
$$(0, -4) \qquad (5, 0)$$

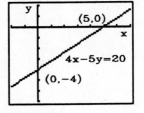

29. y-intercept   x-intercept

$$3x + 4y = 12 \qquad 3x + 4y = 12$$
$$3(0) + 4y = 12 \qquad 3x + 4(0) = 12$$
$$4y = 12 \qquad 3x = 12$$
$$y = 3 \qquad x = 4$$
$$(0, 3) \qquad (4, 0)$$

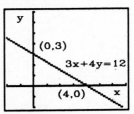

31. y-intercept   x-intercept

$$2x + y = 5 \qquad 2x + y = 5$$
$$2(0) + y = 5 \qquad 2x + 0 = 5$$
$$y = 5 \qquad 2x = 5$$
$$(0, 5) \qquad x = \frac{5}{2} \quad \left(\frac{5}{2}, 0\right)$$

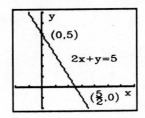

33. $y - 2x = 7$

$\qquad y = 2x + 7$

x	y
-2	3
-1	5

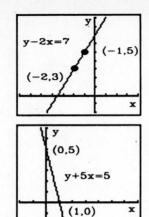

35. $y + 5x = 5$

$\qquad y = -5x + 5$

x	y
0	5
1	0

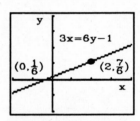

37. $6x - 3y = 10$

$\qquad -3y = -6x + 10$

$\qquad y = 2x - \dfrac{10}{3}$

x	y
0	$-\dfrac{10}{3}$
2	$\dfrac{2}{3}$

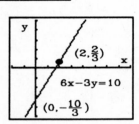

39. $3x = 6y - 1$

$\qquad -6y = -3x - 1$

$\qquad y = \dfrac{1}{2}x + \dfrac{1}{6}$

x	y
0	$\dfrac{1}{6}$
2	$\dfrac{7}{6}$

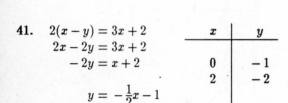

41. $2(x - y) = 3x + 2$

$\quad 2x - 2y = 3x + 2$

$\qquad -2y = x + 2$

$\qquad y = -\dfrac{1}{2}x - 1$

x	y
0	-1
2	-2

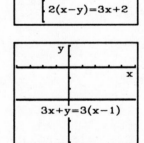

43. $3x + y = 3(x - 1)$

$\quad 3x + y = 3x - 3$

$\qquad y = -3$

(horizontal line with y-coordinate of -3)

45. $y = 3$ {horizontal}

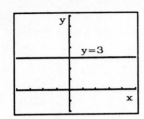

47. $3x + 5 = -1$
$$3x = -6$$
$$x = -2 \quad \text{\{vertical\}}$$

49. $3(y + 2) = y$
$$3y + 6 = y$$
$$2y = -6$$
$$y = -3 \quad \text{\{horizontal\}}$$

51. $3(y + 2x) = 6x + y$
$$3y + 6x = 6x + y$$
$$2y = 0$$
$$y = 0 \quad \text{\{horizontal\}}$$

53. $d = \sqrt{(4-0)^2 + (-3-0)^2}$
$$= \sqrt{16 + 9}$$
$$= 5$$

55. $d = \sqrt{(-3-0)^2 + (2-0)^2}$
$$= \sqrt{9 + 4}$$
$$= \sqrt{13}$$

57. $d = \sqrt{(1-0)^2 + (1-0)^2}$
$$= \sqrt{1 + 1}$$
$$= \sqrt{2}$$

59. $d = \sqrt{(\sqrt{3}-0)^2 + (1-0)^2}$
$$= \sqrt{3 + 1}$$
$$= 2$$

61. $d = \sqrt{(3-6)^2 + (7-3)^2}$
$$= \sqrt{(-3)^2 + 4^2}$$
$$= \sqrt{25}$$
$$= 5$$

63. $d = \sqrt{(4--1)^2 + (-6-6)^2}$
$$= \sqrt{5^2 + (-12)^2}$$
$$= \sqrt{169}$$
$$= 13$$

65. $d = \sqrt{(-2--9)^2 + (-15--39)^2}$
$$= \sqrt{7^2 + 24^2}$$
$$= \sqrt{49 + 576}$$
$$= 25$$

67.
$$d = \sqrt{(3--5)^2 + (-3-5)^2}$$
$$= \sqrt{8^2 + (-8)^2}$$
$$= \sqrt{128}$$
$$= 8\sqrt{2}$$

69.
$$d = \sqrt{(3-5)^2 + (7-7)^2}$$
$$= \sqrt{(-2)^2 + 0^2}$$
$$= \sqrt{4}$$
$$= 2$$

71.
$$d = \sqrt{(\pi-\pi)^2 + (-2-5)^2}$$
$$= \sqrt{0^2 + (-7)^2}$$
$$= \sqrt{49}$$
$$= 7$$

73.
$$(x_m, y_m) = \left(\frac{2+6}{2}, \frac{4+8}{2}\right)$$
$$= \left(\frac{8}{2}, \frac{12}{2}\right)$$
$$= (4, 6)$$

75.
$$(x_m, y_m) = \left(\frac{2+-2}{2}, \frac{-5+7}{2}\right)$$
$$= \left(\frac{0}{2}, \frac{2}{2}\right)$$
$$= (0, 1)$$

77.
$$(x_m, y_m) = \left(\frac{-8+8}{2}, \frac{5+-5}{2}\right)$$
$$= \left(\frac{0}{2}, \frac{0}{2}\right)$$
$$= (0, 0)$$

79.
$$(x_m, y_m) = \left(\frac{0+\sqrt{5}}{2}, \frac{0+\sqrt{5}}{2}\right)$$
$$= \left(\frac{\sqrt{5}}{2}, \frac{\sqrt{5}}{2}\right)$$

81.
$$d = \sqrt{(a-0)^2 + (0-b)^2}$$
$$= \sqrt{a^2 + b^2}$$

83.
$$d = \sqrt{(a-0)^2 + (b-0)^2}$$
$$= \sqrt{a^2 + b^2}$$

85.
$$(x_m, y_m) = \left(\frac{a+0}{2}, \frac{b+0}{2}\right)$$
$$= \left(\frac{a}{2}, \frac{b}{2}\right)$$

87.
$$(x_m, y_m) = \left(\frac{a+b}{2}, \frac{a+b}{2}\right)$$

89. $P(1, 4)$ and $Q(x, y)$

$$x_m = \frac{1+x}{2} \qquad y_m = \frac{4+y}{2}$$
$$3 = \frac{1+x}{2} \qquad 5 = \frac{4+y}{2}$$
$$6 = 1+x \qquad 10 = 4+y$$
$$5 = x \qquad 6 = y$$

Q has coordinates of $(5, 6)$.

91. $P(5, -5)$ and $Q(x, y)$

$$x_m = \frac{5+x}{2} \qquad y_m = \frac{-5+y}{2}$$
$$5 = \frac{5+x}{2} \qquad 5 = \frac{-5+y}{2}$$
$$10 = 5+x \qquad 10 = -5+y$$
$$5 = x \qquad 15 = y$$

Q has coordinates of $(5, 15)$.

93. Call $A(13, -2)$, $B(9, -8)$, $C(5, -2)$.

Then $AB = \sqrt{(13-9)^2 + (-2--8)^2} = \sqrt{4^2 + 6^2} = \sqrt{16+36} = \sqrt{52} = 2\sqrt{13}$

$AC = \sqrt{(13-5)^2 + (-2--2)^2} = \sqrt{8^2 + 0^2} = \sqrt{64} = 8$

$BC = \sqrt{(9-5)^2 + (-8--2)^2} = \sqrt{4^2 + (-6)^2} = \sqrt{16+36} = \sqrt{52} = 2\sqrt{13}$

Since $AB = BC = 2\sqrt{13}$, the triangle is isosceles.

95. Call $A(-3, -1)$, $B(3, 1)$, $C(1, 7)$, $D(-5, 5)$.

$AB = \sqrt{(-3-3)^2 + (-1-1)^2} = \sqrt{(-6)^2 + (-2)^2} = \sqrt{36+4} = \sqrt{40} = 2\sqrt{10}$

$BC = \sqrt{(3-1)^2 + (1-7)^2} = \sqrt{2^2 + (-6)^2} = \sqrt{4+36} = \sqrt{40} = 2\sqrt{10}$

$CD = \sqrt{(1--5)^2 + (7-5)^2} = \sqrt{6^2 + 2^2} = \sqrt{36+4} = \sqrt{40} = 2\sqrt{10}$

$DA = \sqrt{(-5--3)^2 + (5--1)^2} = \sqrt{(-2)^2 + 6^2} = \sqrt{4+36} = \sqrt{40} = 2\sqrt{10}$

Since all four sides have length $2\sqrt{10}$, the figure is a rhombus.

97. The center of a circle is the midpoint of a diameter of the circle. Thus, the center of the circle is the midpoint of the segment between the endpoints of a diameter of a circle.

$$(x_C, y_C) = \left(\frac{-17+9}{2}, \frac{6+-12}{2}\right) = \left(\frac{-8}{2}, \frac{-6}{2}\right) = (-4, -3)$$

99. See the figure below. Since a square is a rhombus, its diagonals bisect each other. Since a square is a rectangle, its diagonals are equal. Thus the distance from the origin to each vertex of the square is the same, say a.

Since the area of the square is 50, the length of a side (s) can be found:

$$s^2 = 50$$
$$s = \sqrt{50}$$

Then the distance between two adjacent vertices {consider $(a, 0)$ and $(0, a)$} must be $\sqrt{50}$:

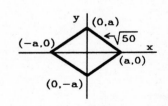

$$d = \sqrt{(a-0)^2 + (0-a)^2}$$

$$\sqrt{50} = \sqrt{a^2 + a^2}$$
$$50 = 2a^2$$
$$25 = a^2$$
$$\pm 5 = a \quad \text{The vertices are } (5, 0), (0, 5), (-5, 0), (0, -5).$$

101. $M(x_M, y_M) = \left(\frac{2+6}{2}, \frac{4+10}{2}\right) = \left(\frac{8}{2}, \frac{14}{2}\right) = (4, 7)$

$N(x_N, y_N) = \left(\frac{4+6}{2}, \frac{10+6}{2}\right) = \left(\frac{10}{2}, \frac{16}{2}\right) = (5, 8)$

$MN = \sqrt{(5-4)^2 + (8-7)^2} = \sqrt{1^2 + 1^2} = \sqrt{2}$

103. $M(x_M, y_M) = \left(\frac{a+0}{2}, \frac{0+b}{2}\right) = \left(\frac{a}{2}, \frac{b}{2}\right).$ Other coordinates can then be found.

L has the same x-coordinate as M: $\; L\left(\frac{a}{2}, 0\right).$

N has the same y-coordinate as M: $\; N\left(0, \frac{b}{2}\right).$

Area$(OLMN) = length \cdot width = OL \cdot ON = \frac{a}{2} \cdot \frac{b}{2} = \frac{ab}{4} = \frac{1}{4}ab.$

Area$(AOB) = \frac{1}{2} \cdot base \cdot height = \frac{1}{2}(a)(b).$

$\frac{1}{2} \cdot$ Area$(AOB) = \frac{1}{2} \cdot \frac{1}{2}ab = \frac{1}{4}ab =$ Area$(OLMN).$

105. Set up a coordinate system with the origin at the home port, the positive x-axis to the east of the port, and the positive y-axis to the north of the port. Then the ship has coordinates $(47 + 23, 84 + 72) = (70, 156)$. The distance from the port to the ship is then the distance from $(0, 0)$ to $(70, 156)$.

$$d = \sqrt{(70-0)^2 + (156-0)^2} = \sqrt{70^2 + 156^2} = \sqrt{4900 + 24336} = \sqrt{29236}$$

$$\approx 171 \text{ miles}$$

Exercise 3.2 (page 154)

1. $m = \frac{5-10}{2-3} = \frac{-5}{-1} = 5$

3. $m = \frac{-2-5}{3--1} = \frac{-7}{4} = -\frac{7}{4}$

5. $m = \frac{-7-1}{8-4} = \frac{-8}{4} = -2$

7. $m = \frac{3--3}{-4--4} = \frac{6}{0}$ undefined slope

9. $m = \dfrac{\frac{2}{3} - \frac{7}{3}}{\frac{3}{2} - \frac{5}{2}} = \dfrac{-\frac{5}{3}}{-\frac{2}{2}} = \dfrac{-\frac{5}{3}}{-1} = \frac{5}{3}$

11. $m = \dfrac{\frac{3}{2} - -\frac{1}{2}}{\frac{3}{7} - \frac{5}{7}} = \dfrac{\frac{4}{2}}{-\frac{2}{7}} = \dfrac{\frac{2}{1}}{-\frac{2}{7}} = \frac{2}{1} \div \left(-\frac{2}{7}\right) = \frac{2}{1} \cdot \left(-\frac{7}{2}\right) = -7$

13. $m = \frac{a-b-0}{a-b} = \frac{a-b}{a-b} = 1$ (as long as $a \neq b$)

15. $m = \dfrac{c-a}{(a+b)-(b+c)} = \dfrac{c-a}{a+b-b-c} = \dfrac{c-a}{a-c} = -1$ (as long as $a \neq c$)

17. $(0, 2)$ and $(1, 5)$ are on the line.

$$m = \frac{2-5}{0-1} = \frac{-3}{-1} = 3$$

19. $\left(0, -\dfrac{3}{10}\right)$ and $\left(\dfrac{3}{5}, 0\right)$ are on the line.

$$m = \frac{-\dfrac{3}{10}-0}{0-\dfrac{3}{5}} = \frac{-\dfrac{3}{10}}{-\dfrac{3}{5}}$$

$$= \left(-\frac{3}{10}\right) \div \left(-\frac{3}{5}\right)$$

$$= \frac{3}{10} \cdot \frac{5}{3}$$

$$= \frac{1}{2}$$

21. $(0, -3)$ and $(3, -1)$ are on the line.

$$m = \frac{-3--1}{0-3} = \frac{-2}{-3} = \frac{2}{3}$$

23. $(0, -2)$ and $(1, -2)$ are on the line.

$$m = \frac{-2--2}{0-1} = \frac{0}{-1} = 0$$

25. $m_1 m_2 = 3\left(-\dfrac{1}{3}\right) = -1$

perpendicular

27. $m_1 = \sqrt{8} = 2\sqrt{2} = m_2$

parallel

29. $m_1 m_2 = -\sqrt{2} \cdot \dfrac{\sqrt{2}}{2} = \dfrac{-2}{2} = -1$

perpendicular

31. $m_1 m_2 = (-0.125)8 = -1$

perpendicular

33. $m_1 m_2 = ab^{-1}(-a^{-1}b) = -aa^{-1}b^{-1}b = -a^0 b^0 = -1$

perpendicular

For **35 – 39**, calculate the slope of $\overleftrightarrow{RS}$: $m = \dfrac{5-7}{-3-2} = \dfrac{-2}{-5} = \dfrac{2}{5}$

35. $m_{PQ} = \dfrac{4-6}{2-7} = \dfrac{-2}{-5} = \dfrac{2}{5}$

parallel

37. $m_{PQ} = \dfrac{6-1}{-4--2} = \dfrac{5}{-2} = -\dfrac{5}{2}$

perpendicular

39. $m_{PQ} = \dfrac{a-6a}{a-3a} = \dfrac{-5a}{-2a} = \dfrac{5}{2}$ neither parallel nor perpendicular

41. $m_{PQ} = \dfrac{8-9}{-2--6} = \dfrac{-1}{4} = -\dfrac{1}{4}$; $m_{PR} = \dfrac{8-5}{-2-2} = \dfrac{3}{-4} = -\dfrac{3}{4}$; not on same line

43. $m_{PQ} = \dfrac{a-0}{-a-0} = \dfrac{a}{-a} = -1$; $m_{PR} = \dfrac{a--a}{-a-a} = \dfrac{2a}{-2a} = -1$; on same line

45. $m_{PQ} = \frac{4 - -5}{5 - 2} = \frac{9}{3} = 3$

$m_{PR} = \frac{4 - -3}{5 - 8} = \frac{7}{-3} = -\frac{7}{3}$

$m_{QR} = \frac{-5 - -3}{2 - 8} = \frac{-2}{-6} = \frac{1}{3}$

No lines are perpendicular.

47. $m_{PQ} = \frac{3 - 9}{1 - 1} = \frac{-6}{0}$ undefined slope

$m_{PR} = \frac{3 - 3}{1 - 7} = \frac{0}{-6} = 0$

$m_{QR} = \frac{9 - 3}{1 - 7} = \frac{6}{-6} = -1$

Lines PQ and PR are perpendicular.

49. $m_{PQ} = \frac{0 - b}{0 - a} = \frac{-b}{-a} = \frac{b}{a}$

$m_{QR} = \frac{b - a}{a - -b} = \frac{b - a}{a + b}$

$m_{PR} = \frac{0 - a}{0 - -b} = \frac{-a}{b} = -\frac{a}{b}$

Lines PQ and PR are perpendicular.

51.

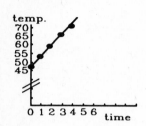

$\frac{\Delta T}{\Delta t}$ represents the change in T in one unit of t, or the hourly temperature change.

53. Let $y =$ distance in miles and $x =$ time in hours.

distance $=$ rate $\times$ time

$$y = 590x$$

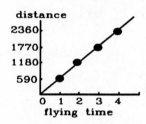

The slope represents the distance traveled in 1 hour.

55. $m_{AB} = \frac{-1 - 4}{-1 - -3} = -\frac{5}{2}$; $m_{AC} = \frac{-1 - 1}{-1 - 4} = \frac{2}{5}$;   Since $-\frac{5}{2} \cdot \frac{2}{5} = -1$, the lines are perpendicular. The triangle is right.

57. To show that the figure is a square, it is sufficient to show that all 4 sides are the same length, **and** that one of the angles is right.

$d(AB) = \sqrt{(1 - 3)^2 + (-1 - 0)^2} = \sqrt{(-2)^2 + (-1)^2} = \sqrt{5}$

$d(BC) = \sqrt{(3 - 2)^2 + (0 - 2)^2} = \sqrt{(1^2) + (-2)^2} = \sqrt{5}$

$d(CD) = \sqrt{(2 - 0)^2 + (2 - 1)^2} = \sqrt{(2)^2 + (1)^2} = \sqrt{5}$

$d(DA) = \sqrt{(0 - 1)^2 + (1 - -1)^2} = \sqrt{(-1)^2 + (2)^2} = \sqrt{5}$

$m_{AB} = \frac{-1 - 0}{1 - 3} = \frac{1}{2}$; $m_{BC} = \frac{0 - 2}{3 - 2} = -2$;   Since $\frac{1}{2} \cdot (-2) = -1$, the lines are perpendicular. The figure is thus a square.

59. To show that the figure is a parallelogram, it is sufficient to show that both pairs of opposite sides are parallel ($AB \parallel CD$ and $BC \parallel DA$).

$$m_{AB} = \frac{-2-3}{-2-3} = 1; \quad m_{BC} = \frac{3-6}{3-2} = -3; \quad m_{CD} = \frac{6-1}{2--3} = 1;$$

$$m_{DA} = \frac{1--2}{-3--2} = -3. \text{ Since } m_{AB} = m_{CD}, \; AB \parallel CD. \text{ Since } m_{BC} = m_{DA},$$

$BC \parallel DA$. So both pairs of opposite sides are parallel.

61. $x_M = \frac{5+7}{2} = 6, \; y_M = \frac{9+5}{2} = 7.$ M has coordinates $(6, 7)$.

$x_N = \frac{1+7}{2} = 4, \; y_N = \frac{3+5}{2} = 4.$ N has coordinates $(4, 4)$.

$$m_{MN} = \frac{7-4}{6-4} = \frac{3}{2}; \quad m_{AC} = \frac{9-3}{5-1} = \frac{3}{2}$$

Since the slopes are equal, the lines are parallel.

63. If both slopes are 0, then both lines have changes in $y = 0$, so both are horizontal, and are certainly parallel. Similarly, if both slopes are undefined, the lines are both vertical and then parallel.

If the equal slopes are not 0 or undefined, then that equal slope can be written as $\frac{\Delta y}{\Delta x}$, where $\Delta y \neq 0$ and $\Delta x \neq 0$. Call the lines L and M and draw triangles ABC and DEF as indicated in the below drawing. Obviously, by side-angle-side, ABC is congruent to DEF. Then $\angle CAB \cong \angle FDE$. Thus L and M have the same angle of inclination and are parallel.

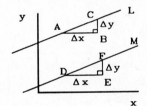

Exercise 3.3   (page 163)

1.
$$y - y_1 = m(x - x_1)$$
$$y - 4 = 2(x - 2)$$

3.
$$y - y_1 = m(x - x_1)$$
$$y - \frac{1}{2} = 2\left(x - -\frac{3}{2}\right)$$
$$y - \frac{1}{2} = 2\left(x + \frac{3}{2}\right)$$

5.
$$y - y_1 = m(x - x_1)$$
$$y - -5 = -5(x - 5)$$

7.
$$y - y_1 = m(x - x_1)$$
$$y - 0 = \pi(x - \pi)$$
$$y = \pi(x - \pi)$$

9.
$$y - y_1 = m(x - x_1)$$
$$y - 4 = 3(x - 1)$$
$$y - 4 = 3x - 3$$
$$y = 3x + 1$$

11.
$$y - y_1 = m(x - x_1)$$
$$y - \frac{3}{2} = -2\left(x - \frac{5}{2}\right)$$
$$y - \frac{3}{2} = -2x + \frac{10}{2}$$
$$y = -2x + \frac{13}{2}$$

13.
$$y - y_1 = m(x - x_1)$$
$$y - 8 = -4(x - 0)$$
$$y - 8 = -4x + 0$$
$$y = -4x + 8$$

15.
$$y - y_1 = m(x - x_1)$$
$$y - 0 = \frac{1}{2}(x - 0)$$
$$y = \frac{1}{2}x$$

17.
$$y = mx + b$$
$$y = 3x - 2$$

19.
$$y = mx + b$$
$$y = 5x - \frac{1}{5}$$

21.
$$y = mx + b$$
$$y = ax + \frac{1}{a}$$

23.
$$y = mx + b$$
$$y = ax + a$$

25. Substitute $(0, 0)$ for (x, y).
$$y = mx + b$$
$$0 = \frac{3}{2}(0) + b$$
$$0 = b$$
$$y = \frac{3}{2}x + 0$$
$$2y = 3x$$
$$3x - 2y = 0$$

27. Substitute $(-3, 5)$ for (x, y).
$$y = mx + b$$
$$5 = -3(-3) + b$$
$$5 = 9 + b$$
$$-4 = b$$
$$y = -3x - 4$$
$$3x + y = -4$$

29. Substitute $(0, \sqrt{2})$ for (x, y).
$$y = mx + b$$
$$\sqrt{2} = \sqrt{2}(0) + b$$
$$\sqrt{2} = b$$
$$y = \sqrt{2}\, x + \sqrt{2}$$
$$-\sqrt{2}\, x + y = \sqrt{2}$$
$$\sqrt{2}\, x - y = -\sqrt{2}$$

31. Substitute (r, s) for (x, y).
$$y = mx + b$$
$$s = \frac{s}{r}(r) + b$$
$$0 = b$$
$$y = \frac{s}{r}x$$
$$-\frac{s}{r}x + y = 0$$
$$\frac{s}{r}x - y = 0$$
$$sx - ry = 0$$

33. $m_{PQ} = \dfrac{2 - 3}{3 - 2} = -1$
$$y - y_1 = m(x - x_1)$$
$$y - 2 = -1(x - 3)$$
$$y - 2 = -x + 3$$
$$x + y = 5$$

35. $m_{PQ} = \dfrac{-2 - -8}{3 - 4} = -6$
$$y - y_1 = m(x - x_1)$$
$$y - -2 = -6(x - 3)$$
$$y + 2 = -6x + 18$$
$$6x + y = 16$$

37. $m_{PQ} = \dfrac{0-13}{0--9} = -\dfrac{13}{9}$

$$y - y_1 = m(x - x_1)$$
$$y - 0 = -\frac{13}{9}(x - 0)$$
$$y = -\frac{13}{9}x$$
$$9y = -13x$$
$$13x + 9y = 0$$

39. $m_{PQ} = \dfrac{\frac{1}{3} - -\frac{1}{3}}{\frac{1}{2} - \frac{3}{2}} = \dfrac{\frac{2}{3}}{-\frac{2}{2}} = -\dfrac{2}{3}$

$$y - y_1 = m(x - x_1)$$
$$y - \frac{1}{3} = -\frac{2}{3}\left(x - \frac{1}{2}\right)$$
$$3\left(y - \frac{1}{3}\right) = 3\left[-\frac{2}{3}\left(x - \frac{1}{2}\right)\right]$$
$$3y - 1 = -2\left(x - \frac{1}{2}\right)$$
$$3y - 1 = -2x + 1$$
$$2x + 3y = 2$$

41.
$$y = -7x + 12$$
$$y = mx + b$$
$$\text{slope} = m = -7$$
$$y\text{-intercept} = b = 12$$

43.
$$2y = 7x - 3$$
$$y = \frac{7}{2}x - \frac{3}{2}$$
$$y = mx + b$$
$$\text{slope} = \frac{7}{2}; \ y\text{-intercept} = -\frac{3}{2}$$

45.
$$3(x - 1) = y$$
$$3x - 3 = y$$
$$y = 3x - 3$$
$$y = mx + b$$
$$\text{slope} = m = 3$$
$$y\text{-intercept} = b = -3$$

47.
$$2(y - 3) + x = 2y - 7$$
$$2y - 6 + x = 2y - 7$$
$$-6 + x = -7$$
$$x = -1$$
This is a vertical line, so there is an undefined slope and no y-intercept.

49. $m_{PQ} = \dfrac{5 - -5}{-3 - 7} = -1$

51. First, find the slope of the line
$$y = 3(x - 7) = 3x - 21$$
This line has a slope of 3, and so does any parallel line.

53. First, find the slope of the line
$$y + 3x = 8$$
$$y = -3x + 8$$
This line's slope is -3, so any perpendicular line has a slope of $\frac{1}{3}$.

55. The x-axis is horizontal, and therefore has a slope of 0. Any parallel line also has a slope of 0.

57. First, find the slope of line PQ.
$$m = \frac{-5 - 5}{3 - -3} = \frac{-10}{6} = -\frac{5}{3}$$
A perpendicular line would then have a slope of $\frac{3}{5}$.

59.
$$y - y_1 = m(x - x_1)$$
$$y - 4 = \frac{1}{2}(x - 3)$$
$$2y - 8 = x - 3$$
$$x - 2y = -5$$

61. $m = \frac{4 - -4}{-2 - 5} = -\frac{8}{7}$

$$y - y_1 = m(x - x_1)$$
$$y - 4 = -\frac{8}{7}(x - -2)$$
$$7y - 28 = -8(x + 2)$$
$$7y - 28 = -8x - 16$$
$$8x + 7y = 12$$

63. $m = \frac{0 - 11}{0 - -2} = -\frac{11}{2}$

$$y - y_1 = m(x - x_1)$$
$$y - 0 = -\frac{11}{2}(x - 0)$$
$$2y = -11x$$
$$11x + 2y = 0$$

65. y-intercept of $7 \Rightarrow (0, 7)$ on line.

$$m = \frac{7 - 4}{0 - -3} = 1$$

$$y - y_1 = m(x - x_1)$$
$$y - 7 = 1(x - 0)$$
$$x - y = -7$$

67.
$$y = mx + b$$
$$y = -\frac{2}{3}x + 10$$
$$3y = -2x + 30$$
$$2x + 3y = 30$$

69. If the line is parallel to the x-axis then it has a constant y coordinate. That y coordinate must be the same as the y coordinate of the point the line goes through. The equation is
$$y = -5$$

71. The slope of the line $y = 3x - 17$ is 3, and so the slope of any parallel is also 3.
$$y - y_1 = m(x - x_1)$$
$$y - -5 = 3(x - 0)$$
$$3x - y = 5$$

73. If the line is parallel to the x-axis, then it has a constant x coordinate. Since the x-intercept is -4, the x-coordinate must then be -4. The equation is $x = -4$.

75. The line desired has the same slope as the line $3y - 5x = 0$ and goes through the x-intercept of the line $5x - 2y = 15$.

$$3y - 5x = 0$$
$$3y = 5x$$

$$y = \frac{5}{3}x$$

The slope is $\frac{5}{3}$.

$$5x - 2y = 15$$
$$5x - 2(0) = 15$$
$$5x = 15$$
$$x = 3$$
$$(3, 0)$$

$$y - y_1 = m(x - x_1)$$

$$y - 0 = \frac{5}{3}(x - 3)$$

$$3y = 5(x - 3)$$
$$5x - 3y = 15$$

77. Find the slope of the line
$$3x + 2y = 6$$
$$y = -\frac{3}{2}x + 3$$

This line has a slope of $-\frac{3}{2}$.

The desired line has a slope of $-\frac{3}{2}$.
$$y - y_1 = m(x - x_1)$$
$$y - 0 = -\frac{3}{2}(x - 0)$$

$$y = -\frac{3}{2}x$$

79. Find the slope of the line
$$3x + 5y = 25$$
$$y = -\frac{3}{5}x + 5$$

This line has a slope of $-\frac{3}{5}$.

The desired line has a slope of $\frac{5}{3}$.
$$y - y_1 = m(x - x_1)$$
$$y - 3 = \frac{5}{3}(x - -2)$$

$$y = \frac{5}{3}x + \frac{19}{3}$$

81. $m = \dfrac{s-0}{r-0} = \dfrac{s}{r}$

$$y - y_1 = m(x - x_1)$$
$$y - 0 = \tfrac{s}{r}(x - 0)$$
$$y = \tfrac{s}{r}x + 0$$
$$y = \tfrac{s}{r}x$$

83. Let $x =$ the Fahrenheit temperature and $y =$ the Celsius temperature. Two points: $(32, 0)$ and $(212, 100)$.

$$m = \frac{100 - 0}{212 - 32} = \frac{100}{180} = \frac{5}{9}$$
$$y - y_1 = m(x - x_1)$$
$$y - 0 = \tfrac{5}{9}(x - 32)$$
$$y = \tfrac{5}{9}(x - 32) \quad \text{or}$$
$$C = \tfrac{5}{9}(F - 32)$$

85. Let T replace y on the coordinate plane, and also in the equation of a line. We then have ordered pairs of the form (x, T), specifically in this problem $(12{,}000, 30)$ and $(30{,}000, -30)$. Find the equation of the line through these two points.

$$m = \frac{30 - -30}{12{,}000 - 30{,}000} = -\frac{1}{300}$$
$$T - T_1 = m(x - x_1)$$
$$T - 30 = -\frac{1}{300}(x - 12{,}000)$$
$$T = -\frac{1}{300}x + 70$$

At sea level, $x = 0$:

$$T = -\frac{1}{300}(0) + 70 = 70$$

At sea level, the temperature is $70°$.

87. Certainly zero U.S. gallons is equal to zero liters. Let $x =$ U.S. gallons and $y =$ liters. Then find the equation of the line through the ordered pairs $(0, 0)$ and $(1, 3.785)$.

$$m = \frac{3.785 - 0}{1 - 0} = 3.785$$
$$y - y_1 = m(x - x_1)$$
$$y - 0 = 3.785(x - 0)$$
$$y = 3.785x$$
$$\# \text{ liters} = 3.785 \ (\# \text{ U.S. gallons})$$

89. When $x = 0$, $y = 300$. After 1 week $x = 1$ and $y = 291$. Use these two ordered pairs to find the equation of a line:

$$m = \frac{300 - 291}{0 - 1} = -9$$
$$y - y_1 = m(x - x_1)$$
$$y - 300 = -9(x - 0)$$
$$y - 300 = -9x$$
$$y = -9x + 300$$

91. Let $y =$ the value of the bed and $x =$ the number of years from now. Two points: $(3, 2750)$ and $(7, 3298)$.

$$m = \frac{3298 - 2750}{7 - 3} = 137$$
$$y - y_1 = m(x - x_1)$$
$$y - 2750 = 137(x - 3)$$
$$y - 2750 = 137x - 411$$
$$y = 137x + 2339$$

Its value now can be found with $x = 0$.

$$y = 137(0) + 2339$$
$$y = 2339$$

The value now is \$2339.

93. Let January $= 1$, February $= 2$, March $= 3$ and April $= 4$. Then let $x =$ the month and $y =$ the inventory. There are then two ordered pairs: $(1, 375)$ and $(4, 264)$. Find the line.

$$m = \frac{375 - 264}{1 - 4} = -37$$

$$y - y_1 = m(x - x_1)$$
$$y - 375 = -37(x - 1)$$
$$y - 375 = -37x + 37$$
$$y = -37x + 412$$

The inventory in March can be found by letting $x = 3$.

$$y = -37x + 412$$

$$y = -37(3) + 412$$

$$y = -111 + 412$$
$$y = 301$$

The inventory will be 301 units in March.

95. Let $y =$ the number of barrels and $x =$ the number of years. We then can find two ordered pairs: $(0, 1900)$ and $(1, 1830)$. Find the line.

$$m = \frac{1900 - 1830}{0 - 1} = -70$$
$$y - y_1 = m(x - x_1)$$

$$y - 1900 = -70(x - 0)$$
$$y - 1900 = -70x$$
$$y = -70x + 1900$$

The production after $3\frac{1}{2}$ years can be found by letting $x = 3\frac{1}{2}$

$$y = -70x + 1900$$

$$y = -70\left(\frac{7}{2}\right) + 1900$$

$$y = -\frac{490}{2} + 1900$$

$$y = -245 + 1900 = 1655$$

The production will be 1655 barrels.

97. Let $x =$ the number of years the piping has been used and $y =$ the value of the piping. Then there are two ordered pairs: $(0, 137{,}000)$ and $(12, -33{,}000)$. Find the line.

$$m = \frac{137{,}000 - -33{,}000}{0 - 12}$$

$$= -\frac{42{,}500}{3}$$

$$y - y_1 = m(x - x_1)$$

$$y - 137{,}000 = -\frac{42{,}500}{3}(x - 0)$$

$$y = -\frac{42{,}500}{3}x + 137{,}000$$

99. If the x-intercept is a, then that is the point $(a, 0)$. Similarly, the y-intercept is the point $(0, b)$. Then find an equation for the line through the points.

$$m = \frac{0 - b}{a - 0} = -\frac{b}{a}$$

$$y - y_1 = m(x - x_1)$$
$$y - 0 = -\frac{b}{a}(x - a)$$

$$ay = -b(x - a)$$
$$ay = -bx + ab$$
$$bx + ay = ab$$
$$\frac{bx}{ab} + \frac{ay}{ab} = \frac{ab}{ab}$$

$$\frac{x}{a} + \frac{y}{b} = 1$$

101. If $B = 0$, then:
$$Ax + By = C$$
$$Ax + (0)y = C$$
$$Ax = C$$

$$x = \frac{C}{A}$$

This is the equation of a vertical line with a constant x-coordinate

of $\frac{C}{A}$, and thus with a x-intercept

which is also $\frac{C}{A}$.

103. Find the slope of
$$Ax + By = C$$
$$By = -Ax + C$$

$$y = -\frac{A}{B}x + \frac{C}{B}$$

The slope of this line is $-\frac{A}{B}$.

Find the slope of
$$Bx - Ay = C$$
$$-Ay = -Bx + C$$

$$y = \frac{B}{A}x - \frac{C}{A}$$

The slope of this line is $\frac{B}{A}$. Since the product of the slopes is -1, the lines are perpendicular.

105. The x-intercept has a y-coordinate of 0.
$$y = mx + b$$
$$0 = mx + b$$
$$-mx = b$$

$$x = -\frac{b}{m}$$

The x-intercept is $-\frac{b}{m}$.

107. The x-axis is a horizontal line with a constant y-coordinate of 0. Thus the equation is $y = 0$.

Exercise 3.4 (page 175)

1. x-intercept(s): set $y = 0$
$$y = x^2 - 4$$
$$0 = x^2 - 4$$
$$4 = x^2$$
$$\pm 2 = x$$
$(2, 0)$ and $(-2, 0)$ are the x-intercepts.

y-intercept(s): set $x = 0$
$$y = x^2 - 4$$
$$y = 0^2 - 4$$
$$y = -4$$
$(0, -4)$ is the y-intercept.

3. x-intercept(s): set $y = 0$
$$y = 4x^2 - 2x$$
$$0 = 4x^2 - 2x$$
$$0 = 2x(2x - 1)$$

$2x = 0$ or $2x - 1 = 0$
$x = 0$ $2x = 1$

$$x = \frac{1}{2}$$

$(0, 0)$ and $\left(\frac{1}{2}, 0\right)$ are the x-intercepts.

y-intercept(s): set $x = 0$
$$y = 4x^2 - 2x$$
$$y = 4(0)^2 - 2(0)$$
$$y = 0$$
$(0, 0)$ is the y-intercept.

5. x-intercept(s): set $y = 0$

$$y = x^2 - 4x - 5$$
$$0 = x^2 - 4x - 5$$
$$0 = (x - 5)(x + 1)$$

$x = 5$ $x = -1$

$(5, 0)$ and $(-1, 0)$ are the x-intercepts.

y-intercept(s): set $x = 0$

$$y = x^2 - 4x - 5$$
$$y = 0^2 - 4(0) - 5$$
$$y = -5$$

$(0, -5)$ is the y-intercept.

7. x-intercept(s): set $y = 0$

$$y = x^2 + x - 2$$
$$0 = x^2 + x - 2$$
$$0 = (x + 2)(x - 1)$$

$x = -2$ $x = 1$

$(-2, 0)$ and $(1, 0)$ are the x-intercepts.

y-intercept(s): set $x = 0$

$$y = x^2 + x - 2$$
$$y = 0^2 + 0 - 2$$
$$y = -2$$

$(0, -2)$ is the y-intercept.

9. x-intercept(s): set $y = 0$

$$y = x^3 - 9x$$
$$0 = x^3 - 9x$$
$$0 = x(x + 3)(x - 3)$$

$x = 0$ or $x = -3$ or $x = 3$

$(0, 0)$, $(-3, 0)$ and $(3, 0)$

y-intercept(s): set $x = 0$

$$y = x^3 - 9x$$
$$y = 0^3 - 9(0)$$
$$y = 0$$

$(0, 0)$ is the y-intercept.

11. x-intercept(s): set $y = 0$

$$y = x^4 - 1$$
$$0 = x^4 - 1$$
$$y = (x^2 + 1)(x + 1)(x - 1)$$

$x = -1$ or $x = 1$. $(-1, 0)$ and $(1, 0)$

y-intercept(s): set $x = 0$

$$y = x^4 - 1$$
$$y = 0^4 - 1$$
$$y = -1$$

$(0, -1)$ is the y-intercept.

13. x-intercept: $(0, 0)$
y-intercept: $(0, 0)$

15. x-intercepts: $(\sqrt{2}, 0)$, $(-\sqrt{2}, 0)$
y-intercept: $(0, 2)$

17. x-intercepts: $(4, 0)$, $(0, 0)$
y-intercept: $(0, 0)$

19. x-intercepts: $(4, 0)$, $(0, 0)$
y-intercept: $(0, 0)$

21. $y = x^2 + 2$

about *x*-axis

$-y = x^2 + 2$
$y = -x^2 - 2$
not identical to original
NO SYMMETRY

about origin

$-y = (-x)^2 + 2$
$y = -x^2 - 2$
not identical to original
NO SYMMETRY

about *y*-axis

$y = (-x)^2 + 2$
$y = x^2 + 2$
identical to original
SYMMETRY

23. $y^2 + 1 = x$

about *x*-axis

$(-y)^2 + 1 = x$
$y^2 + 1 = x$
identical to original
SYMMETRY

about origin

$(-y)^2 + 1 = -x$
$-y^2 - 1 = x$
not identical to original
NO SYMMETRY

about *y*-axis

$y^2 + 1 = -x$
$-y^2 - 1 = x$
not identical to original
NO SYMMETRY

25. $y^2 = x^2$

about *x*-axis

$(-y)^2 = x^2$
$y^2 = x^2$
identical to original
SYMMETRY

about origin

$(-y)^2 = (-x)^2$
$y^2 = x^2$
identical to original
SYMMETRY

about *y*-axis

$y^2 = (-x)^2$
$y^2 = x^2$
identical to original
SYMMETRY

27. $y = 3x^2 + 7$

about *x*-axis

$-y = 3x^2 + 7$
$y = -3x^2 - 7$
not identical to original
NO SYMMETRY

about origin

$-y = 3(-x)^2 + 7$
$y = -3x^2 - 7$
not identical to original
NO SYMMETRY

about *y*-axis

$y = 3(-x)^2 + 7$
$y = 3x^2 + 7$
identical to original
SYMMETRY

29. $y = 3x^3 + 7$

about *x*-axis

$-y = 3x^3 + 7$
$y = -3x^3 - 7$
not identical to origin
NO SYMMETRY

about origin

$-y = 3(-x)^3 + 7$
$-y = -3x^3 + 7$
not identical to original
NO SYMMETRY

about *y*-axis

$y = 3(-x)^3 + 7$
$y = -3x^3 + 7$
not identical to original
NO SYMMETRY

31. $y^2 = 3x$

about *x*-axis

$(-y)^2 = 3x$
$y^2 = 3x$
identical to original
SYMMETRY

about origin

$(-y)^2 = 3(-x)$
$y^2 = -3x$
not identical to original
NO SYMMETRY

about *y*-axis

$y^2 = 3(-x)$
$y^2 = -3x$
not identical to original
NO SYMMETRY

33. $y = |x|$

about x-axis	about origin	about y-axis
$-y = \|x\|$	$-y = \|-x\|$	$y = \|-x\|$
not identical to original	$-y = \|x\|$	$y = \|x\|$
NO SYMMETRY	not identical to original	identical to original
	NO SYMMETRY	SYMMETRY

35. $|y| = x$

about x-axis	about origin	about y-axis
$\|-y\| = x$	$\|-y\| = -x$	$\|y\| = -x$
$\|y\| = x$	$\|y\| = -x$	not identical to original
identical to original	not identical to original	NO SYMMETRY
SYMMETRY	NO SYMMETRY	

37. x-intercepts: $(0, 0)$, $(4, 0)$
y-intercept: $(0, 0)$
symmetry: none

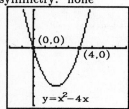

39. x-intercept: $(0, 0)$
y-intercept: $(0, 0)$
symmetry: about origin

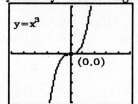

41. x-intercept: $(2, 0)$
y-intercept: $(0, 2)$
symmetry: none

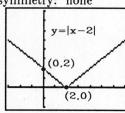

43. x-intercepts: $(3, 0)$, $(-3, 0)$
y-intercept: $(0, 3)$
symmetry: about y-axis

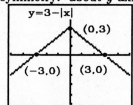

45. x-intercept: $(0, 0)$
y-intercept: $(0, 0)$
symmetry: about x-axis

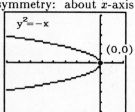

47. x-intercept: $(0, 0)$
y-intercept: $(0, 0)$
symmetry: about x-axis

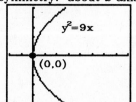

49. x-intercept: $(1, 0)$
y-intercept: $(0, -1)$
symmetry: none

(graph: $y=\sqrt{x}-1$, points $(1,0)$ and $(0,-1)$)

51. x-intercept: none
y-intercept: none
symmetry: about origin

(graph: $xy=4$, points $(2,2)$ and $(-2,-2)$)

53. $(h, k) = (0, 0); \quad r = 1$
$$(x - h)^2 + (y - k)^2 = r^2$$
$$(x - 0)^2 + (y - 0)^2 = 1^2$$
$$x^2 + y^2 = 1$$
$$x^2 + y^2 - 1 = 0$$

55. $(h, k) = (6, 8); \quad r = 4$
$$(x - h)^2 + (y - k)^2 = r^2$$
$$(x - 6)^2 + (y - 8)^2 = 4^2$$
$$x^2 - 12x + 36 + y^2 - 16y + 64 = 16$$
$$x^2 + y^2 - 12x - 16y + 84 = 0$$

57. $(h, k) = (3, -4); \quad r = \sqrt{2}$
$$(x - h)^2 + (y - k)^2 = r^2$$
$$(x - 3)^2 + (y + 4)^2 = (\sqrt{2})^2$$
$$x^2 - 6x + 9 + y^2 + 8y + 16 = 2$$
$$x^2 + y^2 - 6x + 8y + 23 = 0$$

59. The center is the midpoint of the ends of a diameter.

$$h = \frac{3+3}{2} = 3; \quad k = \frac{-2+8}{2} = 3$$

The radius is the distance from the center to an end of a diameter.

$$r = \sqrt{(3-3)^2 + (3--2)^2} = 5$$
$$(x - h)^2 + (y - k)^2 = r^2$$
$$(x - 3)^2 + (y - 3)^2 = 5^2$$
$$x^2 - 6x + 9 + y^2 - 6y + 9 = 25$$
$$x^2 + y^2 - 6x - 6y - 7 = 0$$

61. The radius is the distance from the center to a point on the circle (the origin).

$$r = \sqrt{(-3-0)^2 + (4-0)^2} = \sqrt{9+16} = 5 \qquad (h, k) = (-3, 4)$$
$$(x - h)^2 + (y - k)^2 = r^2$$
$$(x + 3)^2 + (y - 4)^2 = 5^2$$
$$x^2 + 6x + 9 + y^2 - 8y + 16 = 25$$
$$x^2 + y^2 + 6x - 8y = 0$$

63.
$$x^2 + y^2 - 25 = 0$$
$$x^2 + y^2 = 25$$
$$(x - 0)^2 + (y - 0)^2 = 5^2$$
$$(h, k) = (0, 0); \quad r = 5$$

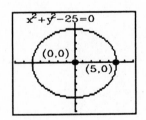

115

65. $(x-1)^2 + (y+2)^2 = 4$

$\quad\quad (x-1)^2 + (y+2)^2 = 2^2$

$\quad\quad (h,\,k) = (1,\,-2); \quad r = 2$

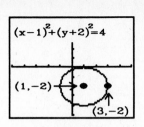

67. $x^2 + y^2 + 2x - 24 = 0$

$\quad\quad x^2 + 2x + y^2 = 24$

$\quad\quad x^2 + 2x + 1 + y^2 = 24 + 1$

$\quad\quad (x+1)^2 + (y-0)^2 = 25$

$\quad\quad (h,\,k) = (-1,\,0); \quad r = \sqrt{25} = 5$

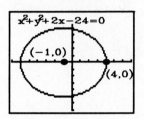

69. $9x^2 + 9y^2 - 12y = 5$

$\quad\quad x^2 + y^2 - \dfrac{12}{9}y = \dfrac{5}{9}$

$\quad\quad x^2 + y^2 - \dfrac{4}{3}y = \dfrac{5}{9}$

$\quad\quad x^2 + y^2 - \dfrac{4}{3}y + \dfrac{4}{9} = \dfrac{5}{9} + \dfrac{4}{9}$

$\quad\quad (x-0)^2 + \left(y - \dfrac{2}{3}\right)^2 = 1$

$\quad\quad (h,\,k) = \left(0,\,\dfrac{2}{3}\right); \quad r = 1$

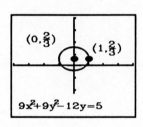

71. $4x^2 + 4y^2 - 4x + 8y + 1 = 0$

$\quad\quad x^2 + y^2 - x + 2y = -\dfrac{1}{4}$

$\quad\quad x^2 - x + y^2 + 2y = -\dfrac{1}{4}$

$\quad\quad x^2 - x + \dfrac{1}{4} + y^2 + 2y + 1 = -\dfrac{1}{4} + \dfrac{1}{4} +$

$\quad\quad \left(x - \dfrac{1}{2}\right)^2 + (y+1)^2 = 1$

$\quad\quad (h,\,k) = \left(\dfrac{1}{2},\,-1\right); \quad r = 1$

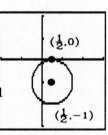

Exercise 3.5 (page 181)

1. $y = x^2 + 1$

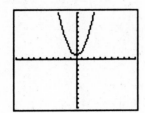

3. $y = 3 - x^2$

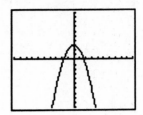

5. $y = x^3 - x$

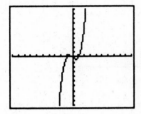

7. $y = \dfrac{x^2 - x}{5}$

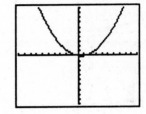

9.

```
RANGE
Xmin=-30
Xmax=30
Xscl=5
Ymin=-30
Ymax=30
Yscl=5
Xres=1
```

$y = x^2 - 10x$

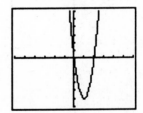

11.

```
RANGE
Xmin=-15
Xmax=15
Xscl=2
Ymin=-3
Ymax=27
Yscl=2
Xres=1
```

$x = 9 + x^2$

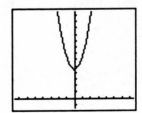

13. $y = |x|$

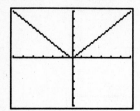

15. $y = |x| - 3$

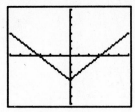

17. $y^2 = x$
  $y = \pm \sqrt{x}$

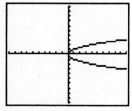

19. $x^2 - y^2 = 4$
  $y = \pm \sqrt{x^2 - 4}$

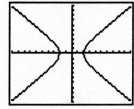

21. $y = 2x^2 - x + 1$
  vertex $(0.25, 0.88)$

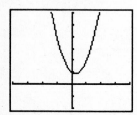

23. $y = 7 + x - x^2$
  vertex $(0.50, 7.25)$

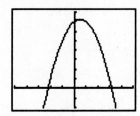

25. Graph $y = x^2 - 7$, and find any points with a y-coordinate of 0.
  x-intercepts $(\pm 2.65, 0)$
  Solutions: $x \approx \pm 2.65$

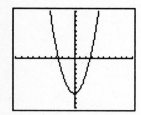

27. Graph $y = x^3 - 3$, and find any points with a y-coordinate of 0.
  x-intercept: $(1.44, 0)$
  Solution: $x \approx 1.44$

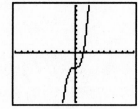

29. Graph $y = x^2 + x - 6$, and find all points where y is less than 0.

Solution: $-3 < x < 2$, or $x \in (-3, 2)$.

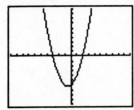

Exercise 3.6 (page 187)

1. $\frac{4}{x} = \frac{2}{7}$

$28 = 2x$

$14 = x$

3. $\frac{x}{2} = \frac{3}{x+1}$

$x(x+1) = 6$

$x^2 + x - 6 = 0$

$(x+3)(x-2) = 0$

$x = -3 \ \text{or} \ x = 2$

5. Let $w = $ the number of women.

$\frac{3}{5} = \frac{w}{30}$

$90 = 5w$

$18 = w$

There are 18 women in the class.

7. $y = kx$

$15 = k(30)$

$\frac{15}{30} = k$

$k = \frac{1}{2}$

9. $I = \frac{k}{R}$

$50 = \frac{k}{20}$

$1000 = k$

11. $E = kIR$

$125 = k(5)(25)$

$125 = 125k$

$1 = k$

13. $y = kx$

$15 = k(4)$

$\frac{15}{4} = k$

$y = \frac{15}{4}x$

$y = \frac{15}{4} \cdot \frac{7}{5}$

$y = \frac{21}{4}$

15. $P = krs$

$16 = k(5)(-8)$

$-\frac{16}{40} = k$

$k = -\frac{2}{5}$

$P = -\frac{2}{5}rs$

$P = -\frac{2}{5}(2)(10)$

$P = -8$

17. direct variation

19. neither type of variation

119

21.

$$V = k\frac{T}{P}$$

$$20 = k\frac{330}{40}$$

$$800 = 330k$$

$$\frac{800}{330} = k$$

$$V = \frac{80}{33} \cdot \frac{T}{P}$$

$$V = \frac{80}{33} \cdot \frac{300}{50}$$

$$V = \frac{160}{11} = 14\frac{6}{11}$$

The volume will be $14\frac{6}{11}$ cubic feet.

23.

$$d = kt^2$$

$$16 = k \cdot 1^2$$

$$16 = k$$

$$d = 16t^2$$

$$144 = 16t^2$$

$$9 = t^2$$

$$\pm 3 = t$$

It will take 3 seconds (throw out the negative answer).

25.

$$P = k\frac{V^2}{R}$$

$$20 = k\frac{20^2}{20}$$

$$400 = 400k$$

$$1 = k$$

$$P = \frac{V^2}{R}$$

$$40 = \frac{V^2}{10}$$

$$400 = V^2$$

$$\pm 20 = V$$

The voltage would be 20 volts.

27.

$$F = k\sqrt{T}$$

$$144 = k\sqrt{2}$$

$$\frac{144}{\sqrt{2}} = k$$

$$F = \frac{144}{\sqrt{2}} \cdot \sqrt{T}$$

$$F = \frac{144}{\sqrt{2}} \cdot \sqrt{18}$$

$$F = 144\sqrt{9}$$

$$F = 432$$

The frequency would be 432 hertz.

29.

$$A = kl^2$$

$$\frac{A}{l^2} = k$$

To find k then, we need to find the area of an equilateral triangle with sides of a given length. Consider an equilateral triangle with sides of length 2, like the one pictured to the right. From geometry, $AX = 1$ and $BX = \sqrt{3}$. To find the area of $\triangle ABC$:

$$A = \tfrac{1}{2}bh = \tfrac{1}{2}(2)(\sqrt{3}) = \sqrt{3}$$

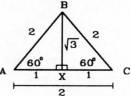

$$A = kl^2$$

$$\sqrt{3} = k(2)^2$$

$$\boxed{\frac{\sqrt{3}}{4} = k}$$

Chapter 3 Review Exercises (page 189)

1. $A\,(2, 0)$

3. $C\,(0, -1)$

5. Quadrant II

7. y-axis

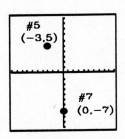

9. x-intercept $(5, 0)$
y-intercept $(0, -3)$

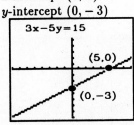

11. x-intercept $(-7, 0)$
y-intercept $(0, -7)$

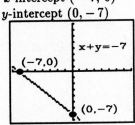

13. $d = \sqrt{(-3-3)^2 + (7 - -1)^2} = \sqrt{(-6)^2 + 8^2} = \sqrt{100} = 10$

$x_M = \dfrac{-3+3}{2} = 0; \quad y_M = \dfrac{7+ -1}{2} = 3; \quad$ midpoint $M\,(0, 3)$

15. $d = \sqrt{(\sqrt{3} - \sqrt{3})^2 + (9-7)^2} = \sqrt{0^2 + 2^2} = 2$

$x_M = \dfrac{\sqrt{3} + \sqrt{3}}{2} = \sqrt{3}; \quad y_M = \dfrac{9+7}{2} = 8; \quad$ midpoint $M\,(\sqrt{3}, 8)$

17. $m = \dfrac{-5-7}{3-1} = \dfrac{-12}{2} = -6$

19. $m = \dfrac{a-b}{b-a} = -1$

21. $m = \dfrac{0-7}{0 - -5} = -\dfrac{7}{5}$

$y - y_1 = m(x - x_1)$

$y - 0 = -\dfrac{7}{5}(x - 0)$

$y = -\dfrac{7}{5}x$

23. $m = \dfrac{-5-1}{7-4} = -2$

$y - y_1 = m(x - x_1)$

$y - 1 = -2(x - 4)$

$y - 1 = -2x + 8$

$y = -2x + 9$

121

25.
$$y - y_1 = m(x - x_1)$$
$$y - 17 = 0(x - -5)$$
$$y - 17 = 0$$
$$y = 17$$

27. Find the slope of the segment:
$$m = \frac{4 - -10}{2 - 4} = -7$$
Use this slope for the parallel line:
$$y - y_1 = m(x - x_1)$$
$$y - -2 = -7(x - 8)$$
$$y + 2 = -7x + 56$$
$$y = -7x + 54$$

29. Find the slope of
$$3x - 4y = 7$$
$$-4y = -3x + 7$$
$$y = \frac{3}{4}x - \frac{7}{4}$$

slope $= \frac{3}{4}$

$$y - y_1 = m(x - x_1)$$
$$y - 0 = \frac{3}{4}(x - 2)$$
$$y = \frac{3}{4}x - \frac{3}{2}$$

31. $y = x^2 + 2$

x-intercept:
$$0 = x^2 + 2$$
$$-2 = x^2$$
none

y-intercept:
$$y = 0^2 + 2$$
$$y = 2$$
$$(0, 2)$$

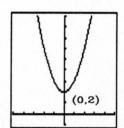

SYMMETRY

about y-axis
$$y = (-x)^2 + 2$$
$$y = x^2 + 2$$
SYMMETRY

about origin
$$-y = (-x)^2 + 2$$
$$-y = x^2 + 2$$
NO SYMMETRY

about x-axis
$$-y = x^2 + 2$$
NO SYMMETRY

33. $y = \frac{1}{2}|x|$

x-intercept:
$$0 = \frac{1}{2}|x|$$
$$0 = x$$
$$(0, 0)$$

y-intercept:
$$y = \frac{1}{2}|0|$$
$$y = 0$$
$$(0, 0)$$

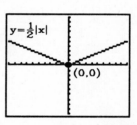

SYMMETRY

about y-axis
$$y = \frac{1}{2}|-x|$$
$$y = \frac{1}{2}|x|$$
SYMMETRY

about origin
$$-y = \frac{1}{2}|-x|$$
$$-y = \frac{1}{2}|x|$$
NO SYMMETRY

about x-axis
$$-y = \frac{1}{2}|x|$$
NO SYMMETRY

35. $y = \sqrt{x} + 2$

x-intercept:
$0 = \sqrt{x} + 2$
$-2 = \sqrt{x}$
none

y-intercept:
$y = \sqrt{0} + 2$
$y = 2$
$(0, 2)$

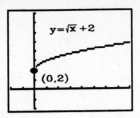

SYMMETRY

about y-axis

8 $\quad y = \sqrt{-x} + 2$

NO SYMMETRY

about origin

$-y = \sqrt{-x} + 2$

NO SYMMETRY

about x-axis

$-y = \sqrt{x} + 2$

NO SYMMETRY

37. $y = x^3 + 1$

x-intercept:
$0 = x^3 + 1$
$-1 = x^3$
$-1 = x$
$(-1, 0)$

y-intercept:
$y = 0^3 + 1$
$y = 1$
$(0, 1)$

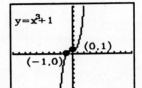

SYMMETRY

about y-axis

$y = (-x)^3 + 1$
$y = -x^3 + 1$
NO SYMMETRY

about origin

$-y = (-x)^3 + 1$
$-y = -x^3 + 1$
$y = x^3 - 1$
NO SYMMETRY

about x-axis

$-y = x^3 + 1$
$y = -x^3 - 1$
NO SYMMETRY

39. $(x - h)^2 + (y - k)^2 = r^2$
$(x + 3)^2 + (y - 4)^2 = 144$

41. $x^2 + y^2 - 2y = 15$
$x^2 + y^2 - 2y + 1 = 16$
$(x - 0)^2 + (y - 1)^2 = 4^2$
center at $(0, 1)$; radius $= 4$

123

43. $y = (x - 2)(x + 5)$

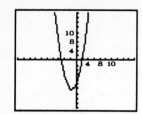

45. $y^2 = x - 3$
$y = \pm\sqrt{x - 3}$

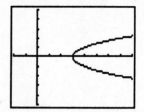

47. Graph $y = x^2 - 11$, and find all points with a y-coordinate of 0.
$x \approx \pm 3.32$

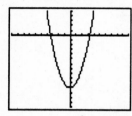

49. Graph $y = |x^2 - 2| - 1$, and find all points with a y-coordinate of 0.
$x \approx \pm 1$ or ± 1.73

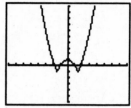

51.
$$\frac{x + 3}{10} = \frac{x - 1}{x}$$
$$x(x + 3) = 10(x - 1)$$
$$x^2 + 3x = 10x - 10$$
$$x^2 - 7x + 10 = 0$$
$$(x - 2)(x - 5) = 0$$
$$x = 2 \text{ or } x = 5$$

53.
$$\frac{3x - 5}{x} = \frac{x - 3}{1}$$
$$1(3x - 5) = x(x - 3)$$
$$3x - 5 = x^2 - 3x$$
$$0 = x^2 - 6x + 5$$
$$0 = (x - 1)(x - 5)$$
$$x = 1 \text{ or } x = 5$$

55.
$$F = ks$$
$$3 = k(5)$$
$$\frac{3}{5} = k$$
$$F = \frac{3}{5}s$$
$$F = \frac{3}{5}(3)$$
$$F = \frac{9}{5} = 1.8$$

The force would be 1.8 pounds.

57.
$$E = kv^2$$
At 30 mph, $E = k(30)^2 = 900k$

At 50 mph, $E = k(50)^2 = 2500k$

The increase from $900k$ to $2500k$ is a factor of
$$\frac{2500k}{900k} = \frac{25}{9}.$$

It increases by a factor of $\frac{25}{9}$.

59. Let x = fixed amount per job, y = amount per roll, and n = number of rolls. The difference in bills between the jobs is $294 - 177 = \$117$. Both bills include the same fixed amount for each job, so the difference of \$117 must be accounted for by the difference in number of rolls hung. In other words, 9 rolls cost \$117.

cost per roll = $y = \dfrac{117}{9} = 13$

So, $\qquad$ Bill = $x + ny$ $\qquad\qquad$ Then, $\qquad\qquad$ Bill = $34 + 13n$
$\qquad\qquad\quad 294 = x + 20y$ $\qquad\qquad\qquad\qquad\qquad\qquad = 34 + 13(27)$
$\qquad\qquad\quad 294 = x + 20(13)$ $\qquad\qquad\qquad\qquad\qquad\qquad = 34 + 351$
$\qquad\qquad\quad 294 = x + 260$ $\qquad\qquad\qquad\qquad\qquad\qquad = 385$
$\qquad\qquad\quad\;\; 34 = x$ $\qquad\qquad\qquad$ The bill would be \$385.

Chapter 3 Test $\quad$ (page 191)

1. quadrant II

3. $x + 3y = 6$

x-intercept: $\qquad$ y-intercept:
$\quad x + 3y = 6$ $\qquad\quad x + 3y = 6$
$\quad x + 3(0) = 6$ $\qquad\quad 0 + 3y = 6$
$\qquad\qquad x = 6$ $\qquad\qquad\quad y = 2$

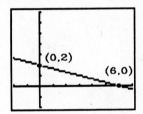

5. $\qquad 2(x + y) = 3x + 5$

$\qquad\qquad 2x + 2y = 3x + 5$
$\qquad\qquad -x + 2y = 5$

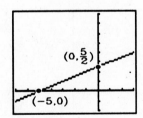

7. $\qquad \frac{1}{2}(x - 2y) = y - 1$

$\qquad\qquad x - 2y = 2y - 2$
$\qquad\qquad x - 4y = -2$

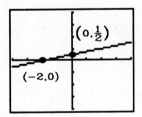

9. $d = \sqrt{(1 - -3)^2 + (-1 - 4)^2} = \sqrt{4^2 + (-5)^2} = \sqrt{16 + 25} = \sqrt{41}$

11. $(x_M, y_M) = \left(\dfrac{3 + -3}{2}, \dfrac{-7 + 7}{2}\right) = (0, 0)$

13. $m = \dfrac{\Delta y}{\Delta x} = \dfrac{-9 - 1}{3 - -5} = \dfrac{-10}{8} = -\dfrac{5}{4}$

15. $y = 3x - 2$: $m = 3$; $y = 2x - 3$: $m = 2$; Since the slopes are not equal, and the product of the slopes is not -1, the lines are neither parallel nor perpendicular.

17.
$$y - y_1 = m(x - x_1)$$
$$y - -5 = 2(x - 3)$$
$$y + 5 = 2x - 6$$
$$-2x + y = -11$$
$$2x - y = 11$$
$$y = 2x - 11$$

19. Find the slope of the line given:
$$2x - y = 3$$
$$-y = -2x + 3$$
$$y = 2x - 3$$
The slope of this line is 2. The line described has a slope of 2 and $b = 5$.
$$y = mx + b$$
$$y = 2x + 5$$

21. $m = \dfrac{-\frac{3}{2} - \frac{1}{2}}{2 - 3} = \dfrac{-2}{-1} = 2$
$$y - y_1 = m(x - x_1)$$
$$y - -\frac{3}{2} = 2(x - 2)$$
$$y + \frac{3}{2} = 2x - 4$$
$$4x - 2y = 11$$
$$y = 2x - \frac{11}{2}$$

23.

x-intercept(s)	y-intercept(s)
$y = x^3 - 16x$	$y = x^3 - 16x$
$0 = x^3 - 16x$	$y = 0^3 - 16(0)$
$0 = x(x + 4)(x - 4)$	$y = 0$
$x = 0$ or $x = \pm 4$	
$(0, 0), (\pm 4, 0)$	$(0, 0)$

25. $y^2 = x - 1$

about x-axis	about origin	about y-axis
$(-y)^2 = x - 1$	$(-y)^2 = -x - 1$	$y^2 = -x - 1$
$y^2 = x - 1$	$y^2 = -x - 1$	$y^2 = -x - 1$
identical to original	not identical to original	not identical to original
SYMMETRY	NO SYMMETRY	NO SYMMETRY

27. $y = x^2 - 9$

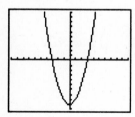

29. $y = 2\sqrt{x}$

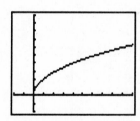

31. $(x - 5)^2 + (y - 7)^2 = 8^2$

33. $x^2 + y^2 = 9$: center $(0, 0)$, $r = 3$

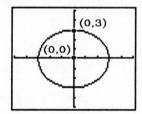

35. $y = kx^2$

37. $P = kQ$
$7 = k(2)$
$\frac{7}{2} = k$

$P = \frac{7}{2}Q$

$P = \frac{7}{2}(5)$

$P = \frac{35}{2}$

39. Graph $y = x^2 - 7$, and find any positive x-intercept(s).
Solution: $x \approx 2.65$

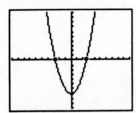

Cumulative Review Exercises (page 193)

1. $A \cap B = \{3, 4, 5, 6\} \cap \{2, 3\} = \{3\}$

3. $\{x \mid -4 \le x < 7\}$:

5. commutative property of addition

7. $(81a^4)^{1/2} = 81^{1/2}(a^4)^{1/2} = 9a^2$

9. $(a^{-3}b^{-2})^{-2} = a^6b^4$

11. $\left(\frac{4x^0y^2}{x^2y}\right)^{-2} = \left(\frac{4y}{x^2}\right)^{-2} = \left(\frac{x^2}{4y}\right)^2 = \frac{x^4}{16y^2}$

13. $(a^{1/2}b)^2(ab^{1/2})^2 = ab^2 \cdot a^2b$
$= a^3b^3$

15. $\frac{3}{\sqrt{3}} = \frac{3}{\sqrt{3}} \cdot \frac{\sqrt{3}}{\sqrt{3}} = \frac{3\sqrt{3}}{3} = \sqrt{3}$

17. $\frac{3}{y - \sqrt{3}} = \frac{3}{(y - \sqrt{3})} \cdot \frac{(y + \sqrt{3})}{(y + \sqrt{3})} = \frac{3(y + \sqrt{3})}{y^2 - 3}$

19. $\sqrt{75} - 3\sqrt{5} = \sqrt{25 \cdot 3} - 3\sqrt{5} = 5\sqrt{3} - 3\sqrt{5}$

21. $(\sqrt{2} - \sqrt{3})^2 = (\sqrt{2} - \sqrt{3})(\sqrt{2} - \sqrt{3}) = 2 - \sqrt{6} - \sqrt{6} + 3 = 5 - 2\sqrt{6}$

23. $(3x^2 - 2x + 5) - 3(x^2 + 2x - 1) = 3x^2 - 2x + 5 - 3x^2 - 6x + 3 = -8x + 8$

25. $(3x - 5)(2x + 7) = 6x^2 + 21x - 10x - 35 = 6x^2 + 11x - 35$

$$
\begin{array}{r}
2x^2 \quad - \quad x \quad + \ 1
\end{array}
$$

27. $3x+2 \,\big)\, 6x^3 \ + \ x^2 \ + \ x \ + \ 2$

$$
\begin{array}{r}
\underline{-(\ 6x^3 \ + \ 4x^2)} \\
-\ 3x^2 \ + \ x \\
\underline{-(-\ 3x^2 \ - \ 2x)} \\
3x \ + \ 2 \\
\underline{-(\quad 3x \ + \ 2)} \\
0
\end{array}
$$

29. $3t^2 - 6t = 3t(t-2)$

31.
$$
\begin{aligned}
x^8 + x^4 + 1 = (x^8 + 2x^4 + 1) - x^4 &= (x^4+1)^2 - (x^2)^2 \\
&= (x^4+1+x^2)(x^4+1-x^2) \\
&= (x^4+x^2+1)(x^4-x^2+1) \\
&= [(x^4+2x^2+1)-x^2](x^4-x^2+1) \\
&= [(x^2+1)^2 - x^2](x^4-x^2+1) \\
&= (x^2+1+x)(x^2+1-x)(x^4-x^2+1) \\
&= (x^2+x+1)(x^2-x+1)(x^4-x^2+1)
\end{aligned}
$$

33.
$$
\frac{x^2-4}{x^2+5x+6} \cdot \frac{x^2-2x-15}{x^2+3x-10} = \frac{(x+2)(x-2)}{(x+2)(x+3)} \cdot \frac{(x-5)(x+3)}{(x+5)(x-2)} = \frac{x-5}{x+5}
$$

35.
$$
\frac{2}{x+3} + \frac{5x}{x-3} = \frac{2(x-3)}{(x+3)(x-3)} + \frac{5x(x+3)}{(x+3)(x-3)} = \frac{2x-6+5x^2+15x}{(x+3)(x-3)}
$$
$$
= \frac{5x^2+17x-6}{(x+3)(x-3)}
$$

37.
$$
\frac{\frac{1}{a}+\frac{1}{b}}{\frac{1}{ab}} = \frac{\frac{1}{a}+\frac{1}{b}}{\frac{1}{ab}} \cdot \frac{\frac{ab}{1}}{\frac{ab}{1}} = \frac{b+a}{1} = a+b
$$

39.
$$
\frac{3x}{x+5} = \frac{x}{x-5}
$$
$$
\begin{aligned}
3x(x-5) &= x(x+5) \\
3x^2 - 15x &= x^2 + 5x \\
2x^2 - 20x &= 0 \\
2x(x-10) &= 0
\end{aligned}
$$
$x = 0$ or $x = 10$ (both check)

41.
$$
\frac{1}{R} = \frac{1}{R_1} + \frac{1}{R_2}
$$
$$
RR_1R_2\frac{1}{R} = RR_1R_2\left(\frac{1}{R_1}+\frac{1}{R_2}\right)
$$
$$
\begin{aligned}
R_1R_2 &= RR_2 + RR_1 \\
R_1R_2 &= R(R_1+R_2)
\end{aligned}
$$
$$
\frac{R_1R_2}{R_1+R_2} = R, \text{ or } R = \frac{R_1R_2}{R_1+R_2}
$$

43. See the figure to the right.

$$x(40 - 2x) = 192$$
$$40x - 2x^2 = 192$$
$$-2x^2 + 40x - 192 = 0$$
$$-2(x^2 - 20x + 96) = 0$$
$$-2(x - 8)(x - 12) = 0$$
$$x = 8 \text{ or } x = 12$$

The dimensions will be 8 feet by 24 feet, or 12 feet by 16 feet.

45. $\dfrac{2+i}{2-i} = \dfrac{(2+i)}{(2-i)} \cdot \dfrac{(2+i)}{(2+i)} = \dfrac{4 + 4i + i^2}{4 - i^2} = \dfrac{4 + 4i - 1}{4 - (-1)} = \dfrac{3 + 4i}{5} = \dfrac{3}{5} + \dfrac{4}{5}i$

47. $|3 + 4i| = \sqrt{3^2 + 4^2} = \sqrt{9 + 16} = \sqrt{25} = 5$

49.
$$\frac{x+3}{x-1} - \frac{6}{x} = 1$$

$$x(x-1) \cdot \left(\frac{x+3}{x-1} - \frac{6}{x} \right) = x(x-1) \cdot 1$$

$$x(x+3) - (x-1)6 = x(x-1)$$
$$x^2 + 3x - 6x + 6 = x^2 - x$$
$$6 = 2x$$
$$3 = x \quad \text{The solution checks.}$$

51.
$$\sqrt{y+2} + \sqrt{11-y} = 5$$
$$\sqrt{y+2} = 5 - \sqrt{11-y}$$
$$(\sqrt{y+2})^2 = (5 - \sqrt{11-y})^2$$
$$y + 2 = 25 - 10\sqrt{11-y} + 11 - y$$
$$10\sqrt{11-y} = -2y + 34$$
$$(10\sqrt{11-y})^2 = (-2y + 34)^2$$
$$100(11 - y) = 4y^2 - 136y + 1156$$
$$0 = 4y^2 - 36y + 56$$
$$0 = 4(y - 2)(y - 7)$$
$$y = 2 \text{ or } y = 7 \qquad \text{Both solutions check.}$$

53.
$$5x - 7 \leq 4$$
$$5x \leq 11$$
$$x \leq \frac{11}{5}$$

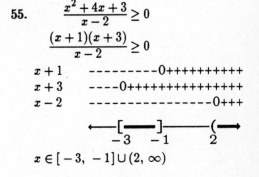

11/5

55.
$$\frac{x^2 + 4x + 3}{x - 2} \geq 0$$
$$\frac{(x + 1)(x + 3)}{x - 2} \geq 0$$

$x + 1 \qquad \text{--------}0{+}{+}{+}{+}{+}{+}{+}{+}{+}{+}$
$x + 3 \qquad \text{----}0{+}{+}{+}{+}{+}{+}{+}{+}{+}{+}{+}{+}{+}{+}$
$x - 2 \qquad \text{----------------}0{+}{+}{+}$

$x \in [-3, -1] \cup (2, \infty)$

57.
$$|2x - 3| \geq 5$$

$$2x - 3 \geq 5 \quad \text{or} \quad 2x - 3 \leq -5$$
$$x \geq 4 \qquad\qquad x \leq -1$$
$$x \in (-\infty, -1] \cup [4, \infty)$$

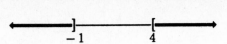

59. $5x - 3y = 15$
x-intercept: $(3, 0)$
y-intercept: $(0, -5)$

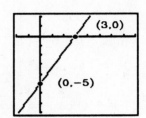

61. length $= \sqrt{(-2-3)^2 + \left(\frac{7}{2} - -\frac{1}{2}\right)^2} = \sqrt{(-5)^2 + 4^2} = \sqrt{25 + 16} = \sqrt{41}$

$$(x_M, y_M) = \left(\frac{-2+3}{2}, \frac{\frac{7}{2} + -\frac{1}{2}}{2}\right) = \left(\frac{1}{2}, \frac{3}{2}\right); \qquad m = \frac{\Delta y}{\Delta x} = \frac{\frac{7}{2} - -\frac{1}{2}}{-2-3} = \frac{4}{-5} = -\frac{4}{5}$$

63. $m = \dfrac{5 - -7}{-3 - 3} = \dfrac{12}{-6} = -2$

$$y - y_1 = m(x - x_1)$$
$$y - 5 = -2(x - -3)$$
$$y - 5 = -2(x + 3)$$
$$y - 5 = -2x - 6$$
$$y = -2x - 1$$

65. $\qquad 3x - 5y = 7$

$$-5y = -3x + 7$$
$$y = \frac{3}{5}x - \frac{7}{5} \quad \left\{\text{slope} = \frac{3}{5}\right\}$$

Use the parallel slope.

$$y - y_1 = m(x - x_1)$$
$$y - 3 = \frac{3}{5}(x - -5)$$
$$y - 3 = \frac{3}{5}x + 3$$
$$y = \frac{3}{5}x + 6$$

67. $x^2 = y - 2$
x-intercept(s): none
y-intercept(s): $(0, 2)$
symmetry: about y-axis

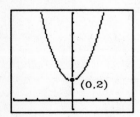

69. $x^2 + y^2 = 100$
x-intercept(s): $(\pm 10, 0)$
y-intercept(s): $(0, \pm 10)$
symmetry: about x-, y-axes and origin

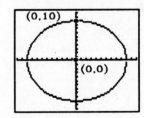

71.
$$\frac{x-2}{x} = \frac{x-6}{5}$$

$$5(x-2) = x(x-6)$$
$$5x - 10 = x^2 - 6x$$
$$0 = x^2 - 11x + 10$$
$$0 = (x-1)(x-10)$$
$$x = 1 \text{ or } x = 10$$

73. Let V = amount per visit.
Let X = amount per x-ray.
Since the totals for 2 x-rays and 4 x-rays both include the office visit cost, the difference is the cost of 2 x-rays.

$$2X = 54 - 37$$
$$X = 8.50$$

$$\boxed{\begin{array}{c}\text{Cost of}\\ \text{2 x-rays}\end{array}} = V + 2X$$
$$37 = V + 2X$$
$$37 = V + 17$$
$$20 = V$$

$$\boxed{\begin{array}{c}\text{Cost of}\\ \text{5 x-rays}\end{array}} = V + 5X$$
$$= 20 + 5(8.50)$$
$$= 20 + 42.50$$
$$= 62.50$$

5 exposures will cost \$62.50.

Exercise 4.1 (page 205)

1. $2x + 3y = 6$

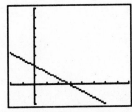

domain = $\{x \mid x \in \mathbb{R}\}$
range = $\{y \mid y \in \mathbb{R}\}$

3. $x^2 + y^2 = 16$

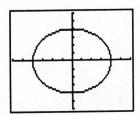

domain = $\{x \mid x \in \mathbb{R}, \ -4 \le x \le 4\}$
range = $\{y \mid y \in \mathbb{R}, \ -4 \le y \le 4\}$

5.

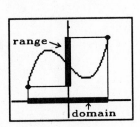

7. function

9. not a function
If $x = 4$, then $y = \pm 2$.

11. function

13. not a function

If $x = 2$, then $y = \pm 3$.

15. not a function

If $x = 2$, then $y = \pm 2$.

17. function

19. domain $= \{x \mid x \in \mathbb{R}\}$

range $= \{y \mid y \in \mathbb{R}\}$

21. domain $= \{x \mid x \in \mathbb{R}\}$

range $= \{y \mid y \in \mathbb{R}, y \geq 0\}$

23. domain $= \{x \mid x \in \mathbb{R}, x \neq -1\}$

range $= \{y \mid y \in \mathbb{R}, y \neq 0\}$

25. domain $= \{x \mid x \in \mathbb{R}, x \geq 0\}$

range $= \{y \mid y \in \mathbb{R}, y \geq 0\}$

27. function

29. function

31. not a function – fails vertical line test

33. $f(x) = 3x - 2$

$$\begin{aligned} f(2) &= 3(2) - 2 \\ &= 6 - 2 \\ &= 4 \end{aligned}$$

$$\begin{aligned} f(-3) &= 3(-3) - 2 \\ &= -9 - 2 \\ &= -11 \end{aligned}$$

$$f(k) = 3k - 2$$

$$\begin{aligned} f(k^2 - 1) &= 3(k^2 - 1) - 2 \\ &= 3k^2 - 3 - 2 \\ &= 3k^2 - 5 \end{aligned}$$

35. $f(x) = \frac{1}{2}x + 3$

$$\begin{aligned} f(2) &= \frac{1}{2}(2) + 3 \\ &= 1 + 3 \\ &= 4 \end{aligned}$$

$$\begin{aligned} f(-3) &= \frac{1}{2}(-3) + 3 \\ &= -\frac{3}{2} + \frac{6}{2} \\ &= \frac{3}{2} \end{aligned}$$

$$f(k) = \frac{1}{2}k + 3$$

$$\begin{aligned} f(k^2 - 1) &= \frac{1}{2}(k^2 - 1) + 3 \\ &= \frac{1}{2}k^2 - \frac{1}{2} + \frac{6}{2} \\ &= \frac{1}{2}k^2 + \frac{5}{2} \end{aligned}$$

37. $f(x) = x^2$

$$\begin{aligned} f(2) &= 2^2 \\ &= 4 \end{aligned}$$

$$\begin{aligned} f(-3) &= (-3)^2 \\ &= 9 \end{aligned}$$

$$f(k) = k^2$$

$$\begin{aligned} f(k^2 - 1) &= (k^2 - 1)^2 \\ &= (k^2 - 1)(k^2 - 1) \\ &= k^4 - 2k^2 + 1 \end{aligned}$$

39. $f(x) = \dfrac{2}{x+4}$

$$f(2) = \frac{2}{2+4}$$
$$= \frac{2}{6}$$
$$= \frac{1}{3}$$

$$f(-3) = \frac{2}{-3+4}$$
$$= \frac{2}{1}$$
$$= 2$$

$$f(k) = \frac{2}{k+4}$$

$$f(k^2 - 1) = \frac{2}{k^2 - 1 + 4}$$
$$= \frac{2}{k^2 + 3}$$

41. $f(x) = \dfrac{1}{x^2 - 1}$

$$f(2) = \frac{1}{2^2 - 1}$$
$$= \frac{1}{3}$$

$$f(-3) = \frac{1}{(-3)^2 - 1}$$
$$= \frac{1}{8}$$

$$f(k) = \frac{1}{k^2 - 1}$$

$$f(k^2 - 1) = \frac{1}{(k^2 - 1)^2 - 1}$$
$$= \frac{1}{k^4 - 2k^2}$$

43. $f(x) = \sqrt{x^2 + 1}$

$$f(2) = \sqrt{2^2 + 1}$$
$$= \sqrt{5}$$

$$f(-3) = \sqrt{(-3)^2 + 1}$$
$$= \sqrt{10}$$

$$f(k) = \sqrt{k^2 + 1}$$

$$f(k^2 - 1) = \sqrt{(k^2 - 1)^2 + 1}$$
$$= \sqrt{k^4 - 2k^2 + 1 + 1}$$
$$= \sqrt{k^4 - 2k^2 + 2}$$

45.

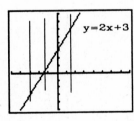

function

47.

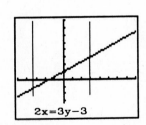

function

49.

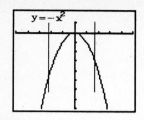

function

51.

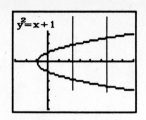

not a function

53.

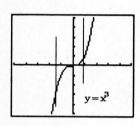

function

55.

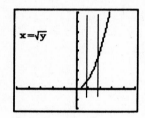

function

57.

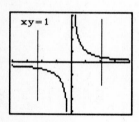

function

59.

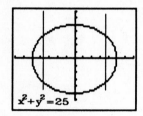

not a function

61. $C = 2\pi r$

$$\frac{C}{2\pi} = r, \text{ or } r = \frac{C}{2\pi}$$

63. $A = s^2$, so $s = \sqrt{A}$
$P = 4s$
$\quad = 4\sqrt{A}$

65. $A = s^2$, so $s = \sqrt{A}$
$V = s^3$

$V = (\sqrt{A})^3$

$V = A^{3/2}$

67. $d^2 = s^2 + s^2$
$\quad = 2s^2$

$\dfrac{d^2}{2} = s^2$

$\sqrt{\dfrac{d^2}{2}} = s$

$A = s^2$

$A = \left(\sqrt{\dfrac{d^2}{2}}\right)^2 = \dfrac{d^2}{2}$

69. points: $(0, 32)$, $(-40, -40)$

$m = \dfrac{32 - -40}{0 - -40} = \dfrac{72}{40} = \dfrac{9}{5}$

$C = mF + b$

$32 = \dfrac{9}{5}(0) + b$

$32 = b$

$C = \dfrac{9}{5}F + 32$

71. points: $(1000, 4.7)$, $(9000, 14.3)$

$m = \dfrac{4.7 - 14.3}{1000 - 9000} = \dfrac{-9.6}{-8000} = 0.0012$

$c = mn + b$

$4.7 = 0.0012(1000) + b$

$4.7 = 1.2 + b$

$3.5 = b$

$c = 0.0012n + 3.5$

73. $y = \sqrt{2x - 5}$

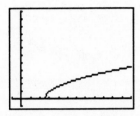

domain $= \{x \mid x \in \mathbb{R}, x \geq 2.5\}$
range $= \{y \mid y \in \mathbb{R}, y \geq 0\}$

75. $y = \sqrt[3]{5x - 1}$

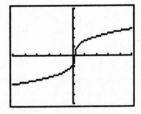

domain $= \{x \mid x \in \mathbb{R}\}$
range $= \{y \mid y \in \mathbb{R}\}$

Exercise 4.2 (page 207)

1. $y = x^2 - x$

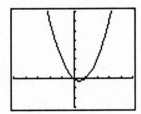

3. $y = -3x^2 + 2$

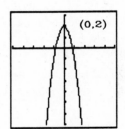

(0,2)

5. $y = -\dfrac{1}{2}x^2 + 3$

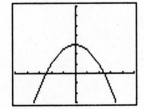

7. $y = x^2 - 4x + 1$

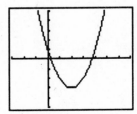

9. $y = x^2 - 1$

$a = 1, b = 0, c = -1$

$h = -\dfrac{b}{2a} = -\dfrac{0}{2(1)} = 0$

$k = c - \dfrac{b^2}{4a} = -1 - \dfrac{0^2}{4(1)} = -1$

vertex $(0, -1)$

11. $y = x^2 - 4x + 4$

$a = 1, b = -4, c = 4$

$h = -\dfrac{b}{2a} = -\dfrac{-4}{2(1)} = -(-2) = 2$

$k = c - \dfrac{b^2}{4a} = 4 - \dfrac{(-4)^2}{4(1)} = 4 - 4 = 0$

vertex $(2, 0)$

13. $y = x^2 + 6x - 3$

$a = 1, b = 6, c = -3$

$h = -\dfrac{b}{2a} = -\dfrac{6}{2(1)} = -3$

$k = c - \dfrac{b^2}{4a} = -3 - \dfrac{6^2}{4(1)} = -12$

vertex $(-3, -12)$

15. $y = -2x^2 + 12x - 17$

$a = -2, b = 12, c = -17$

$h = -\dfrac{b}{2a} = -\dfrac{12}{2(-2)} = 3$

$k = c - \dfrac{b^2}{4a} = -17 - \dfrac{12^2}{4(-2)} = 1$

vertex $(3, 1)$

17.

$$x^2 + 20y - 400 = 0$$
$$20y = -x^2 + 400$$
$$y = -\frac{1}{20}x^2 + 20$$

This graph is a parabola. Since the coefficient of x^2 is negative, the parabola opens down. The maximum y value (and height of the arch) is the y-coordinate of the vertex, or k.

$$k = c - \frac{b^2}{4a} = 20 - \frac{0^2}{4\left(-\dfrac{1}{20}\right)} = 20$$

The arch reaches a maximum height of 20 feet.

19. Consider the figure below. Right triangle ABC is a 45-45-90 right triangle and so is isosceles. To find x, use the Pythagorean Theorem.

$$x^2 + x^2 = 100^2$$
$$2x^2 = 100^2$$
$$x^2 = \frac{100^2}{2}$$
$$x = \pm\frac{100}{\sqrt{2}}$$

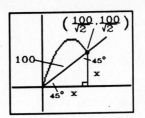

Since x is a distance, use the positive value. Now, the point

$\left(\dfrac{100}{\sqrt{2}}, \dfrac{100}{\sqrt{2}}\right)$ must be on the graph.

$$y = -x^2 + ax$$

$$\frac{100}{\sqrt{2}} = -\left(\frac{100}{\sqrt{2}}\right)^2 + a\left(\frac{100}{\sqrt{2}}\right)$$

$$\frac{100}{\sqrt{2}} = -\frac{100^2}{2} + a\left(\frac{100}{\sqrt{2}}\right)$$

$$100 = -\frac{100^2 \cdot \sqrt{2}}{2} + 100a$$

$$200 = -100^2 \cdot \sqrt{2} + 200a$$

$$200 + 100^2 \cdot \sqrt{2} = 200a$$

$$\frac{200 + 100^2 \cdot \sqrt{2}}{200} = a$$

$$\frac{200}{200} + \frac{100 \cdot 100\sqrt{2}}{200} = a$$

$$1 + 50\sqrt{2} = a$$

21. Let x stand for one side of the enclosure. Then $50 - x$ is the other side. Let y stand for the area of the enclosure. $y = x(50 - x) = 50x - x^2 = -x^2 + 50x$. $y = -x^2 + 50x$ is the equation of a parabola. Since the coefficient of x^2 is negative, the parabola opens down. The vertex is then a maximum. The y-coordinate of the vertex is the maximal area, obtained when x is the x-coordinate of the vertex.

$$\text{vertex} = \left(-\frac{b}{2a}, c - \frac{b^2}{4a}\right) = \left(-\frac{50}{2(-1)}, 0 - \frac{50^2}{4(-1)}\right) = (25, 625)$$

The maximum area is 625 square feet, with dimensions of 25 feet by 25 feet.

137

23. Let x stand for the depth of the trough. Then the width is $24 - 2x$.
Let y stand for the cross-sectional area. $y = x(24 - 2x) = -2x^2 + 24x$.
$y = -2x^2 + 24x$ is the equation of a parabola. Since the coefficient of x^2 is negative, the parabola opens down. The vertex is then a maximum. The y-coordinate of the vertex is the maximal area, obtained when x is the x-coordinate of the vertex.
$$\text{vertex} = \left(-\frac{b}{2a},\ c - \frac{b^2}{4a} \right) = \left(-\frac{24}{2(-2)},\ 0 - \frac{24^2}{4(-2)} \right) = (6,\ 72)$$
The maximum area is 72 square inches, with dimensions of 6 inches by 12 inches.

25. Let x stand for one number. Then $6 - x$ stands for the other number.
Let y stand for the sum of their squares.
$y = x^2 + (6 - x)^2 = x^2 + 36 - 12x + x^2 = 2x^2 - 12x + 36$
This is the equation of a parabola. Since the coefficient of x^2 is positive, the parabola opens up. The vertex is then a minimum. The y-coordinate of the vertex is the minimum sum of squares, obtained when the numbers are x, from the x-coordinate of the vertex, and $6 - x$.
$$\text{vertex} = \left(-\frac{b}{2a},\ c - \frac{b^2}{4a} \right) = \left(-\frac{-12}{2(2)},\ 36 - \frac{(-12)^2}{4(2)} \right) = (3,\ 18)$$

The maximum sum of squares is 18, occurring when the numbers are both 3.

27. Let y stand for revenue. $y = p(1200 - p) = -p^2 + 1200p$.
This is the equation of a parabola (with p instead of x). Since the coefficient of p^2 is negative, the parabola opens down. The vertex is then a maximum. The y-coordinate of the vertex is the maximum revenue, occurring when the price is the p-coordinate of the vertex.
$$\text{vertex} = \left(-\frac{b}{2a},\ c - \frac{b^2}{4a} \right) = \left(-\frac{1200}{2(-1)},\ 0 - \frac{1200^2}{4(-1)} \right) = (600,\ 360{,}000)$$
When the price is \$600, there is a maximum revenue of \$360,000.

29. If the hotel is two-thirds full, then there are two-thirds of 300 (or 200) rooms occupied. Every \$5 over \$90 means that 10 fewer rooms will be occupied.
Let $x =$ the number of \$5 increases in room rates. Then $90 + 5x$ will equal the room rate, and $200 - 10x$ will equal the number of occupied rooms.
Let $y =$ the income. $\quad y = (90 + 5x)(200 - 10x) = 18{,}000 - 900x + 1000x - 50x^2$
$$= -50x^2 + 100x + 18{,}000$$
This is the equation of a parabola. Since the coefficient of x^2 is negative, the parabola opens down and the vertex is a maximum. The y-coordinate of the vertex is the maximum income, obtained when the number of \$5 increases in the room rate is the x-coordinate of the vertex.
$$\text{vertex} = \left(-\frac{b}{2a},\ c - \frac{b^2}{4a} \right) = \left(-\frac{100}{2(-50)},\ 18{,}000 - \frac{100^2}{4(-50)} \right) = (1,\ 18{,}050)$$
The maximum income occurs with one \$5 increase, or when the room rate is \$95.

31. $s = 80t - 16t^2 = -16t^2 + 80t$ is the equation of a parabola. Since the coefficient of t^2 is negative, the parabola opens down, and the vertex is a maximum. The maximum height will be the y-coordinate (s) of the vertex. The x-coordinate (t) of the vertex will be the time needed to get to this height.

$$\text{vertex} = \left(-\frac{b}{2a},\, c - \frac{b^2}{4a}\right) = \left(-\frac{80}{2(-16)},\, 0 - \frac{80^2}{4(-16)}\right) = \left(\tfrac{5}{2},\, 100\right)$$

The maximum height is reached after 2.5 seconds.

33. See the work for **#31**. The maximum height is 100 feet.

35. Since the point (x, y) is on line $\overleftrightarrow{AB}$, its coordinates must satisfy the equation of the line. Find the equation:

$m = \dfrac{\Delta y}{\Delta x} = \dfrac{9 - 0}{0 - 12} = -\dfrac{3}{4}$

$y - y_1 = m(x - x_1)$

$y - 9 = -\dfrac{3}{4}(x - 0)$

$y = -\dfrac{3}{4}x + 9$

Then the point (x, y) can be considered the point $\left(x,\, -\dfrac{3}{4}x + 9\right)$. The length of the rectangle is x, while the height of the rectangle is $-\dfrac{3}{4}x + 9$.

$\text{Area} = \text{length} \cdot \text{height} = x\left(-\dfrac{3}{4}x + 9\right) = -\dfrac{3}{4}x^2 + 9x$. This is a parabola which opens down, so the vertex is a maximum.

$$\text{vertex} = \left(-\frac{b}{2a},\, c - \frac{b^2}{4a}\right) = \left(-\frac{9}{2\left(-\frac{3}{4}\right)},\, 0 - \frac{9^2}{4\left(-\frac{3}{4}\right)}\right) = (6, 27)$$

The maximum area is 27, with dimensions of 6 by 4.5.

37. $y = 2x^2 + 9x - 56$
$(-2.25, 66.13)$

39. $y = (x - 7)(5x + 2)$
$(3.3, -68.5)$

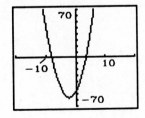

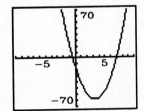

Exercise 4.3 (page 220)

1. $y = x^3$

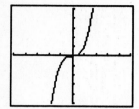

3. $y = x^3 + x^2$

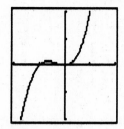

5. $y = -x^3$

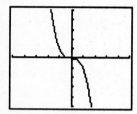

7. $y = x^4 - 2x^2 + 1$

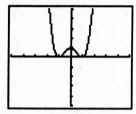

9. $y = x^4 + x^2$
       $y = (-x)^4 + (-x)^2$
       $y = x^4 + x^2$
   even

11. $y = x^3 + x^2$
       $y = (-x)^3 + (-x)^2$
       $y = -x^3 + x^2$
   neither even nor odd

13. $y = x^5 + x^3$
       $y = (-x)^5 + (-x)^3$
       $y = -x^5 - x^3$
   odd

15. $y = 2x^3 - 3x$
       $y = 2(-x)^3 - 3(-x)$
       $y = -2x^3 + 3x$
   odd

17. decreasing on $(-\infty, 0)$
      increasing on $[0, \infty)$

19. increasing on $(-\infty, 0)$
      decreasing on $(0, \infty)$

21. $y = x^2 - 4x + 4$

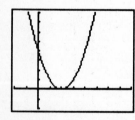

decreasing on $(-\infty, 2)$
increasing on $[2, \infty)$

23.

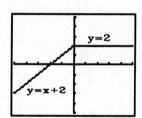

25.

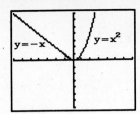

27.

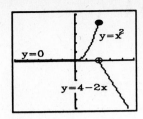

29. $y = [\![2x]\!]$

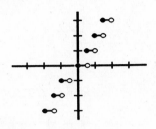

31. $y = [\![x]\!] - 1$

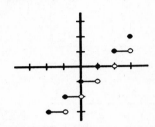

33.

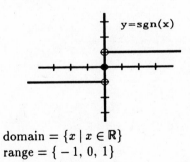

cost of 275 miles = $26

35.

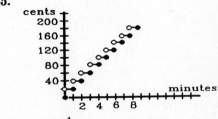

cost of $7\frac{1}{2}$ minutes = $1.60

37.

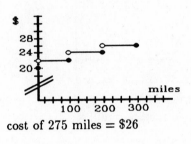

domain = $\{x \mid x \in \mathbb{R}\}$
range = $\{-1, 0, 1\}$

39. The vertex moves along the lines $y = \pm x$.

41. x-intercepts are a and b.

Exercise 4.4 (page 225)

1. $y = x^2 - 2$
   shifted 2 units down

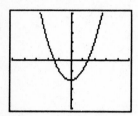

3. $y = (x+3)^2$
   shifted 3 units left

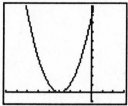

5. $y = (x+1)^2 + 2$
   shifted 2 units up, 1 unit left

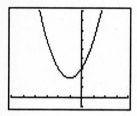

7. $y = \left(x + \frac{1}{2}\right)^2 - \frac{1}{2}$

   shifted 1/2 unit down, 1/2 unit left

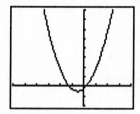

9. $y = x^3 + 1$
   shifted 1 unit up

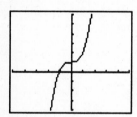

11. $y = (x-2)^3$
    shifted 2 units right

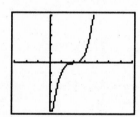

13. $y = (x-2)^3 - 3$
    shifted 3 units down, 2 units right

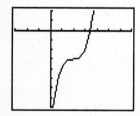

15. $y + 2 = x^3 \implies y = x^3 - 2$
    shifted 2 units down

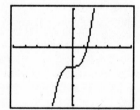

142

17. $y = 2x^2$
stretched vertically by 2

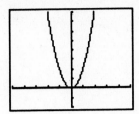

19. $y = -3x^2$ {reversed}
stretched vertically by 3

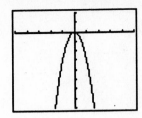

21. $y = \left(\frac{1}{2}x\right)^3$
stretched horizontally by 2

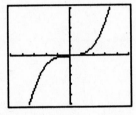

23. $y = -8x^3$ {reversed}
stretched vertically a factor of 8

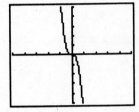

25. $y = |x-2| + 1$
$y = |x|$ shifted 1 unit up and
2 units right.

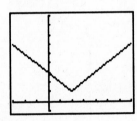

27. $y = \sqrt{x-2} + 1$
$y = \sqrt{x}$ shifted 1 unit up and
2 units right.

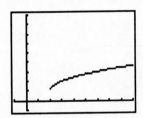

29. $y = |3x|$
$y = |x|$ compressed horizontally
by a factor of 3.

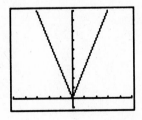

31. $y = 2\sqrt{x} + 3$
$y = \sqrt{x}$ stretched vertically by a factor
of 2, then shifted 3 units up.

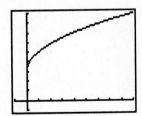

33. $y = -2|x| + 3$:   same as $y = |x|$, except reversed, stretched by a factor of 2, and then shifted up 3 units.

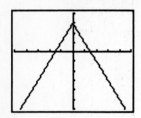

35, 37. Answers will vary.

Exercise 4.5   (page 236)

1. $\{x \mid x \in \mathbf{R}, x \neq 2\}$

3. $\{x \mid x \in \mathbf{R}, x \neq \pm 5\}$

5. $\{x \mid x \in \mathbf{R}, x \neq 0, x \neq \pm 1\}$

7. $\{x \mid x \in \mathbf{R}\}$

9. $y = \dfrac{1}{x - 2}$
   vertical: $x = 2$
   horizontal: $y = 0$
   slant: none
   x-intercept(s): none
   y-intercept(s): $\left(0, -\frac{1}{2}\right)$
   symmetry: none

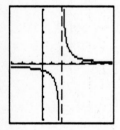

11. $y = \dfrac{x}{x - 1}$
   vertical: $x = 1$
   horizontal: $y = 1$
   slant: none
   x-intercept(s): $(0, 0)$
   y-intercept(s): $(0, 0)$
   symmetry: none

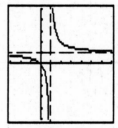

13. $y = \dfrac{x+1}{x+2}$

vertical: $x = -2$
horizontal: $y = 1$
slant: none
x-intercept(s): $(-1, 0)$

y-intercept(s): $\left(0, \frac{1}{2}\right)$

symmetry: none

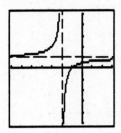

15. $y = \dfrac{2x-1}{x-1}$

vertical: $x = 1$
horizontal: $y = 2$
slant: none
x-intercept(s): $\left(\frac{1}{2}, 0\right)$

y-intercept(s): $(0, 1)$

symmetry: none

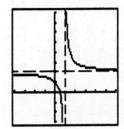

17. $y = \dfrac{x^2 - 9}{x^2 - 4} = \dfrac{(x+3)(x-3)}{(x+2)(x-2)}$

vertical: $x = -2$, $x = 2$
horizontal: $y = 1$
slant: none
x-intercept(s): $(-3, 0)$, $(3, 0)$

y-intercept(s): $\left(0, \frac{9}{4}\right)$

symmetry: about y-axis

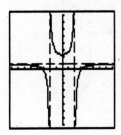

19. $y = \dfrac{x^2 - x - 2}{x^2 - 4x + 3} = \dfrac{(x-2)(x+1)}{(x-3)(x-1)}$

vertical: $x = 3$, $x = 1$
horizontal: $y = 1$
slant: none
x-intercept(s): $(2, 0)$, $(-1, 0)$

y-intercept(s): $\left(0, -\frac{2}{3}\right)$

symmetry: none

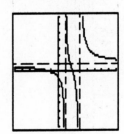

21. $y = \dfrac{x^2 + 2x - 3}{x^3 - 4x} = \dfrac{(x+3)(x-1)}{x(x+2)(x-2)}$

    vertical: $x = 0$, $x = 2$, $x = -2$
    horizontal: $y = 0$
    slant: none
    x-intercept(s): $(-3, 0)$, $(1, 0)$
    y-intercept(s): none
    symmetry: none

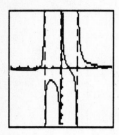

23. $y = \dfrac{x^2 - 9}{x^2} = \dfrac{(x+3)(x-3)}{x^2}$

    vertical: $x = 0$
    horizontal: $y = 1$
    slant: none
    x-intercept(s): $(3, 0)$, $(-3, 0)$
    y-intercept(s): none
    symmetry: about y-axis

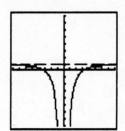

25. $y = \dfrac{x}{(x+3)^2}$

    vertical: $x = -3$
    horizontal: $y = 0$
    slant: none
    x-intercept(s): $(0, 0)$
    y-intercept(s): $(0, 0)$
    symmetry: none

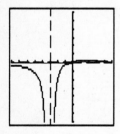

27. $y = \dfrac{x + 1}{x^2(x - 2)}$

    vertical: $x = 0$, $x = 2$
    horizontal: $y = 0$
    slant: none
    x-intercept(s): $(-1, 0)$
    y-intercepts: none
    symmetry: none

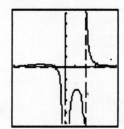

29. $y = \dfrac{x}{x^2 + 1}$

    vertical: none

    horizontal: $y = 0$

    slant: none

    x-intercept(s): $(0, 0)$

    y-intercept(s): $(0, 0)$

    symmetry: about origin

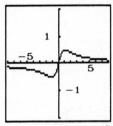

31. $y = \dfrac{3x^2}{x^2 + 1}$

    vertical: none

    horizontal: $y = 3$

    slant: none

    x-intercept(s): $(0, 0)$

    y-intercept(s): $(0, 0)$

    symmetry: about y-axis

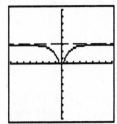

33. $y = \dfrac{x^2 - 2x - 8}{x - 1} = \dfrac{(x - 4)(x + 2)}{x - 1}$

    vertical: $x = 1$

    horizontal: none

    slant: $y = x$

    x-intercept(s): $(4, 0)$, $(-2, 0)$

    y-intercept(s): $(0, 8)$

    symmetry: none

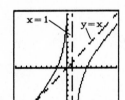

35. $y = \dfrac{x^3 + x^2 + 6x}{x^2 - 1} = \dfrac{x(x^2 + x + 6)}{(x + 1)(x - 1)}$

    vertical: $x = 1$, $x = -1$

    horizontal: none

    slant: $y = x$

    x-intercept(s): $(0, 0)$

    y-intercept(s): $(0, 0)$

    symmetry: none

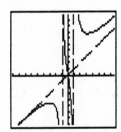

37. $y = \dfrac{x^2}{x}$

    If $x \neq 0$, then $y = x$.

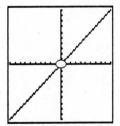

39. $y = \dfrac{x^3 + x}{x} = \dfrac{x(x^2 + 1)}{x}$

    If $x \neq 0$, then $y = x^2 + 1$.

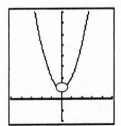

41. $y = \dfrac{x^2 - 2x + 1}{x - 1} = \dfrac{(x-1)(x-1)}{x-1}$

If $x \neq 1$, then $y = x - 1$.

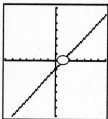

43. $y = \dfrac{x^3 - 1}{x - 1} = \dfrac{(x-1)(x^2 + x + 1)}{x - 1}$

If $x \neq 1$, then $y = x^2 + x + 1$

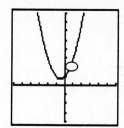

45. No. The horizontal asymptote is determined from the quotient

$\dfrac{P(x)}{Q(x)}$, which is unique. There is

at most one horizontal asymptote.

47. $y = \dfrac{ax + b}{cx^2 + d} = \dfrac{\dfrac{ax}{x^2} + \dfrac{b}{x^2}}{\dfrac{cx^2}{x^2} + \dfrac{d}{x^2}} = \dfrac{\dfrac{a}{x} + \dfrac{b}{x^2}}{c + \dfrac{d}{x^2}}$

As x gets large, $y \approx \dfrac{0 + 0}{c + 0} = 0$

The horizontal asymptote is $y = 0$.

49. $y = \dfrac{ax^3 + b}{cx^2 + d} = \dfrac{\dfrac{ax^3}{x^2} + \dfrac{b}{x^2}}{\dfrac{cx^2}{x^2} + \dfrac{d}{x^2}} = \dfrac{ax + \dfrac{b}{x^2}}{c + \dfrac{d}{x^2}}.$

As x gets large, $y \approx \dfrac{ax + 0}{c + 0} = \frac{a}{c}x.$ The slant asymptote is $y = \frac{a}{c}x.$

51-53. Answers will vary.

Exercise 4.6   (page 244)

1. $\begin{aligned} f + g &= (2x + 1) + (3x - 2) \\ &= 5x - 1 \end{aligned}$

domain $= \{x \mid x \in \mathbb{R}\}$

3. $\begin{aligned} f \cdot g &= (2x + 1)(3x - 2) \\ &= 6x^2 - x - 2 \end{aligned}$

domain $= \{x \mid x \in \mathbb{R}\}$

5. $\begin{aligned} f - g &= (x^2 + x) - (x^2 - 1) \\ &= x + 1 \end{aligned}$

domain $= \{x \mid x \in \mathbb{R}\}$

7. $f/g = \dfrac{x^2 + x}{x^2 - 1}$

domain $= \{x \mid x \in \mathbb{R}, x \neq \pm 1\}$

9. $\begin{aligned} (f + g)(2) &= f(2) + g(2) \\ &= [2^2 - 1] + [3(2) - 2] \\ &= 3 + 4 \\ &= 7 \end{aligned}$

11. $\begin{aligned} (f - g)(0) &= f(0) - g(0) \\ &= [0^2 - 1] - [3(0) - 2] \\ &= -1 - -2 \\ &= 1 \end{aligned}$

13. $(f \cdot g)(2) = f(2) \cdot g(2)$
$= [2^2 - 1] \cdot [3(2) - 2)]$
$= 3 \cdot 4$
$= 12$

15. $(f/g)\left(\frac{2}{3}\right) = \dfrac{f\left(\frac{2}{3}\right)}{g\left(\frac{2}{3}\right)}$

$= \dfrac{\left(\frac{2}{3}\right)^2 - 1}{3\left(\frac{2}{3}\right) - 2}$

$= \dfrac{\frac{4}{9} - \frac{9}{9}}{2 - 2} = \dfrac{-\frac{5}{9}}{0}$

The value is undefined.

17. $f(x) = 3x^2$, $g(x) = 2x$

19. $f(x) = 3x^2$, $g(x) = x^2 - 1$

21. $y = x(3x^2 + 1) = 3x^3 + x$
$f(x) = 3x^3$, $g(x) = -x$

23. $y = x^2 + 7x - 18 = (x - 2)(x + 9)$
$f(x) = x - 2$, $g(x) = x + 9$

25. $(f \circ g)(2) = f(g(2))$
$= f(5(2) - 2)$
$= f(8)$
$= 2(8) - 5$
$= 11$

27. $(f \circ f)\left(-\frac{1}{2}\right) = f\left(f\left(-\frac{1}{2}\right)\right)$
$= f\left(2\left(-\frac{1}{2}\right) - 5\right)$
$= f(-6)$
$= 2(-6) - 5$
$= -17$

29. $(f \circ g)(-3) = f((g(-3))$
$= f(4(-3) + 4)$
$= f(-8)$
$= 3(-8)^2 - 2$
$= 190$

31. $(f \circ f)(\sqrt{3}) = f(f(\sqrt{3}))$
$= f(3(\sqrt{3})^2 - 2)$
$= f(7)$
$= 3(7)^2 - 2$
$= 145$

33. $(f \circ g)(x) = f(g(x))$
$= f(x + 1)$
$= 3(x + 1)$
$= 3x + 3$
domain $= \{x \mid x \in \mathbb{R}\}$

35. $(f \circ f)(x) = f(f(x))$
$= f(3x)$
$= 3(3x)$
$= 9x$
domain $= \{x \mid x \in \mathbb{R}\}$

37. $(g \circ f)(x) = g(f(x))$
$= g(x^2)$
$= 2(x^2)$
$= 2x^2$
domain $= \{x \mid x \in \mathbb{R}\}$

39. $(g \circ g)(x) = g(g(x))$
$= g(2x)$
$= 2(2x)$
$= 4x$
domain $= \{x \mid x \in \mathbb{R}\}$

41.
$$(f \circ g)(x) = f(g(x))$$
$$= f(x^2 + 1)$$
$$= \sqrt{x^2 + 1}$$

domain $= \{x \mid x \in \mathbb{R}\}$

43.
$$(f \circ f)(x) = f(f(x))$$
$$= f(\sqrt{x})$$
$$= \sqrt{\sqrt{x}}$$
$$= \sqrt[4]{x}$$

domain $= \{x \mid x \in \mathbb{R}, x \geq 0\}$

45.
$$(g \circ f)(x) = g(f(x))$$
$$= g(\sqrt{x+1})$$
$$= (\sqrt{x+1})^2 - 1$$
$$= x$$

domain $= \{x \mid x \in \mathbb{R}, x \geq -1\}$

47.
$$(g \circ g)(x) = g(g(x))$$
$$= g(x^2 - 1)$$
$$= (x^2 - 1)^2 - 1$$
$$= x^4 - 2x^2$$

domain $= \{x \mid x \in \mathbb{R}\}$

49. $f(x) = x - 2, g(x) = 3x$

51. $f(x) = x - 2, g(x) = x^2$

53. $f(x) = x^2, g(x) = x - 2$

55. $f(x) = \sqrt{x}, g(x) = x + 2$

57. $f(x) = x + 2, g(x) = \sqrt{x}$

59. $f(x) = x, g(x) = x$

61. $(f + f)(x) = f(x) + f(x) = 3x + 3x = 6x; \quad f(x + x) = f(2x) = 3(2x) = 6x$

63.
$$(f \circ f)(x) = f(f(x))$$
$$= f\left(\frac{x-1}{x+1}\right) = \frac{\frac{x-1}{x+1} - 1}{\frac{x-1}{x+1} + 1} = \frac{\frac{x-1}{x+1} - \frac{x+1}{x+1}}{\frac{x-1}{x+1} + \frac{x+1}{x+1}} = \frac{\frac{-2}{x+1}}{\frac{2x}{x+1}} = -\frac{1}{x}$$

65. $f(x) = x^2 - x$
$(f \circ g)(x) = (x-3)^2 - (x-3)$

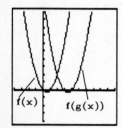

67. $f(x) = x^2 - x$
$(f \circ h)(x) = (3x)^2 - 3x$

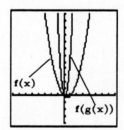

Exercise 4.7 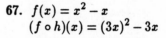 (page 251)

1. one-to-one

3. not one-to-one: $x = \pm 1 \Rightarrow y = 4$.

5. not one-to-one: $x = \pm 1 \Rightarrow y = 0$.

7. not one-to-one: $x = \pm 1 \Rightarrow y = 1$.

9. not one-to-one: $x = \pm 2 \Rightarrow y = 1$.

11. one-to-one

13. one-to-one

15. not one-to-one (not a function)

17.
$$(f \circ g)(x) = f(g(x))$$
$$= f\left(\tfrac{1}{5}x\right)$$
$$= 5\left(\tfrac{1}{5}x\right)$$
$$= x$$

$$(g \circ f)(x) = g(f(x))$$
$$= g(5x)$$
$$= \tfrac{1}{5}(5x)$$
$$= x$$

19.
$$(f \circ g)(x) = f(g(x))$$
$$= f\left(\frac{1}{x-1}\right)$$
$$= \frac{\dfrac{1}{x-1}+1}{\dfrac{1}{x-1}}$$
$$= \frac{\dfrac{1}{x-1}+\dfrac{x-1}{x-1}}{\dfrac{1}{x-1}}$$
$$= \frac{\dfrac{x}{x-1}}{\dfrac{1}{x-1}}$$
$$= x$$

$$(g \circ f)(x) = g(f(x))$$
$$= g\left(\frac{x+1}{x}\right)$$
$$= \frac{1}{\dfrac{x+1}{x}-1}$$
$$= \frac{1}{\dfrac{x+1}{x}-\dfrac{x}{x}}$$
$$= \frac{1}{\dfrac{1}{x}}$$
$$= x$$

21.
$$y = 3x$$
$$x = 3y$$
$$y = \tfrac{x}{3}$$
$$f^{-1}(x) = \tfrac{x}{3}$$

23.
$$y = 3x + 2$$
$$x = 3y + 2$$
$$x - 2 = 3y$$
$$y = \frac{x-2}{3}$$
$$f^{-1}(x) = \frac{x-2}{3}$$

25.
$$y = \frac{1}{x+3}$$
$$x = \frac{1}{y+3}$$
$$(y+3)x = 1$$
$$y + 3 = \tfrac{1}{x}$$
$$y = \tfrac{1}{x} - 3$$
$$f^{-1}(x) = \tfrac{1}{x} - 3$$

27.
$$y = \frac{1}{2x}$$
$$x = \frac{1}{2y}$$
$$2yx = 1$$
$$y = \frac{1}{2x}$$
$$f^{-1}(x) = \frac{1}{2x}$$

29.
$$y = 5x$$
$$x = 5y$$
$$y = \frac{x}{5}$$
$$f^{-1}(x) = \frac{x}{5}$$

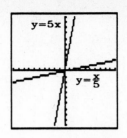

31.
$$y = 2x - 4$$
$$x = 2y - 4$$
$$x + 4 = 2y$$
$$y = \frac{x + 4}{2}$$
$$f^{-1}(x) = \frac{x + 4}{2}$$

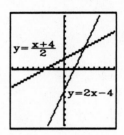

33.
$$x - y = 2$$
$$-y = -x + 2$$
$$y = x - 2$$
$$x = y - 2$$
$$x + 2 = y$$
$$f^{-1}(x) = x + 2$$

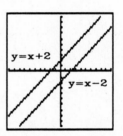

35.
$$2x + y = 4$$
$$y = -2x + 4$$
$$x = -2y + 4$$
$$x - 4 = -2y$$
$$y = \frac{x - 4}{-2}$$
$$f^{-1}(x) = \frac{x - 4}{-2}$$

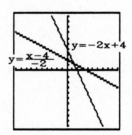

37.
$$y = \frac{1}{2x}$$
$$x = \frac{1}{2y}$$
$$2yx = 1$$
$$y = \frac{1}{2x}$$
$$f^{-1}(x) = \frac{1}{2x}$$

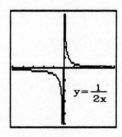

39.
$$y = \frac{x+1}{x-1}$$

$$x = \frac{y+1}{y-1}$$

$$x(y-1) = y+1$$
$$xy - x = y+1$$
$$xy - y = x+1$$
$$y(x-1) = x+1$$

$$y = \frac{x+1}{x-1}$$

$$f^{-1}(x) = \frac{x+1}{x-1}$$

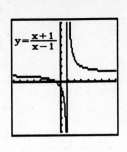

41.
$$f(x) = x^2 - 3,\ x \le 0$$
$$y = x^2 - 3,\ x \le 0$$
$$x = y^2 - 3,\ y \le 0$$
$$x + 3 = y^2,\ y \le 0$$
$$y^2 = x + 3,\ y \le 0$$

$$y = \pm\sqrt{x+3}\ ,\ y \le 0$$

$$y = -\sqrt{x+3}$$

$$f^{-1}(x) = -\sqrt{x+3}$$

43.
$$f(x) = x^4 - 8,\ x \ge 0$$
$$y = x^4 - 8,\ x \ge 0$$
$$x = y^4 - 8,\ y \ge 0$$
$$x + 8 = y^4,\ y \ge 0$$
$$y^4 = x + 8,\ y \ge 0$$

$$y = \pm\ ^4\!\sqrt{x+8},\ y \ge 0$$

$$y = \ ^4\!\sqrt{x+8}$$

$$f^{-1}(x) = \ ^4\!\sqrt{x+8}$$

45.
$$f(x) = \sqrt{4-x^2},\ 0 \le x \le 2$$

$$y = \sqrt{4-x^2},\ 0 \le x \le 2$$

$$x = \sqrt{4-y^2},\ 0 \le y \le 2$$

$$x^2 = 4 - y^2,\ 0 \le y \le 2$$

$$y^2 = 4 - x^2,\ 0 \le y \le 2$$

$$y = \pm\sqrt{4-x^2},\ 0 \le y \le 2$$

$$y = \sqrt{4-x^2}$$

$$f^{-1}(x) = \sqrt{4-x^2}$$

47.
$$f(x) = \frac{x}{x-1}$$

$$y = \frac{x}{x-1}$$

$$x = \frac{y}{y-1}$$

$$(y-1)x = y$$
$$yx - x = y$$
$$yx - y = x$$
$$y(x-1) = x$$

$$y = \frac{x}{x-1}$$

$$f^{-1}(x) = \frac{x}{x-1}$$

domain of $f = \{x \mid x \in \mathbb{R}, x \neq 1\}$
range of f = domain of f^{-1}
$\qquad = \{x \mid x \in \mathbb{R}, x \neq 1\}$

49.
$$f(x) = \frac{1}{x} - 2$$

$$y = \frac{1}{x} - 2$$

$$x = \frac{1}{y} - 2$$

$$x + 2 = \frac{1}{y}$$

$$y(x+2) = 1$$

$$y = \frac{1}{x+2}$$

$$f^{-1}(x) = \frac{1}{x+2}$$

domain of $f = \{x \mid x \in \mathbb{R}, x \neq 0\}$
range of f = domain of f^{-1}
$\qquad = \{x \mid x \in \mathbb{R}, x \neq -2\}$

Chapter 4 Review Exercises (page 254)

1. function

3. not a function: $y = \pm 1 \Rightarrow x = -4$.

5. domain $= \{x \mid x \in \mathbb{R}\}$
range $= \{y \mid y \in \mathbb{R}, y \geq -2\}$

7. domain $= \{x \mid x \in \mathbb{R}, x \geq 1\}$
range $= \{y \mid y \in \mathbb{R}, y \geq 0\}$

9. $f(x) = 5x - 2$
$$\begin{aligned} f(2) &= 5(2) - 2 & f(-3) &= 5(-3) - 2 & f(0) &= 5(0) - 2 \\ &= 8 & &= -17 & &= -2 \end{aligned}$$

11. $f(x) = |x - 2|$
$$\begin{aligned} f(2) &= |2 - 2| & f(-3) &= |-3 - 2| & f(0) &= |0 - 2| \\ &= 0 & &= 5 & &= 2 \end{aligned}$$

13. $y = x^2 - x$
$a = 1, \ b = -1, \ c = 0$

$$h = -\frac{b}{2a} = -\frac{-1}{2(1)} = \frac{1}{2}$$

$$k = c - \frac{b^2}{4a} = 0 - \frac{(-1)^2}{4(1)} = -\frac{1}{4}$$

Vertex $\left(\frac{1}{2}, \ -\frac{1}{4}\right)$

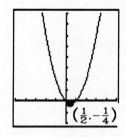
$\left(\frac{1}{2}, -\frac{1}{4}\right)$

15. $y = x^2 - 3x - 4$

$a = 1, b = -3, c = -4$

$h = -\dfrac{b}{2a} = -\dfrac{-3}{2(1)} = \dfrac{3}{2}$

$k = c - \dfrac{b^2}{4a} = -4 - \dfrac{(-3)^2}{4(1)} = -\dfrac{25}{4}$

Vertex $\left(\dfrac{3}{2}, -\dfrac{25}{4}\right)$

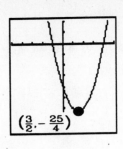

$\left(\dfrac{3}{2}, -\dfrac{25}{4}\right)$

17. $3x^2 + y - 300 = 0$

$\qquad\qquad y = -3x^2 + 300$

$a = -3, b = 0, c = 300$

$h = -\dfrac{b}{2a} = -\dfrac{0}{2(-3)} = 0$

$k = c - \dfrac{b^2}{4a} = 300 - \dfrac{0^2}{4(-3)} = 300$

Since the graph is a parabola with a negative coefficient of x^2, the vertex is a maximum. The maximum height is the y-coordinate, or 300.

19. $y = x^3 - x$: odd

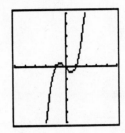

21. $y = x^3 - x^2$: neither

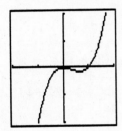

23. increasing on $(-\infty, 0]$

decreasing on $(0, \infty)$

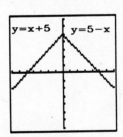

25. $y = [x - 2]$

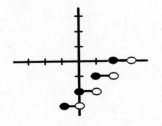

27. $y = \sqrt{x + 2} + 3$: Start with $y = \sqrt{x}$ and shift 2 units to the left and 3 up.

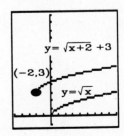

29. $y = \frac{1}{3}x^3$: Start with $y = x^3$ and compress vertically by a factor of $\frac{1}{3}$.

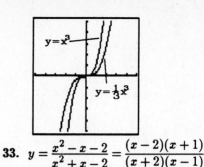

31. $y = \dfrac{x}{(x-1)^2}$

vertical: $x = 1$
horizontal: $y = 0$
slant: none
x-intercept(s): $(0, 0)$
y-intercept(s): $(0, 0)$
symmetry: none

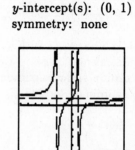

33. $y = \dfrac{x^2 - x - 2}{x^2 + x - 2} = \dfrac{(x-2)(x+1)}{(x+2)(x-1)}$

vertical: $x = -2$, $x = 1$
horizontal: $y = 1$
slant: none
x-intercept(s): $(2, 0)$, $(-1, 0)$
y-intercept(s): $(0, 1)$
symmetry: none

35. $\begin{aligned}(f + g)(x) &= f(x) + g(x) \\ &= x^2 - 1 + 2x + 1 \\ &= x^2 + 2x\end{aligned}$

37. $\begin{aligned}(f \circ g)(x) &= f(g(x)) \\ &= f(2x + 1) \\ &= (2x + 1)^2 - 1 \\ &= 4x^2 + 4x + 1 - 1 \\ &= 4x^2 + 4x\end{aligned}$

39. $\begin{aligned}y &= 7x - 1 \\ x &= 7y - 1 \\ x + 1 &= 7y \\[4pt] y &= \frac{x+1}{7} \\[4pt] f^{-1}(x) &= \frac{x+1}{7}\end{aligned}$

41. $\begin{aligned}y &= \frac{x}{1-x} \\[4pt] x &= \frac{y}{1-y} \\[4pt] (1-y)x &= y \\ x - yx &= y \\ -y - yx &= -x \\ y(-1-x) &= -x \\[4pt] y &= \frac{-x}{-1-x} \\[4pt] f^{-1}(x) &= \frac{x}{1+x}\end{aligned}$

Chapter 4 Test (page 255)

1. domain $= \{x \mid x \in \mathbf{R}, x \neq 5\}$
range $= \{y \mid y \in \mathbf{R}, y \neq 0\}$

3. $f(-1) = \dfrac{-1}{-1-1} = \dfrac{-1}{-2} = \dfrac{1}{2}$

$f(2) = \dfrac{2}{2-1} = \dfrac{2}{1} = 2$

5. $y = 3(x-7)^2 - 3$
$h = 7, y = -3$
vertex $(7, -3)$

7. $y = 3x^2 - 24x + 38$
$a = 3, b = -24, c = 38$

$h = -\dfrac{b}{2a} = -\dfrac{-24}{2(3)} = 4$

$k = c - \dfrac{b^2}{4a} = 38 - \dfrac{(-24)^2}{4(3)} = -10$

vertex: $(4, -10)$

9. $y = x^4 - x^2$

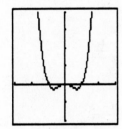

11. $h = 100t - 16t^2$
$h = -16t^2 + 100t$
$a = -16, b = 100, c = 0$
The maximum height occurs at the vertex of the parabola. This will occur when the number of seconds is the same as the x-coordinate of the vertex, which is

$$-\dfrac{b}{2a} = -\dfrac{100}{2(-16)} = \dfrac{25}{8} \text{ seconds.}$$

13. The roadway is at $y = 0$, so to find the distance to the cable's lowest point, the y-coordinate of the vertex is needed.

$x^2 - 2500y + 25,000 = 0$

$y = \dfrac{1}{2500}x^2 + 10$

$c - \dfrac{b^2}{4a} = 10 - \dfrac{0^2}{4a} = 10$

The distance is 10 feet.

15. $y = \dfrac{x-1}{x^2-9} = \dfrac{x-1}{(x+3)(x-3)}$

vertical: $x = -3, x = 3$

horizontal: $y = 0$

17. $y = \dfrac{x^2}{x^2 - 9} = \dfrac{x^2}{(x+3)(x-3)}$

    vertical: $x = -3,\ x = 3$
    horizontal: $y = 1$
    x-intercept(s): $(0,\ 0)$
    y-intercept(s): $(0,\ 0)$
    symmetry: about y-axis

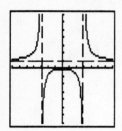

19. $y = \dfrac{2x^2 - 3x - 2}{x - 2} = \dfrac{(2x+1)(x-2)}{x-2}$

    Like $y = 2x + 1$, except $x \neq 2$.

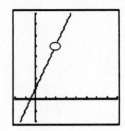

21. $f + g = f(x) + g(x) = 3x + x^2 + 2 = x^2 + 3x + 2$

23. $f/g = \dfrac{f(x)}{g(x)} = \dfrac{3x}{x^2 + 2}$

25.

$$f(x) = \frac{x+1}{x-1}$$

$$y = \frac{x+1}{x-1}$$

$$x = \frac{y+1}{y-1}$$

$$x(y-1) = y+1$$

$$xy - x = y + 1$$

$$xy - y = x + 1$$

$$y(x-1) = x + 1$$

$$y = \frac{x+1}{x-1}$$

$$f^{-1}(x) = \frac{x+1}{x-1}$$

27. $y = (x - 3)^2 + 1$

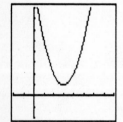

29.

$$y = \frac{3}{x} - 2$$

$$x = \frac{3}{y} - 2$$

$$x + 2 = \frac{3}{y}$$

$$y(x + 2) = 3$$

$$y = \frac{3}{x + 2}$$

$$\text{range} = \{x \mid x \in \mathbb{R}, x \neq -2\}$$

Exercise 5.1 (page 258)

1. $P(1) = 2(1)^4 - 2(1)^3 + 5(1)^2 - 1 = 2(1) - 2(1) + 5(1) - 1 = 2 - 2 + 5 - 1 = \boxed{4}$

$$
\begin{array}{r}
2x^3 \qquad\quad + 5x \;\; + 5 \\[2pt]
\hline
x-1 \;)\; 2x^4 \;-\; 2x^3 \;+\; 5x^2 \;+\; 0x \;-\; 1 \\[2pt]
\underline{-(\;2x^4 \;-\; 2x^3)} \\[2pt]
0 \;\;+\; 5x^2 \;+\; 0x \\[2pt]
\underline{-(\; 5x^2 \;-\; 5x)} \\[2pt]
5x \;\;-\; 1 \\[2pt]
\underline{-(\; 5x \;-\; 5)} \\[2pt]
\boxed{4}
\end{array}
$$

3. $P(-2) = 2(-2)^4 - 2(-2)^3 + 5(-2)^2 - 1 = 32 + 16 + 20 - 1 = \boxed{67}$

$$
\begin{array}{r}
2x^3 \;-\; 6x^2 \;+\; 17x \;-\; 34 \\[2pt]
\hline
x+2 \;)\; 2x^4 \;-\; 2x^3 \;+\; 5x^2 \;+\; 0x \;-\; 1 \\[2pt]
\underline{-(\; 2x^4 \;+\; 4x^3)} \\[2pt]
-6x^3 \;+\; 5x^2 \\[2pt]
\underline{-(-6x^3 \;-\; 12x^2)} \\[2pt]
17x^2 \;+\; 0x \\[2pt]
\underline{-(\; 17x^2 \;+\; 34x)} \\[2pt]
-34x \;-\; 1 \\[2pt]
\underline{-(-34x \;-\; 68)} \\[2pt]
\boxed{67}
\end{array}
$$

5. $P(-3) = 2(-3)^4 - 2(-3)^3 + 5(-3)^2 - 1 = 162 + 54 + 45 - 1 = \boxed{260}$

$$
\begin{array}{r}
2x^3 \;-\; 8x^2 \;+\; 29x \;-\; 87 \\[2pt]
\hline
x+3 \;)\; 2x^4 \;-\; 2x^3 \;+\; 5x^2 \;+\; 0x \;-\; 1 \\[2pt]
\underline{-(\; 2x^4 \;+\; 6x^3)} \\[2pt]
-8x^3 \;+\; 5x^2 \\[2pt]
\underline{-(-8x^3 \;-\; 24x^2)} \\[2pt]
29x^2 \;+\; 0x \\[2pt]
\underline{-(\; 29x^2 \;+\; 87x)} \\[2pt]
-87x \;-\; 1 \\[2pt]
\underline{-(-87x \;-\; 261)} \\[2pt]
\boxed{260}
\end{array}
$$

7. $P\left(\frac{1}{2}\right) = 2\left(\frac{1}{2}\right)^4 - 2\left(\frac{1}{2}\right)^3 + 5\left(\frac{1}{2}\right)^2 - 1 = \frac{1}{8} - \frac{1}{4} + \frac{5}{4} - 1 = \frac{1}{8} + 1 - 1 = \boxed{\frac{1}{8}}$

$$
\begin{array}{r}
2x^3 \;-\; x^2 \;+\; \frac{9}{2}x \;-\; \frac{9}{4} \\[4pt]
x - \frac{1}{2} \overline{\bigg)\; 2x^4 \;-\; 2x^3 \;+\; 5x^2 \;+\; 0x \;-\; 1} \\
-(\;2x^4 \;-\; x^3\,) \\
\hline
-\,x^3 \;+\; 5x^2 \\
-(-\,x^3 \;+\; \tfrac{1}{2}x^2\,) \\
\hline
\tfrac{9}{2}x^2 \;+\; 0x \\
-(\;\tfrac{9}{2}x^2 \;-\; \tfrac{9}{4}x\,) \\
\hline
-\,\tfrac{9}{4}x \;-\; 1 \\
-(-\,\tfrac{9}{4}x \;+\; \tfrac{9}{8}\,) \\
\hline
\boxed{\tfrac{1}{8}}
\end{array}
$$

9. $P(1) = 3(1)^4 + 5(1)^3 - 4(1)^2 - 2(1) + 1 = 3 + 5 - 4 - 2 + 1 = 3$

11. $P(-2) = 3(-2)^4 + 5(-2)^3 - 4(-2)^2 - 2(-2) + 1 = 48 - 40 - 16 + 4 + 1 = -3$

13. $P(12) = 3(12)^4 + 5(12)^3 - 4(12)^2 - 2(12) + 1 = 70{,}249$

15. $P(-3.25) = 3(-3.25)^4 + 5(-3.25)^3 - 4(-3.25)^2 - 2(-3.25) + 1 \approx 128.308$

17.

$$
\begin{array}{r}
x^4 \;+\; 2x^3 \;+\; 4x^2 \;+\; 8x \;+\; 16 \\[4pt]
x - 2 \overline{\bigg)\; x^5 \;+\; 0x^4 \;+\; 0x^3 \;+\; 0x^2 \;+\; 0x \;-\; 1} \\
-(\;x^5 \;-\; 2x^4\,) \\
\hline
2x^4 \;+\; 0x^3 \\
-(\;2x^4 \;-\; 4x^3\,) \\
\hline
4x^3 \;+\; 0x^2 \\
-(\;4x^3 \;-\; 8x^2\,) \\
\hline
8x^2 \;+\; 0x \\
-(\;8x^2 \;-\; 16x\,) \\
\hline
16x \;-\; 1 \\
-(\;16x \;-\; 32\,) \\
\hline
\boxed{31} = P(2)
\end{array}
$$

19.

$$-x^4 + 3x^3 - 3x^2 + 9x - 47$$

$$
x+3 \enclose{longdiv}{\;-x^5 + 0x^4 + 6x^3 + 0x^2 - 20x + 1}
$$

$$\underline{-(-x^5 - 3x^4)}$$

$$3x^4 + 6x^3$$

$$\underline{-(\;\;3x^4 + 9x^3)}$$

$$-3x^3 + 0x^2$$

$$\underline{-(-3x^3 - 9x^2)}$$

$$9x^2 - 20x$$

$$\underline{-(\;\;9x^2 + 27x)}$$

$$-47x + 1$$

$$\underline{-(-47x - 141)}$$

$$\boxed{142} = P(-3)$$

21. $x - 1$ is a factor if $P(1) = 0$.
$P(x) = x^7 - 1$
$P(1) = 1^7 - 1 = 0$
$x - 1$ is a factor of $x^7 - 1$.

23. $x - 1$ is a factor if $P(1) = 0$.
$P(x) = 3x^5 + 4x^2 - 7$
$P(1) = 3(1)^5 + 4(1)^2 - 7 = 0$
$x - 1$ is a factor of $3x^5 + 4x^2 - 7$.

25. $x + 3$ is a factor if $P(-3) = 0$.
$P(x) = 2x^3 - 2x^2 + 1$
$\quad = 2(-3)^3 - 2(-3)^2 + 1$
$\quad = -54 - 18 + 1 = -71$
$x + 3$ is not a factor of $2x^3 - 2x^2 + 1$.

27. $x - 1$ is a factor if $P(1) = 0$.
$P(x) = x^{1984} - x^{1776} + x^{1492} - x^{1066}$
$\quad = 1^{1984} - 1^{1776} + 1^{1492} - 1^{1066}$
$\quad = 1 - 1 + 1 - 1 = 0$
$x - 1$ is a factor of the polynomial.

29. If -1 is a root of the equation, then $x + 1$ is a factor of $x^3 + 3x^2 - 13x - 15$.

$$
x^2 + 2x - 15
$$

$$
x+1 \enclose{longdiv}{\;x^3 + 3x^2 - 13x - 15}
$$

$$\underline{-(\;x^3 + x^2)}$$

$$2x^2 - 13x$$

$$\underline{-(\;2x^2 + 2x)}$$

$$-15x - 15$$

$$\underline{-(-15x - 15)}$$

$x^3 + 3x^2 - 13x - 15 = 0$
$(x + 1)(x^2 + 2x - 15) = 0$
$(x + 1)(x + 5)(x - 3) = 0$

The solution set is $\{-1, -5, 3\}$.

31. If 1 is a double root of the equation, then $x - 1$ is a double factor of $x^4 - 2x^3 - 2x^2 + 6x - 3$. Then $(x - 1)(x - 1) = x^2 - 2x + 1$ is a factor.

$$
\begin{array}{r}
x^2 \qquad\quad - \ 3 \\
x^2 - 2x + 1 \overline{\big)\ x^4 \quad - \ 2x^3 \quad - \ 2x^2 \quad + \ 6x \quad - \ 3} \\
\underline{-(\ x^4 \quad - \ 2x^3 \quad + \ x^2)} \\
- \ 3x^2 \quad + \ 6x \quad - \ 3 \\
\underline{-(- \ 3x^2 \quad + \ 6x \quad - \ 3)}
\end{array}
$$

$$
\begin{aligned}
x^4 - 2x^3 - 2x^2 + 6x - 3 &= 0 \quad \bullet \\
(x^2 - 2x + 1)(x^2 - 3) &= 0
\end{aligned}
$$

$$
\begin{array}{ll}
x^2 - 2x + 1 = 0 \quad \text{or} & x^2 - 3 = 0 \\
(x - 1)(x - 1) = 0 & \quad x^2 = 3 \\
x = 1 & \quad x = \pm\sqrt{3} \quad \text{The solution set is } \{1, \ \sqrt{3}, \ -\sqrt{3}\}.
\end{array}
$$

33. If 2 and 3 are roots of the equation, then $x - 2$ and $x - 3$ are factors of $x^4 - 5x^3 + 7x^2 - 5x + 6$. Then $(x - 2)(x - 3) = x^2 - 5x + 6$ is a factor.

$$
\begin{array}{r}
x^2 \qquad\quad + \ 1 \\
x^2 - 5x + 6 \overline{\big)\ x^4 \quad - \ 5x^3 \quad + \ 7x^2 \quad - \ 5x \quad + \ 6} \\
\underline{-(\ x^4 \quad - \ 5x^3 \quad + \ 6x^2)} \\
x^2 \quad - \ 5x \quad + \ 6 \\
\underline{-(\ x^2 \quad - \ 5x \quad + \ 6)}
\end{array}
$$

$$
\begin{aligned}
x^4 - 5x^3 + 7x^2 - 5x + 6 &= 0 \\
(x^2 - 5x + 6)(x^2 + 1) &= 0
\end{aligned}
$$

$$
\begin{array}{ll}
x^2 - 5x + 6 = 0 \quad \text{or} & x^2 + 1 = 0 \\
(x - 2)(x - 3) = 0 & \quad x^2 = -1 \\
x = 2 \text{ or } x = 3 & \quad x = \pm i \quad \text{The solution set is } \{2, \ 3, \ i, -i\}.
\end{array}
$$

35. $P(x) = (x - 4)(x - 5)$
$\qquad = x^2 - 9x + 20$

37. $P(x) = (x - 1)(x - 1)(x - 1)$
$\qquad = (x^2 - 2x + 1)(x - 1)$
$\qquad = x^3 - 3x^2 + 3x - 1$

39. $P(x) = (x - 2)(x - 4)(x - 5) = (x^2 - 6x + 8)(x - 5) = x^3 - 11x^2 + 38x - 40$

41. $P(x) = (x - 1)(x + 1)(x - \sqrt{2})(x + \sqrt{2}) = (x^2 - 1)(x^2 - 2) = x^4 - 3x^2 + 2$

43. $P(x) = (x - \sqrt{2})(x - i)(x + i) = (x - \sqrt{2})(x^2 - i^2) = x^3 - \sqrt{2}\,(x^2) + x - \sqrt{2}$

45. $P(x) = (x - 0)[x - (1 + i)][x - (1 - i)]$
$\qquad = x[x^2 - (1 - i)x - (1 + i)x + (1 + i)(1 - i)]$
$\qquad = x(x^2 - x + ix - x - ix + 1 - i^2)$
$\qquad = x^2 - 2x^2 + 2x$

47. $x^3 = 1 \Rightarrow x^3 - 1 = 0$. Since 1 is a solution of this equation, $x - 1$ is a factor.

$$
\begin{array}{r}
x^2 \quad + \quad x \quad + \quad 1 \\
x - 1 \;\overline{)\; x^3 \quad + \quad 0x^2 \quad + \quad 0x \quad - \quad 1} \\
-(\; x^3 \quad - \quad x^2\;) \\
\hline
x^2 \quad + \quad 0x \\
-(\quad x^2 \quad - \quad x) \\
\hline
x \quad - \quad 1 \\
-(\quad x \quad - \quad 1) \\
\end{array}
$$

$$x^3 - 1 = 0$$
$$(x - 1)(x^2 + x + 1) = 0$$

$x - 1 = 0$ or $x^2 + x + 1 = 0$

$x = 1$

$$x = \frac{-1 \pm \sqrt{1^2 - 4(1)(1)}}{2(1)}$$

$$= \frac{-1 \pm i\sqrt{3}}{2}$$

The three cube roots are 1, $-\frac{1}{2} + \frac{\sqrt{3}}{2}i$ and $-\frac{1}{2} - \frac{\sqrt{3}}{2}i$.

49. Solve $x^3 + 125 = 0$. Note: -5 is a root, so $x + 5$ is a factor.

$$
\begin{array}{r}
x^2 \quad - \quad 5x \quad + \quad 25 \\
x + 5 \;\overline{)\; x^3 \quad + \quad 0x^2 \quad + \quad 0x \quad + \quad 125} \\
-(\; x^3 \quad + \quad 5x^2\;) \\
\hline
- 5x^2 \quad + \quad 0x \\
-(- 5x^2 \quad - \quad 25x) \\
\hline
25x \quad + \quad 125 \\
-(\quad 25x \quad + \quad 125) \\
\end{array}
$$

$$x^3 + 125 = 0$$
$$(x + 5)(x^2 - 5x + 25) = 0$$

$x + 5 = 0$ or $x^2 - 5x + 25 = 0$

$x = -5$

$$x = \frac{5 \pm \sqrt{(-5)^2 - 4(1)(25)}}{2(1)}$$

$$= \frac{5 \pm 5i\sqrt{3}}{2}$$

The three cube roots are $-5, \frac{5}{2} + \frac{5\sqrt{3}}{2}i$ and $\frac{5}{2} - \frac{5\sqrt{3}}{2}i$

51. If 0 is a zero of $P(x)$, then $P(0) = 0$.

$$P(0) = a_n(0)^n + a_{n-1}(0)^{n-1} + \cdots + a_1(0) + a_0$$
$$= 0 + 0 + 0 + \cdots 0 + a_0 \quad \Rightarrow \quad \text{If } P(0) = 0, \text{ then } a_0 = 0.$$

Exercise 5.2 (page 263)

1.

$$
\begin{array}{r|rrrr}
2 & 5 & 2 & -1 & 1 \\
  &   & 10 & 24 & 46 \\
\hline
  & 5 & 12 & 23 & 47 \\
\end{array}
$$

$P(2) = 47$

3.

$$
\begin{array}{r|rrrr}
-5 & 5 & 2 & -1 & 1 \\
   &   & -25 & 115 & -570 \\
\hline
   & 5 & -23 & 114 & -569 \\
\end{array}
$$

$P(-5) = -569$

5.

$$
\begin{array}{r|rrrr}
0 & 5 & 2 & -1 & 1 \\
  &   & 0 & 0 & 0 \\
\hline
  & 5 & 2 & -1 & 1 \\
\end{array}
$$

$P(0) = 1$

7.

$$
\begin{array}{r|rrrr}
-i & 5 & 2 & -1 & 1 \\
   &   & -5i & -5-2i & -2+6i \\
\hline
   & 5 & 2-5i & -6-2i & -1+6i \\
\end{array}
$$

$P(-i) = -1 + 6i$

9.

$$
\begin{array}{r|rrrrr}
1 & 2 & 0 & -1 & 0 & 2 \\
  &   & 2 & 2 & 1 & 1 \\
\hline
  & 2 & 2 & 1 & 1 & 3 \\
\end{array}
$$

$P(1) = 3$

11.

$$
\begin{array}{r|rrrrr}
-2 & 2 & 0 & -1 & 0 & 2 \\
   &   & -4 & 8 & -14 & 28 \\
\hline
   & 2 & -4 & 7 & -14 & 30 \\
\end{array}
$$

$P(-2) = 30$

13.

$$
\begin{array}{r|rrrrr}
\frac{1}{2} & 2 & 0 & -1 & 0 & 2 \\
   &   & 1 & \frac{1}{2} & -\frac{1}{4} & -\frac{1}{8} \\
\hline
   & 2 & 1 & -\frac{1}{2} & -\frac{1}{4} & \frac{15}{8} \\
\end{array}
$$

$P(1) = \frac{15}{8}$

15.

$$
\begin{array}{r|rrrrr}
i & 2 & 0 & -1 & 0 & 2 \\
  &   & 2i & -2 & -3i & 3 \\
\hline
  & 2 & 2i & -3 & -3i & 5 \\
\end{array}
$$

$P(i) = 5$

17.

$$
\begin{array}{r|rrrrr}
1 & 1 & -8 & 14 & 8 & -15 \\
  &   & 1 & -7 & 7 & 15 \\
\hline
  & 1 & -7 & 7 & 15 & 0 \\
\end{array}
$$

$P(1) = 0$

19.

$$
\begin{array}{r|rrrrr}
-3 & 1 & -8 & 14 & 8 & -15 \\
   &   & -3 & 33 & -141 & 399 \\
\hline
   & 1 & -11 & 47 & -133 & 384 \\
\end{array}
$$

$P(-3) = 384$

21.

$$
\begin{array}{r|rrrrr}
3 & 1 & -8 & 14 & 8 & -15 \\
  &   & 3 & -15 & -3 & 15 \\
\hline
  & 3 & -5 & -1 & 5 & 0 \\
\end{array}
$$

$P(3) = 0$

23.

$$
\begin{array}{r|rrrrr}
2 & 1 & -8 & 14 & 8 & -15 \\
  &   & 2 & -12 & 4 & 24 \\
\hline
  & 1 & -6 & 2 & 12 & 9 \\
\end{array}
$$

$P(2) = 9$

25.

$$
\begin{array}{r|rrrrrr}
0 & 1 & 0 & -1 & -8 & 0 & 8 \\
  &   & 0 & 0 & 0 & 0 & 0 \\
\hline
  & 1 & 0 & -1 & -8 & 0 & 8
\end{array}
$$

$P(0) = 8$

27.

$$
\begin{array}{r|rrrrrr}
2 & 1 & 0 & -1 & -8 & 0 & 8 \\
  &   & 2 & 4 & 6 & -4 & -8 \\
\hline
  & 1 & 2 & 3 & -2 & -4 & 0
\end{array}
$$

$P(2) = 0$

29.

$$
\begin{array}{r|rrrrrr}
i & 1 & 0 & -1 & -8 & 0 & 8 \\
  &   & i & -1 & -2i & 2-8i & 8+2i \\
\hline
  & 1 & i & -2 & -8-2i & 2-8i & 16+2i
\end{array}
$$

$P(i) = 16 + 2i$

31.

$$
\begin{array}{r|rrrrrr}
-2i & 1 & 0 & -1 & -8 & 0 & 8 \\
    &   & -2i & -4 & 10i & 20+16i & 32-40i \\
\hline
    & 1 & -2i & -5 & -8+10i & 20+16i & 40-40i
\end{array}
$$

$P(i) = 40 - 40i$

33.

$$
\begin{array}{r|rrrr}
1 & 3 & -2 & -6 & -4 \\
  &   & 3 & 1 & -5 \\
\hline
  & 3 & 1 & -5 & -9
\end{array}
$$

$(x - 1)(3x^2 + x - 5) - 9$

35.

$$
\begin{array}{r|rrrr}
3 & 3 & -2 & -6 & -4 \\
  &   & 9 & 21 & 45 \\
\hline
  & 3 & 7 & 15 & 41
\end{array}
$$

$(x - 3)(3x^2 + 7x + 15) + 41$

37.

$$
\begin{array}{r|rrrr}
-1 & 3 & -2 & -6 & -4 \\
   &   & -3 & 5 & 1 \\
\hline
   & 3 & -5 & -1 & -3
\end{array}
$$

$(x + 1)(3x^2 - 5x - 1) - 3$

39.

$$
\begin{array}{r|rrrr}
-3 & 3 & -2 & -6 & -4 \\
   &   & -9 & 33 & -81 \\
\hline
   & 3 & -11 & 27 & -85
\end{array}
$$

$(x + 3)(3x^2 - 11x + 27) - 85$

41.

$$
\begin{array}{r|rrrr}
1 & 1 & 1 & 1 & -3 \\
  &   & 1 & 2 & 3 \\
\hline
  & 1 & 2 & 3 & 0
\end{array}
$$

$x^2 + 2x + 3$

43.

$$
\begin{array}{r|rrrr}
-1 & 7 & -3 & -5 & 1 \\
   &   & -7 & 10 & -5 \\
\hline
   & 7 & -10 & 5 & -4
\end{array}
$$

$7x^2 - 10x + 5 + \dfrac{-4}{x + 1}$

45.

$$
\begin{array}{r|rrrrr}
3 & 4 & -3 & 0 & -1 & 5 \\
  &   & 12 & 27 & 81 & 240 \\
\hline
  & 4 & 9 & 27 & 80 & 245
\end{array}
$$

$4x^3 + 9x^2 + 27x + 80 + \dfrac{245}{x - 3}$

165

47.

$$4 \,\big|\begin{array}{cccccc} 3 & 0 & 0 & 0 & -768 & 0 \\ & 12 & 48 & 192 & 768 & 0 \end{array}$$

$$\begin{array}{ccccccc} \hline 3 & 12 & 48 & 192 & 0 & 0 \end{array} \qquad 3x^4 + 12x^3 + 48x^2 + 192x$$

49.

$$2 \,\big|\begin{array}{ccccc} 1 & 4 & 1 & 0 & 5 \\ & 2 & 12 & 26 & 52 \end{array}$$

$$\begin{array}{ccccc} \hline 1 & 6 & 13 & 26 & 57 \end{array} \qquad x^3 + 6x^2 + 13x + 26 + \dfrac{57}{x-2}$$

51.

$$-2 \,\big|\begin{array}{cccccc} 5 & 0 & 3 & 1 & 0 & 5 \\ & -10 & 20 & -46 & 90 & -180 \end{array}$$

$$\begin{array}{cccccc} \hline 5 & -10 & 23 & -45 & 90 & -175 \end{array} \qquad 5x^4 - 10x^3 + 23x^2 - 45x + 90 + \dfrac{-175}{x+2}$$

53.
$$\begin{aligned} P(x) &= 3x^3 + 4x^2 + 5x + 12 \\ &= x(3x^2 + 4x + 5) + 12 \\ &= x[x(3x+4)+5] + 12 \end{aligned}$$

55.
$$\begin{aligned} P(x) &= 4x^4 - 2x^3 - 8x^2 + 3x + 5 \\ &= x(4x^3 - 2x^2 - 8x + 3) + 5 \\ &= x[x(4x^2 - 2x - 8) + 3] + 5 \\ &= x\{x[x(4x-2)-8]+3\} + 5 \end{aligned}$$

57.
$$\begin{aligned} P(x) &= 5x^5 + 3x^3 + 9x + 2 \\ &= x(5x^4 + 3x^2 + 9) + 2 \\ &= x[x(5x^3 + 3x) + 9] + 2 \\ &= x\{x[x(5x^2 + 3)] + 9\} + 2 \end{aligned}$$

59.

$$1.3 \,\big|\begin{array}{cccc} 2.5 & -0.78 & -2.7 & 4.3 \\ & 3.25 & 3.211 & 0.6643 \end{array}$$

$$\begin{array}{cccc} \hline 2.5 & 2.47 & 0.511 & 4.9643 \end{array}$$

$$P(1.3) = 4.9643$$

Exercise 5.3 (page 269)

1. 10 roots

3. 4 roots

5. $x(3x^4 - 2) = 12x$
$3x^5 - 2x = 12x$
There are 5 total roots. Thus there are 4 roots other than 0.

7. roots: $3, -i$ **and** i
$$\begin{aligned} P(x) &= 0 \\ (x-3)(x+i)(x-i) &= 0 \\ (x-3)(x^2+1) &= 0 \\ x^3 - 3x^2 + x - 3 &= 0 \end{aligned}$$

9. roots: $2, 2+i$ **and** $2-i$
$$\begin{aligned} (x-2)[x-(2+i)][x-(2-i)] &= 0 \\ (x-2)[x^2 - (2-i)x - (2+i)x + (2+i)(2-i)] &= 0 \\ (x-2)(x^2 - 2x + ix - 2x - ix + 4 - i^2) &= 0 \\ x^3 - 6x^2 + 13x - 10 &= 0 \end{aligned}$$

11. roots: 3, 2, i **and** $-i$
$$(x-3)(x-2)(x-i)(x+i) = 0$$
$$(x^2 - 5x + 6)(x^2 + 1) = 0$$
$$x^4 - 5x^3 + 7x^2 - 5x + 6 = 0$$

13. roots: $i, 1-i$ **and** $-i$ **and** $1+i$
$$(x-i)(x+i)[x-(1-i)][x-(1+i)] = 0$$
$$(x^2+1)[x^2 - (1+i)x - (1-i)x + (1-i)(1+i)] = 0$$
$$(x^2+1)(x^2 - x - ix - x + ix + 1 - i^2) = 0$$
$$x^4 - 2x^3 + 3x^2 - 2x + 2 = 0$$

15. $P(x) = 3x^3 + 5x^2 - 4x + 3 \Rightarrow 2$ sign changes $\Rightarrow 2$ or 0 positive real roots
$P(-x) = -3x^3 + 5x^2 + 4x + 3 \Rightarrow 1$ sign change $\Rightarrow 1$ negative real root

# positive real roots	# negative real roots	# non-real roots
2	1	0
0	1	2

17. $P(x) = 2x^3 + 7x^2 + 5x + 5 \Rightarrow 0$ sign changes $\Rightarrow 0$ positive real roots
$P(-x) = -2x^3 + 7x^2 - 5x + 5 \Rightarrow 2$ sign changes $\Rightarrow 1$ or 3 negative real roots

# positive real roots	# negative real roots	# non-real roots
0	1	2
0	3	0

19. $8x^4 = -5 \Rightarrow 8x^4 + 5 = 0$
$P(x) = 8x^4 + 5 \Rightarrow 0$ sign changes $\Rightarrow 0$ positive real roots
$P(-x) = 8x^4 + 5 \Rightarrow 0$ sign changes $\Rightarrow 0$ negative real roots

# positive real roots	# negative real roots	# non-real roots
0	0	4

21. $P(x) = x^4 + 8x^2 - 5x - 10 \Rightarrow 1$ sign change $\Rightarrow 1$ positive real root
$P(-x) = x^4 + 8x^2 + 5x - 10 \Rightarrow 1$ sign change $\Rightarrow 1$ negative real root

# positive real roots	# negative real roots	# non-real roots
1	1	2

23. $P(x) = -x^{10} - x^8 - x^6 - x^4 - x^2 - 1 \Rightarrow 0$ sign changes $\Rightarrow 0$ positive real roots
$P(-x) = -x^{10} - x^8 - x^6 - x^4 - x^2 - 1 \Rightarrow 0$ sign changes $\Rightarrow 0$ negative real roots

# positive real roots	# negative real roots	# non-real roots
0	0	10

25. $P(x) = x^9 + x^7 + x^5 + x^3 + x = x(x^8 + x^6 + x^4 + x^2 + 1)$
$P'(x) = x^8 + x^6 + x^4 + x^2 + 1 \Rightarrow 0$ sign changes $\Rightarrow 0$ positive real roots
$P'(-x) = x^8 + x^6 + x^4 + x^2 + 1 \Rightarrow 0$ sign changes $\Rightarrow 0$ negative real root

# positive real roots	# negative real roots	# non-real roots	
0	0	8	(and 0)

27. $P(x) = -2x^4 - 3x^2 + 2x + 3 \Rightarrow$ 1 sign change $\Rightarrow$ 1 positive real root

$P(-x) = -2x^4 - 3x^2 - 2x + 3 \Rightarrow$ 1 sign change $\Rightarrow$ 1 negative real root

# positive real roots	# negative real roots	# non-real roots
1	1	2

29.

$$
\begin{array}{r|rrr}
4 & 1 & -2 & -4 \\
  &   & 4 & 8 \\
\hline
  & 1 & 2 & 4
\end{array}
\qquad
\begin{array}{r|rrr}
-2 & 1 & -2 & -4 \\
   &   & -2 & 8 \\
\hline
   & 1 & -4 & 4
\end{array}
$$

4 is an upper bound and -2 is a lower bound.

31.

$$
\begin{array}{r|rrr}
1 & 18 & -6 & -1 \\
  &   & 18 & 12 \\
\hline
  & 18 & 12 & 11
\end{array}
\qquad
\begin{array}{r|rrr}
-1 & 18 & -6 & -1 \\
   &   & -18 & 24 \\
\hline
   & 18 & -24 & 23
\end{array}
$$

1 is an upper bound and -1 is a lower bound.

33.

$$
\begin{array}{r|rrrr}
6 & 6 & -13 & -110 & 0 \\
  &   & 36 & 138 & 168 \\
\hline
  & 6 & 23 & 28 & 168
\end{array}
\qquad
\begin{array}{r|rrrr}
-4 & 6 & -13 & -110 & 0 \\
   &   & -24 & 148 & -152 \\
\hline
   & 6 & -37 & 38 & -152
\end{array}
$$

6 is an upper bound and -4 is a lower bound.

35.

$$
\begin{array}{r|rrrrrr}
3 & 1 & 1 & -8 & -8 & 15 & 15 \\
  &   & 3 & 12 & 12 & 12 & 81 \\
\hline
  & 1 & 4 & 4 & 4 & 27 & 96
\end{array}
$$
3 is an upper bound.

$$
\begin{array}{r|rrrrrr}
-4 & 1 & 1 & -8 & -8 & 15 & 15 \\
   &   & -4 & 12 & -16 & 96 & -444 \\
\hline
   & 1 & -3 & 4 & -24 & 111 & -429
\end{array}
$$
-4 is a lower bound.

37.

$$
\begin{array}{r|rrrrrr}
4 & 3 & -11 & -2 & 38 & -21 & -15 \\
  &   & 12 & 4 & 8 & 184 & 652 \\
\hline
  & 3 & 1 & 2 & 46 & 163 & 637
\end{array}
$$
4 is an upper bound.

$$
\begin{array}{r|rrrrrr}
-2 & 3 & -11 & -2 & 38 & -21 & -15 \\
   &   & -6 & 34 & -64 & 52 & -62 \\
\hline
   & 3 & -17 & 32 & -26 & 31 & -77
\end{array}
$$
-2 is a lower bound.

39. Any polynomial equation can be written in the form $P(x) = 0$, where $P(x)$ is a polynomial. Then the fundamental theorem of algebra guarantees that $P(x)$ has at least one zero, and thus $P(x) = 0$ has at least one root.

41. If an equation has all non-real roots, then it has an even number of roots, since non-real roots occur in conjugate pairs. But an odd degree equation has an odd number of roots, so all roots cannot be non-real. There is at least one real root.

Exercise 5.4 (page 277)

1. From the rational roots theorem, the only possible rational roots are:

$$\pm \tfrac{1}{1}, \ \pm \tfrac{5}{1}$$

Descartes' rule of signs indicates:

# pos	#neg	#non-real
2	1	0
0	1	2

Try $x = -1$:

$$
\begin{array}{r|rrrr}
-1 & 1 & -5 & -1 & 5 \\
   &   & -1 & 6 & -5 \\
\hline
   & 1 & -6 & 5 & 0
\end{array}
$$

$(x+1)(x^2 - 6x + 5) = 0$
$(x+1)(x-1)(x-5) = 0$

rational roots: $-1, 1, 5$

3. From the rational roots theorem, the only possible rational roots are:

$$\pm \tfrac{18}{1}, \ \pm \tfrac{9}{1}, \ \pm \tfrac{6}{1}, \ \pm \tfrac{3}{1}, \ \pm \tfrac{2}{1}, \ \pm \tfrac{1}{1}$$

Descartes' rule of signs indicates:

# pos	# neg	#non-real
2	1	0
0	1	2

Try $x = 2$:

$$
\begin{array}{r|rrrr}
2 & 1 & -2 & -9 & 18 \\
  &   & 2 & 0 & -18 \\
\hline
  & 1 & 0 & -9 & 0
\end{array}
$$

$(x-2)(x^2 - 9) = 0$
$(x-2)(x+3)(x-3) = 0$

rational roots: $2, -3, 3$

5. From the rational roots theorem, the only possible rational roots are:

$$\pm\tfrac{1}{1}, \pm\tfrac{2}{1}$$

Descartes' rule of signs indicates:

# pos	#neg	#non-real
2	1	0
0	1	2

Try $x = -1$:

$$\begin{array}{r|rrrr} -1 & 1 & -2 & -1 & 2 \\ & & -1 & 3 & -2 \\ \hline & 1 & -3 & 2 & \boxed{0} \end{array}$$

$$(x+1)(x^2 - 3x + 2) = 0$$
$$(x+1)(x-1)(x-2) = 0$$

rational roots: $-1, 1, 2$

7. From the rational roots theorem, the only possible rational roots are:

$$\pm\tfrac{24}{1}, \pm\tfrac{12}{1}, \pm\tfrac{8}{1}, \pm\tfrac{6}{1}, \pm\tfrac{4}{1}, \pm\tfrac{3}{1}, \pm\tfrac{2}{1}, \pm\tfrac{1}{1}$$

Descartes' rule of signs indicates:

# pos	# neg	#non-real
4	0	0
2	0	2
0	0	4

Try $x = 1$:

$$\begin{array}{r|rrrrr} 1 & 1 & -10 & 35 & -50 & 24 \\ & & 1 & -9 & 26 & -24 \\ \hline & 1 & -9 & 26 & -24 & \boxed{0} \end{array}$$

Try $x = 2$:

$$\begin{array}{r|rrrr} 2 & 1 & -9 & 26 & -24 \\ & & 2 & -14 & 24 \\ \hline & 1 & -7 & 12 & \boxed{0} \end{array}$$

$$(x-1)(x-2)(x^2 - 7x + 12) = 0$$
$$(x-1)(x-2)(x-3)(x-4) = 0$$

rational roots: 1, 2, 3, 4

9. From the rational roots theorem, the only possible rational roots are:

$$\pm\tfrac{30}{1}, \pm\tfrac{15}{1}, \pm\tfrac{10}{1}, \pm\tfrac{6}{1}, \pm\tfrac{5}{1}, \pm\tfrac{3}{1}, \pm\tfrac{2}{1}, \pm\tfrac{1}{1}$$

Try $x = 2$:

$$\begin{array}{r|rrrrr} 2 & 1 & 3 & -13 & -9 & 30 \\ & & 2 & 10 & -6 & -30 \\ \hline & 1 & 5 & -3 & -15 & \boxed{0} \end{array}$$

$$(x-2)(x+5)(x^2 + 3) = 0$$

$\{x^2 + 3$ has only irrational roots$\}$

Descartes' rule of signs indicates:

# pos	# neg	# non-real
2	2	0
2	0	2
0	2	2
0	0	4

Try $x = -5$:
{new possible roots: $\pm 1, 3, 5, 15$}

$$\begin{array}{r|rrrr} -5 & 1 & 5 & -3 & -15 \\ & & -5 & 0 & 15 \\ \hline & 1 & 0 & -3 & \boxed{0} \end{array}$$

rational roots: 2, -5

11. From the rational roots theorem, the only possible rational roots are:

$$\pm\frac{12}{1}, \pm\frac{6}{1}, \pm\frac{4}{1}, \pm\frac{3}{1}, \pm\frac{2}{1}, \pm\frac{1}{1}$$

Descartes' rule of signs indicates:

# pos	# neg	# non-real
2	3	0
2	1	2
0	3	2
0	1	4

$\boxed{1}$ Try $x = 1$:

```
1 | 1   3   -5   -15    4    12
          1    4   -1  -16   -12
    -----------------------------
    1   4   -1  -16  -12 |  0
```

$\boxed{2}$ Try $x = 2$:

```
2 | 1   4   -1   -16   -12
         2   12    22    12
    -----------------------
    1   6   11    6 |  0
```

$\boxed{3}$ Try $x = -2$:
{new possible roots: $\pm 1, 2, 3, 6$}

```
-2 | 1   6    11    6
        -2   -8   -6
     ----------------
     1   4    3 |  0
```

$$x^5 + 3x^4 - 5x^3 - 15x^2 + 4x + 12 = 0$$
$$(x-1)(x-2)(x+2)(x^2 + 4x + 3) = 0$$
$$(x-1)(x-2)(x+2)(x+1)(x+3) = 0$$

$\boxed{\text{rational roots: } 1, -1, 2, -2, -3}$

13. Observe that $x = 0$ is a root. Then find the solutions of the reduced polynomial.
$$x^7 - 12x^5 + 48x^3 - 64x = 0$$
$$x(x^6 - 12x^4 + 48x^2 - 64) = 0$$
From the rational roots theorem, the only possible rational roots are:

$$\pm\frac{64}{1}, \pm\frac{32}{1}, \pm\frac{16}{1}, \pm\frac{8}{1}, \pm\frac{4}{1}, \pm\frac{2}{1}, \pm\frac{1}{1}$$

Descartes' rule of signs indicates:

# pos	# neg	# non-real
3	3	0
3	1	2
1	3	2
1	1	4

$\boxed{1}$ Try $x = 2$:

```
2 | 1   0   -12    0    48     0   -64
         2    4  -16   -32    32    64
    ---------------------------------
    1   2   -8  -16    16    32 |  0
```

$\boxed{2}$ Try $x = 2$: {from $\pm 1, 2, 4, 8, 16$}

```
2 | 1   2   -8   -16    16    32
         2    8     0   -32   -32
    ----------------------------
    1   4    0  -16   -16 |  0
```

$\boxed{3}$ Try $x = 2$:

```
2 | 1   4    0   -16   -16
         2   12    24    16
    ---------------------
    1   6   12     8 |  0
```

$\boxed{4}$ Try $x = -2$: {from $\pm 1, 2, 4, 8$}

```
-2 | 1   6   12    8
        -2  -8   -8
     ---------------
     1   4    4 |  0
```

$$x(x^6 - 12x^4 + 48x^2 - 64) = 0$$
$$x(x-2)(x-2)(x-2)(x+2)(x^2 + 4x + 4) = 0$$
$$x(x-2)(x-2)(x-2)(x+2)(x+2)(x+2) = 0$$

$\boxed{\text{rational roots: } 0, 2 \text{ \{with multiplicity 3\}}, -2 \text{ \{with multiplicity 3\}}}$

15. From the rational roots theorem, the only possible rational roots are:

$$\pm\frac{8}{1}, \pm\frac{8}{3}, \pm\frac{4}{1}, \pm\frac{4}{3}, \pm\frac{2}{1}, \pm\frac{2}{3}, \pm\frac{1}{1}, \pm\frac{1}{3}$$

Try $x = \frac{2}{3}$:

$$\frac{2}{3} \, \bigg| \quad \begin{array}{rrrr} 3 & -2 & 12 & -8 \\ & 2 & 0 & 8 \end{array}$$

$$\begin{array}{rrrr} \hline 3 & 0 & 12 & \big| \; 0 \end{array}$$

roots: $\frac{2}{3}, \; \pm 2i$

Descartes' rule of signs indicates:

# pos	# neg	# non-real
3	0	0
1	0	2

$$3x^3 - 2x^2 + 12x - 8 = 0$$

$$\left(x - \frac{2}{3}\right)(3x^2 + 12) = 0$$

$$3x^2 + 12 = 0$$
$$3(x^2 + 4) = 0$$
$$x^2 = -4$$
$$x = \pm 2i$$

17. From the rational roots theorem, the only possible rational roots are:

$$\pm\frac{12}{1}, \pm\frac{12}{3}, \pm\frac{6}{1}, \pm\frac{6}{3}, \pm\frac{4}{1}, \pm\frac{4}{3}, \pm\frac{3}{1}, \pm\frac{3}{3}, \pm\frac{2}{1}, \pm\frac{2}{3}, \pm\frac{1}{1}, \pm\frac{1}{3}$$

Descartes' rule of signs indicates:

# pos	# neg	# non-real
3	1	0
1	1	2

Try $x = 2$:

$$2 \, \bigg| \quad \begin{array}{rrrrr} 3 & -14 & 11 & 16 & -12 \\ & 6 & -16 & -10 & 12 \end{array}$$

$$\begin{array}{rrrrr} \hline 3 & -8 & -5 & 6 & \big| \; 0 \end{array}$$

Try $x = -1$

$$-1 \, \bigg| \quad \begin{array}{rrrr} 3 & -8 & -5 & 6 \\ & -3 & 11 & -6 \end{array}$$

$$\begin{array}{rrrr} \hline 3 & -11 & 6 & \big| \; 0 \end{array}$$

$$3x^4 - 14x^3 + 11x^2 + 16x - 12 = 0$$
$$(x - 2)(x + 1)(3x^2 - 11x + 6) = 0$$
$$(x - 2)(x + 1)(3x - 2)(x - 3) = 0$$

roots: $2, 3, \frac{2}{3}, -1$

19. From the rational roots theorem, the only possible rational roots are:

$$\pm\tfrac{3}{1}, \pm\tfrac{3}{2}, \pm\tfrac{3}{3}, \pm\tfrac{3}{4}, \pm\tfrac{3}{6}, \pm\tfrac{3}{12}, \pm\tfrac{1}{1}, \pm\tfrac{1}{2}, \pm\tfrac{1}{3}, \pm\tfrac{1}{4}, \pm\tfrac{1}{6}, \pm\tfrac{1}{12}$$

Descartes' rule of signs indicates:

# pos	# neg	# non-real
3	1	0
1	1	2

Try $x = \tfrac{1}{3}$:

$$
\begin{array}{r|rrrrr}
\tfrac{1}{3} & 12 & 20 & -41 & 20 & -3 \\
 & & 4 & 8 & -11 & 3 \\
\hline
 & 12 & 24 & -33 & 9 & 0
\end{array}
$$

Try $x = \tfrac{1}{2}$:

$$
\begin{array}{r|rrrrr}
\tfrac{1}{2} & 12 & 24 & -33 & 9 \\
 & & 6 & 15 & -9 \\
\hline
 & 12 & 30 & -18 & 0
\end{array}
$$

$$12x^4 + 20x^3 - 41x^2 + 20x - 3 = 0$$

$$\left(x - \tfrac{1}{3}\right)\left(x - \tfrac{1}{2}\right)(12x^2 + 30x - 18) = 0$$

$$6\left(x - \tfrac{1}{3}\right)\left(x - \tfrac{1}{2}\right)(2x^2 + 5x - 3) = 0$$

$$6\left(x - \tfrac{1}{3}\right)\left(x - \tfrac{1}{2}\right)(2x - 1)(x + 3) = 0$$

roots: $\tfrac{1}{2}$ {multiplicity 2}, $\tfrac{1}{3}, -3$

21. From the rational roots theorem, the only possible rational roots are:

$$\pm\left(\tfrac{12}{6}, \tfrac{6}{6}, \tfrac{4}{6}, \tfrac{3}{6}, \tfrac{2}{6}, \tfrac{1}{6}, \tfrac{12}{3}, \tfrac{6}{3}, \tfrac{4}{3}, \tfrac{3}{3}, \tfrac{2}{3}, \tfrac{1}{3}, \tfrac{12}{2}, \tfrac{6}{2}, \tfrac{4}{2}, \tfrac{3}{2}, \tfrac{2}{2}, \tfrac{1}{2}, \tfrac{12}{1}, \tfrac{6}{1}, \tfrac{4}{1}, \tfrac{3}{1}, \tfrac{2}{1}, \tfrac{1}{1}\right)$$

Descartes' rule of signs indicates:

# pos	# neg	# non-real
3	2	0
1	2	2
3	0	2
1	0	4

Try $x = 2$:

$$
\begin{array}{r|rrrrrr}
2 & 6 & -7 & -48 & 81 & -4 & -12 \\
 & & 12 & 10 & -76 & 10 & 12 \\
\hline
 & 6 & 5 & -38 & 5 & 6 & 0
\end{array}
$$

Try $x = \tfrac{1}{2}$:

$$
\begin{array}{r|rrrr}
\tfrac{1}{2} & 6 & 17 & -4 & -3 \\
 & & 3 & 10 & 3 \\
\hline
 & 6 & 20 & 6 & 0
\end{array}
$$

roots: $2, 2, \tfrac{1}{2}, -\tfrac{1}{3}, -3$

Try $x = 2$:

$$
\begin{array}{r|rrrrr}
2 & 6 & 5 & -38 & 5 & 6 \\
 & & 12 & 34 & -8 & -6 \\
\hline
 & 6 & 17 & -4 & -3 & 0
\end{array}
$$

$$6x^5 - 7x^4 - 48x^3 + 81x^2 - 4x - 12 = 0$$

$$(x - 2)(x - 2)\left(x - \tfrac{1}{2}\right)(6x^2 + 20x + 6) = 0$$

$$6(x - 2)(x - 2)\left(x - \tfrac{1}{2}\right)(3x^2 + 10x + 3) = 0$$

$$6(x - 2)(x - 2)\left(x - \tfrac{1}{2}\right)(3x + 1)(x + 3) = 0$$

23. From the rational roots theorem, the only possible rational roots are:

$$\pm\left(\frac{18}{30}, \frac{9}{30}, \frac{6}{30}, \frac{3}{30}, \frac{2}{30}, \frac{1}{30}, \frac{18}{15}, \frac{9}{15}, \frac{6}{15}, \frac{3}{15}, \frac{2}{15}, \frac{1}{15}, \frac{18}{10}, \frac{9}{10}, \frac{6}{10}, \frac{3}{10}, \frac{2}{10}, \frac{1}{10}, \frac{18}{6}, \frac{9}{6}, \frac{6}{6}\right)$$

$$\pm\left(\frac{3}{6}, \frac{2}{6}, \frac{1}{6}, \frac{18}{5}, \frac{9}{5}, \frac{6}{5}, \frac{3}{5}, \frac{2}{5}, \frac{1}{5}, \frac{18}{3}, \frac{9}{3}, \frac{6}{3}, \frac{3}{3}, \frac{2}{3}, \frac{1}{3}, \frac{18}{2}, \frac{9}{2}, \frac{6}{2}, \frac{3}{2}, \frac{2}{2}, \frac{1}{2}, \frac{18}{1}, \frac{9}{1}, \frac{6}{1}, \frac{3}{1}, \frac{2}{1}, \frac{1}{1}\right)$$

Descartes' rule of signs indicates:

# pos	# neg	# non-real
2	1	0
0	1	2

Try $x = \frac{2}{3}$:

$$\frac{2}{3} \bigg|\ \begin{array}{rrrr} 30 & -47 & -9 & 18 \\ & 20 & -18 & -18 \\ \hline 30 & -27 & -27 & \big|\ 0 \end{array}$$

roots: $\dfrac{2}{3}, \dfrac{3}{2}, -\dfrac{3}{5}$

$$30x^3 - 47x^2 - 9x + 18 = 0$$

$$\left(x - \frac{2}{3}\right)(30x^2 - 27x - 27) = 0$$

$$3\left(x - \frac{2}{3}\right)(10x^2 - 9x - 9) = 0$$

$$3\left(x - \frac{2}{3}\right)(2x - 3)(5x + 3) = 0$$

---

25. From the rational roots theorem, the only possible rational roots are:

$$\pm\left(\frac{24}{15}, \frac{12}{15}, \frac{8}{15}, \frac{6}{15}, \frac{4}{15}, \frac{3}{15}, \frac{2}{15}, \frac{1}{15}, \frac{24}{5}, \frac{12}{5}, \frac{8}{5}, \frac{6}{5}, \frac{4}{5}, \frac{3}{5}, \frac{2}{5}, \frac{1}{5}, \frac{24}{3}, \frac{12}{3}, \frac{8}{3}, \frac{6}{3}, \frac{4}{3}, \frac{3}{3}, \frac{2}{3}, \frac{1}{3}\right)$$

$$\pm\left(\frac{24}{1}, \frac{12}{1}, \frac{8}{1}, \frac{6}{1}, \frac{4}{1}, \frac{3}{1}, \frac{2}{1}, \frac{1}{1}\right)$$

Descartes' rule of signs indicates:

# pos	# neg	# non-real
2	1	0
0	1	2

Try $x = \frac{2}{3}$

$$\frac{2}{3} \bigg|\ \begin{array}{rrrr} 15 & -61 & -2 & 24 \\ & 10 & -34 & -24 \\ \hline 15 & -51 & -36 & \big|\ 0 \end{array}$$

roots: $\dfrac{2}{3}, 4, -\dfrac{3}{5}$

$$15x^3 - 61x^2 - 2x + 24 = 0$$

$$\left(x - \frac{2}{3}\right)(15x^2 - 51x - 36) = 0$$

$$3\left(x - \frac{2}{3}\right)(5x^2 - 17x - 12) = 0$$

$$3\left(x - \frac{2}{3}\right)(5x + 3)(x - 4) = 0$$

27. From the rational roots theorem, the only possible rational roots are:

$$\pm\left(\frac{18}{20}, \frac{9}{20}, \frac{6}{20}, \frac{3}{20}, \frac{2}{20}, \frac{1}{20}, \frac{18}{10}, \frac{9}{10}, \frac{6}{10}, \frac{3}{10}, \frac{2}{10}, \frac{1}{10}, \frac{18}{5}, \frac{9}{5}, \frac{6}{5}, \frac{3}{5}, \frac{2}{5}, \frac{1}{5}, \frac{18}{4}, \frac{9}{4}, \frac{6}{4}\right.$$

$$\left.\pm\left(\frac{3}{4}, \frac{2}{4}, \frac{1}{4}, \frac{18}{2}, \frac{9}{2}, \frac{6}{2}, \frac{3}{2}, \frac{2}{2}, \frac{1}{2}, \frac{18}{1}, \frac{9}{1}, \frac{6}{1}, \frac{3}{1}, \frac{2}{1}, \frac{1}{1}\right)\right.$$

Descartes' rule of signs indicates:

# pos	# neg	# non-real
2	1	0
0	1	2

Try $x = \frac{3}{2}$:

$$20x^3 - 44x^2 + 9x + 18 = 0$$

$\frac{3}{2}$ | 20 -44 9   18
        30   -21 -18

$$\underline{20 \quad -14 \quad -12 \;\big|\; 0}$$

roots: $\frac{3}{2}, \frac{6}{5}, -\frac{1}{2}$

$$\left(x - \frac{3}{2}\right)(20x^2 - 14x - 12) = 0$$

$$2\left(x - \frac{3}{2}\right)(10x^2 - 7x - 6) = 0$$

$$2\left(x - \frac{3}{2}\right)(2x + 1)(5x - 6) = 0$$

In #29 and #31, if $1 + i$ is a root, then $1 - i$ is also a root.

$$[x - (1 + i)][x - (1 - i)] = x^2 - (1 - i)x - (1 + i)x + (1 + i)(1 - i)$$
$$= x^2 - 2x + 2$$

Thus $x^2 - 2x + 2$ is a factor of the polynomials in these two problems.

29.

$$\begin{array}{r} x \quad\quad - 3 \\ \hline x^2 - 2x + 2 \,\big)\, x^3 \quad - 5x^2 \;+\; 8x \;-\; 6 \\ -(\; x^3 \quad - 2x^2 \;+\; 2x) \\ \hline - 3x^2 \;+\; 6x \;-\; 6 \\ -(- 3x^2 \;+\; 6x \;-\; 6) \end{array}$$

$$x^3 - 5x^2 + 8x - 6 = (x^2 - 2x + 2)(x - 3) \qquad \boxed{\text{ROOTS: } 3, 1 + i, 1 - i}$$

31.

$$\begin{array}{r} x^2 \quad\quad - 9 \\ \hline x^2 - 2x + 2 \,\big)\, x^4 \quad - 2x^3 \;-\; 7x^2 \;+\; 18x \;-\; 18 \\ -(\; x^4 \quad + 2x^3 \;+\; 2x^2) \\ \hline - 9x^2 \;+\; 18x \;-\; 18 \\ -(- 9x^2 \;+\; 18x \;-\; 18) \end{array}$$

$$x^4 - 2x^3 - 7x^2 + 18x - 18 = (x^2 - 2x + 2)(x^2 - 9)$$
$$= (x^2 - 2x + 2)(x + 3)(x - 3) \qquad \boxed{\text{ROOTS: } \pm 3, 1 \pm i}$$

33. $x^3 - \frac{4}{3}x^2 - \frac{13}{3}x - 2 = 0 \quad \Rightarrow \quad 3x^3 - 4x^2 - 13x - 6 = 0$

$$
\begin{array}{r|rrrr}
-1 & 3 & -4 & -13 & -6 \\
   &   & -3 & 7 & 6 \\
\hline
   & 3 & -7 & -6 & \big|\; 0
\end{array}
$$

$$3x^3 - 4x^2 - 13x - 6 = 0$$
$$(x+1)(3x^2 - 7x - 6) = 0$$
$$(x+1)(3x+2)(x-3) = 0$$

ROOTS: $-1, -\frac{2}{3}, 3$

35. Let $y = x^{-1}$: $\qquad y^5 - 8y^4 + 25y^3 - 38y^2 + 28y - 8 = 0$

Try $y = 1$: $\qquad\qquad\qquad\qquad$ Try $y = 1$:

$$
\begin{array}{r|rrrrrr}
1 & 1 & -8 & 25 & -38 & 28 & -8 \\
  &   & 1 & -7 & 18 & -20 & -8 \\
\hline
  & 1 & -7 & 18 & -20 & 8 & \big|\; 0
\end{array}
\qquad
\begin{array}{r|rrrrr}
1 & 1 & -7 & 18 & -20 & 8 \\
  &   & 1 & -6 & 12 & -8 \\
\hline
  & 1 & -6 & 12 & -8 & \big|\; 0
\end{array}
$$

Try $y = 2$:

$$
\begin{array}{r|rrrr}
2 & 1 & -6 & 12 & -8 \\
  &   & 2 & -8 & 8 \\
\hline
  & 1 & -4 & 4 & \big|\; 0
\end{array}
$$

So $y = 1$ or $y = 2$. Then $x^{-1} = 1$ or $x^{-1} = 2$, and $x = 1$ or $x = \frac{1}{2}$.

$$y^5 - 8y^4 + 25y^3 - 38y^2 + 28y - 8 = 0$$
$$(y-1)(y-1)(y-2)(y^2 - 4y + 4) = 0$$
$$(y-1)(y-1)(y-2)(y-2)(y-2) = 0$$

37. If n is even, then x^n is positive. Since c is positive, Descartes' rule of signs indicates that there are no positive real roots or negative real roots. By substitution, 0 is not a root, so there are no real roots.

39.
$$\frac{1}{R} = \frac{1}{R_1} + \frac{1}{R_2} + \frac{1}{R_3}$$
$$R_1 R_2 R_3 = R R_2 R_3 + R R_1 R_3 + R R_1 R_2$$
$$R_1 R_2 R_3 = R(R_2 R_3 + R_1 R_3 + R_1 R_2)$$

Let $R_2 = R_1 + 10$, $R_3 = R_1 + 50$, and $R = 6$:

$$R_1(R_1 + 10)(R_1 + 50) = 6[(R_1 + 10)(R_1 + 50) + R_1(R_1 + 50) + R_1(R_1 + 10)]$$
$$R_1^3 + 60R_1^2 + 500R_1 = 6(R_1^2 + 60R_1 + 500 + R_1^2 + 50R_1 + R_1^2 + 10R_1)$$
$$R_1^3 + 60R_1^2 + 500R_1 = 18R_1^2 + 720R_1 + 3000$$
$$R_1^3 + 42R_1^2 - 220R_1 - 3000 = 0$$

$$
\begin{array}{r|rrrr}
10 & 1 & 42 & -220 & -3000 \\
   &   & 10 & 520 & 3000 \\
\hline
   & 1 & 52 & 300 & \big|\; 0
\end{array}
$$

$R_1 = 10$, $R_2 = 20$, $R_3 = 60$.

Descartes' rule of signs indicates that the depressed polynomial has no positive real roots.

41.
$$\text{length} \cdot \text{width} = \text{area}$$
$$2x \cdot y = 42$$
$$2x(16 - x^2) = 42$$
$$32x - 2x^3 = 42$$
$$2x^3 - 32x + 42 = 0$$
$$x^3 - 16x + 21 = 0$$

$$
\begin{array}{r|rrrr}
3 & 1 & 0 & -16 & 21 \\
  &   & 3 & 9 & -21 \\
\hline
  & 1 & 3 & -7 & \;\;0
\end{array}
$$

Using the quadratic formula on the depressed polynomial yields

$$x = \frac{-3 \pm \sqrt{3^2 - 4(1)(-7)}}{2(1)} = \frac{-3 \pm \sqrt{37}}{2} \approx \frac{-3 \pm 6.1}{2} \approx 1.54 \text{ or } -4.54$$

So $x = 3$ or $x = \dfrac{-3 + \sqrt{37}}{2}$ (the other answer does not fit the illustration).

Exercise 5.5 (page 282)

1. Let $P(x) = 2x^2 + x - 3$
$P(-2) = 3$, $P(-1) = -2$
The intermediate value theorem guarantees that $P(x)$ will take on every value between -2 and 3 as x varies from -1 to -2, so $P(x)$ will equal 0 in that interval.

3. Let $P(x) = 3x^3 - 11x^2 - 14x$
$P(4) = -40$, $P(5) = 30$
The intermediate value theorem guarantees that $P(x)$ will take on every value between -40 and 30 as x varies from -40 to 30, so $P(x)$ will equal 0 in that interval.

5. $P(x) = x^4 - 8x^2 + 15$
$P(1) = 8$, $P(2) = -1$
The reason is similar to those above.

7. $P(x) = 30x^3 - 61x^2 - 39x + 10$
$P(2) = -72$, $P(3) = 154$
The reason is similar to those above.

9. $P(x) = 30x^3 - 61x^2 - 39x + 10$
$P(0) = 10$, $P(1) = -60$ The reason is similar to those above.

11. Note that $P(1) = -2$ and $P(2) = 1$. Organize the method in a chart:

STEP	x_l	c	x_r	$P(x_l)$	$P(c)$	$P(x_r)$
0	1	1.5	2	$-$	$-$	$+$
1	1.5	1.75	2	$-$	$+$	$+$
2	1.5	1.625	1.75	$-$	$-$	$+$
3	1.625	1.6875	1.75	$-$	$-$	$+$
4	1.6875	1.71875	1.75	$-$	$-$	$+$
5	1.71875	1.734375	1.75	$-$	$+$	$+$

The solution is 1.7, rounded to the nearest tenth.

13. Note that $P(-3) = 4$ and $P(-2) = -1$.

STEP	x_l	c	x_r	$P(x_l)$	$P(c)$	$P(x_r)$
0	-3	-2.5	-2	$+$	$+$	$-$
1	-2.5	-2.25	-2	$+$	$+$	$-$
2	-2.25	-2.125	-2	$+$	$-$	$-$
3	-2.25	-2.1875	-2.125	$+$	$-$	$-$
4	-2.25	-2.21875	-2.1875	$+$	$-$	$-$
5	-2.25	-2.234375	-2.21875	$+$	$-$	$-$

The solution is -2.2, rounded to the nearest tenth.

15. Note that $P(1) = -2$ and $P(2) = 2$.

STEP	x_l	c	x_r	$P(x_l)$	$P(c)$	$P(x_r)$
0	1	1.5	2	$-$	$-$	$+$
1	1.5	1.75	2	$-$	$+$	$+$
2	1.5	1.625	1.75	$-$	$-$	$+$
3	1.625	1.6875	1.75	$-$	$-$	$+$
4	1.6875	1.71875	1.75	$-$	$+$	$+$

The solution is 1.7, rounded to the nearest tenth.

17. Note that $P(-2) = 24$ and $P(-1) = -1$.

STEP	x_l	c	x_r	$P(x_l)$	$P(c)$	$P(x_r)$
0	-2	-1.5	-1	$+$	$+$	$-$
1	-1.5	-1.25	-1	$+$	$+$	$-$
2	-1.25	-1.125	-1	$+$	$-$	$-$
3	-1.25	-1.1875	-1.125	$+$	$+$	$-$
4	-1.1875	-1.15625	-1.125	$+$	$+$	$-$
5	-1.15625	-1.140625	-1.125	$+$	$-$	$-$
6	-1.15625	-1.1484375	-1.140625	$+$	$-$	$-$
7	-1.15625	-1.15234375	-1.1484375	$+$	$-$	$-$
8	-1.15625	$-1.15429...$	$-1.15234...$	$+$	$-$	$-$

The solution is -1.2, rounded to the nearest tenth.

19.

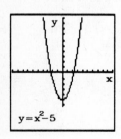

$x \approx \pm 2.24$
The bisection method would find both solutions.

21.

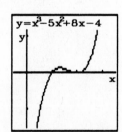

$x = 2,\ x = 1$
The bisection method would not find the solution $x = 2$.

23. Use the distance formula on the point (x, y).

$$d = \sqrt{(x-0)^2 + (y-0)^2}$$
$$1 = \sqrt{x^2 + y^2}$$
$$1 = x^2 + y^2$$
$$1 = x^2 + (x^3)^2$$
$$0 = x^6 + x^2 - 1$$

The x-coordinates are ± 0.82.
$y = x^3 = (\pm 0.82)^3 = \pm 0.55$.

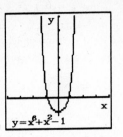

$y = x^6 + x^2 - 1$

Chapter 5 Review Exercises   (page 284)

1.
$$\begin{array}{r|rrrrr} 0 & 4 & 2 & -3 & 0 & -2 \\ & & 0 & 0 & 0 & 0 \\ \hline & 4 & 2 & -3 & 0 & -2 \end{array}$$
$P(0) = -2$

3.
$$\begin{array}{r|rrrrr} -3 & 4 & 2 & -3 & 0 & -2 \\ & & -12 & 30 & -81 & 243 \\ \hline & 4 & -10 & 27 & -81 & 241 \end{array}$$
$P(-3) = 241$

5. $P(2) = 24$, so it is not a factor.

7. $P(5) = 0$, so it is a factor.

9. $P(x) = (x+1)(x-2)(x-\frac{3}{2})$
$$= (x^2 - x - 2)(2x - 3)$$
$$= 2x^3 - 5x^2 - x + 6$$

11. $P(x) = (x-2)(x+5)(x-i)(x+i)$
$$= (x^2 + 3x - 10)(x^2 + 1)$$
$$= x^4 + 3x^3 - 9x^2 + 3x - 10$$

13.
$$\begin{array}{r|rrrrr} 3 & 3 & 0 & 2 & 3 & 7 \\ & & 9 & 27 & 87 & 270 \\ \hline & 3 & 9 & 29 & 90 & 277 \end{array}$$
$3x^3 + 9x^2 + 29x + 90 + \dfrac{277}{x-3}$

15.
$$\begin{array}{r|rrrrrr} -2 & 5 & -4 & 3 & -2 & 1 & -1 \\ & & -10 & 28 & -62 & 128 & -258 \\ \hline & 5 & -14 & 31 & -64 & 129 & -259 \end{array}$$
$5x^4 - 14x^3 + 31x^2 - 64x + 129 + \dfrac{-259}{x+2}$

17. 6 roots

19. 65 roots

21. $P(x) = 3x^4 + 2x^3 - 4x + 2 \Rightarrow 2$ sign changes $\Rightarrow 2$ or 0 positive real roots
$P(-x) = 3x^4 - 2x^3 + 4x + 3 \Rightarrow 2$ sign changes $\Rightarrow 2$ or 0 negative real roots

# positive real roots	# negative real roots	# non-real roots
2	2	0
2	0	2
0	2	2
0	0	4

23. $P(x) = 4x^5 + 3x^4 + 2x^3 + x^2 + x - 7 \Rightarrow 1$ sign change $\Rightarrow 1$ positive real root
$P(-x) = -4x^5 + 3x^4 - 2x^3 + x^2 - 2 - 7 \Rightarrow 4$ sign changes
$\Rightarrow 4$ or 2 or 0 negative real roots

# positive real roots	# negative real roots	# non-real roots
1	4	0
1	2	2
1	0	4

25. $P(x) = x^4 + x^2 + 24567 \Rightarrow 0$ sign changes $\Rightarrow 0$ positive real roots
$P(-x) = x^4 + x^2 + 24567 \Rightarrow 0$ sign changes $\Rightarrow 0$ negative real roots

# positive real roots	# negative real roots	# non-real roots
0	0	4

27. Possible rational roots: $\pm\left(\dfrac{30}{2}, \dfrac{15}{2}, \dfrac{10}{2}, \dfrac{6}{2}, \dfrac{5}{2}, \dfrac{3}{2}, \dfrac{2}{2}, \dfrac{1}{2}, \dfrac{30}{1}, \dfrac{15}{1}, \dfrac{10}{1}, \dfrac{6}{1}, \dfrac{5}{1}, \dfrac{3}{1}, \dfrac{2}{1}, \dfrac{1}{1}\right)$

Descartes' rule of signs indicates 0 positive real roots, 3 or 1 negative real roots.
Try $x = -2$:

$$
\begin{array}{r|rrrr}
-2 & 2 & 17 & 41 & 30 \\
   &   & -4 & -26 & -30 \\
\hline
   & 2 & 13 & 15 & 0
\end{array}
\qquad
\begin{aligned}
2x^3 + 17x^2 + 41x + 30 &= 0 \\
(x+2)(2x^2 + 13x + 15) &= 0 \\
(x+2)(2x+3)(x+5) &= 0
\end{aligned}
$$

ROOTS: -2, $-\dfrac{3}{2}$, -5

29. Possible rational roots: $\pm\left(\dfrac{36}{4}, \dfrac{18}{4}, \dfrac{12}{4}, \dfrac{9}{4}, \dfrac{6}{4}, \dfrac{4}{4}, \dfrac{3}{4}, \dfrac{2}{4}, \dfrac{1}{4}, \dfrac{36}{2}, \dfrac{18}{2}, \dfrac{12}{2}, \dfrac{9}{2}, \dfrac{6}{2}, \dfrac{4}{2}, \dfrac{3}{2}, \dfrac{2}{2}, \dfrac{1}{2}\right)$
$\pm\left(\dfrac{36}{1}, \dfrac{18}{1}, \dfrac{12}{1}, \dfrac{9}{1}, \dfrac{6}{1}, \dfrac{4}{1}, \dfrac{3}{1}, \dfrac{2}{1}, \dfrac{1}{1}\right)$

Descartes' rule of signs: 2 or 0 positive real roots, 2 or 0 negative real roots

Try $x = 2$:
$$
\begin{array}{r|rrrrr}
2 & 4 & 0 & -25 & 0 & 36 \\
  &   & 8 & 16 & -18 & -36 \\
\hline
  & 4 & 8 & -9 & -18 & 0
\end{array}
$$

Try $x = -2$
$$
\begin{array}{r|rrrrr}
-2 & 4 & 8 & -9 & -18 \\
   &   & -8 & 0 & 18 \\
\hline
   & 4 & 0 & -9 & 0
\end{array}
$$

$$4x^4 - 25x^2 + 36 = 0$$
$$(x-2)(x+2)(4x^2 - 9) = 0$$
$$(x-2)(x+2)(2x+3)(2x-3) = 0$$

ROOTS: ± 2, $\pm\dfrac{3}{2}$

31. $P(-1) = -9$ and $P(0) = 18$, so $P = 0$ somewhere between -1 and 0.

33. Note that $P(4) = -6$ and $P(5) = 28$

STEP	x_l	c	x_r	$P(x_l)$	$P(c)$	$P(x_r)$
0	4	4.5	5	−	+	+
1	4	4.25	4.5	−	+	+
2	4	4.125	4.25	−	−	+
3	4.125	4.1875	4.25	−	−	+
4	4.1875	4.21875	4.25	−	−	+
5	4.21875	4.234375	4.25	−	−	+

The solution is 4.2, rounded to the nearest tenth.

35. $x \approx 1.67$

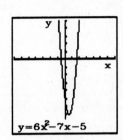

$y = 6x^2 - 7x - 5$

37. Let the width of each collector be the vertical measure of each. Then the length of the rightmost collector is x, the middle length is $x + 2$ and the leftmost length is $x + 5$. To find the width of each, take the area divided by the length. Thus the widths are $\frac{60}{x}$, $\frac{60}{x+2}$ and $\frac{60}{x+5}$. The sum of the widths is 15:

$$\frac{60}{x} + \frac{60}{x+2} + \frac{60}{x+5} = 15$$

$$\frac{x(x+2)(x+5)}{15} \cdot \left(\frac{60}{x} + \frac{60}{x+2} + \frac{60}{x+5}\right) = \frac{x(x+2)(x+5)}{15} \cdot 15$$

$$4(x+2)(x+5) + 4x(x+5) + 4x(x+2) = x(x+2)(x+5)$$
$$x^3 - 5x^2 - 46x - 40 = 0$$

Descartes' rule of signs indicates that there is only 1 positive real root.

$$
\begin{array}{r|rrrr}
10 & 1 & -5 & -46 & -40 \\
   &   & 10 & 50 & 40 \\
\hline
   & 1 & 5 & 4 & 0
\end{array}
$$

Thus $x = 10$ is the only positive real root, and the lengths are 10 feet, 12 feet and 15 feet. The panels are 10×6, 12×5 and 15×4 (all in feet).

Chapter 5 Test   (page 286)

1.
$$
\begin{array}{r|rrrr}
2 & 3 & 0 & -9 & -5 \\
  &   & 6 & 12 & 6 \\
\hline
  & 3 & 6 & 3 & 1
\end{array}
$$
$P(2) = 1$ (the remainder)

3. $P(x) = (x - 5)(x + 1)(x - 0)$
$= (x^2 - 4x - 5)x$
$= x^3 - 4x^2 - 5x$

5.

$$\begin{array}{r|rrrr} 1 & 3 & -2 & 0 & 4 \\ & & 3 & 1 & 1 \\ \hline & 3 & 1 & 1 & \boxed{5} \end{array}$$

$P(1) = 5$

7.

$$\begin{array}{r|rrrr} -\frac{1}{3} & 3 & -2 & 0 & 4 \\ & & -1 & 1 & -\frac{1}{3} \\ \hline & 3 & -3 & 1 & \boxed{\frac{11}{3}} \end{array}$$

$P\left(-\frac{1}{3}\right) = \frac{11}{3}$

9.

$$\begin{array}{r|rrrr} 2 & 2 & -3 & -4 & -1 \\ & & 4 & 2 & -4 \\ \hline & 2 & 1 & -2 & \boxed{-5} \end{array}$$

$(x-2)(2x^2 + x - 2) - 5$

11.

$$\begin{array}{r|rrr} 5 & 2 & -7 & -15 \\ & & 10 & 15 \\ \hline & 2 & 3 & \boxed{0} \end{array}$$

$2x + 3$

13.
$(x-2)(x-i)(x+i) = 0$
$(x-2)(x^2+1) = 0$
$x^3 - 2x^2 + x - 2 = 0$

15.

# pos	# neg	# non-real
3	2	0
3	0	2
1	2	2
1	0	4

17.

$$\begin{array}{r|rrrrrr} 3 & 1 & -1 & -5 & 5 & 4 & -5 \\ & & 3 & 6 & 3 & 24 & 54 \\ \hline & 1 & 2 & 1 & 8 & 18 & 49 \end{array}$$

3 is an upper bound.

$$\begin{array}{r|rrrrrr} -3 & 1 & -1 & -5 & 5 & 4 & -5 \\ & & -3 & 12 & -21 & 48 & -156 \\ \hline & 1 & -4 & 7 & -16 & 52 & -161 \end{array}$$

-3 is a lower bound.

19.

$$\begin{array}{r|rrrr} 2 & 2 & 3 & -11 & -6 \\ & & 4 & 14 & 6 \\ \hline & 2 & 7 & 3 & \boxed{0} \end{array}$$

$2x^3 + 3x^2 - 11x - 6 = 0$
$(x-2)(2x^2 + 7x + 3) = 0$
$(x-2)(2x+1)(x+3) = 0$

ROOTS: $2, -\frac{1}{2}, -3$

Exercise 6.1 (page 295)

1. $4^{\sqrt{3}} \approx 11.0357$

3. $7^{\pi} \approx 451.8079$

5. $y = 3^x$

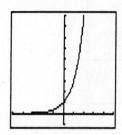

7. $y = \left(\frac{1}{5}\right)^x$

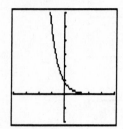

9. $y = -2^x$

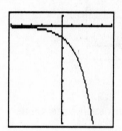

11. $y = \left(\frac{3}{4}\right)^x$

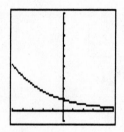

13. $y = 3^x - 1$

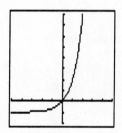

15. $y = 2^x + 1$

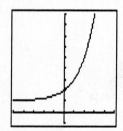

17. $y = 3^{x-1}$

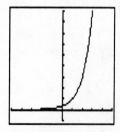

19. $y = 3^{x+1}$

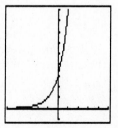

21. $y = 2^{x+1} - 2$

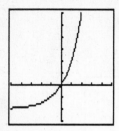

23. $y = 3^{x-2} + 1$

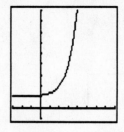

25. $y = 5(2^x)$

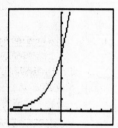

27. $y = 3^{-x}$

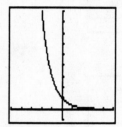

29. $y = b^x$
$25 = b^2$
$5 = b$

31. $y = b^x$
$2 = b^0$
no solution — no such base exists

33. $y = b^x$
$\frac{1}{2} = b^1$

$\frac{1}{2} = b$

35. $y = b^x$
$-1 = b^0$

no solution — no such base exists

37. $A = A_0 2^{-t/h}$
$= 50 \cdot 2^{-100/12.4}$
$\approx 50 \cdot 2^{-8.064516}$
$\approx 50 \cdot 0.003735414$
≈ 0.1868
There will be 0.1868 grams left.

39. $A = A_0\left(\frac{2}{3}\right)^t$

$= A_0\left(\frac{2}{3}\right)^5$

$= A_0 \cdot \frac{32}{243}$

There will be about $0.1317 A_0$ left.

41. If 0.5 grams of ^{253}Cf remain, then half of the amount of ^{253}Cf has decayed, so the amount of time that has passed equals the half-life of ^{253}Cf, or 18 days. Let A stand for the amount of ^{254}Cf which remains.

$A = A_0 2^{-t/h}$
$= 1 \cdot 2^{-18/60}$
$= 2^{-0.3}$
≈ 0.8123 There will be about 0.8123 grams of ^{254}Cf left.

184

43. $A = A_0\left(1 + \dfrac{r}{k}\right)^{kt}$

$= 500\left(1 + \dfrac{0.08}{4}\right)^{4 \cdot 10}$

$= 500(1.02)^{40}$

$= 1104.02$

The account will have \$1104.02.

45. $A = A_0\left(1 + \dfrac{r}{k}\right)^{kt}$

$= 500\left(1 + \dfrac{0.05}{4}\right)^{4 \cdot 5}$

$= 500(1.0125)^{20}$

$= 641.02$

$A = A_0\left(1 + \dfrac{r}{k}\right)^{kt}$

$= 500\left(1 + \dfrac{0.055}{4}\right)^{4 \cdot 5}$

$= 500(1.01375)^{20}$

$= 657.03$

The difference is \$16.01.

47. $A = A_0\left(1 + \dfrac{r}{k}\right)^{kt}$

$= 1\left(1 + \dfrac{0.05}{1}\right)^{1 \cdot 300}$

$= (1.05)^{300}$

$= 2{,}273{,}996.13$

It will be worth \$2,273,996.13.

49. $A = A_0\left(1 + \dfrac{r}{360}\right)^{365t}$

$= 1000\left(1 + \dfrac{0.07}{360}\right)^{365 \cdot 5}$

$= 1000(1.000194444)^{1825}$

$= 1425.93$

It will be worth \$1425.93.

51. $I = 8\left(\dfrac{1}{2}\right)^{x}$

$= 8\left(\dfrac{1}{2}\right)^{2}$

$= 8 \cdot \dfrac{1}{4}$

$= 2$

The intensity is 2 lumens.

53. $I = 6(0.5)^{x}$

$= 6(0.5)^{4.5}$

$= 0.2652 \quad [A]$

$I = 8(0.55)^{x}$

$= 8(0.55)^{4.5}$

$= 0.5429 \quad [B]$

It will be brighter at B.

55. $A = A_0 2^{-t/h}$

$= 1 \cdot 2^{-12/8}$

$= 0.3536$

35.36% of the dose is retained.

57. $A = A_0 2^{-t/h}$

$= 1 \cdot 2^{-2.5/1.5}$

$= 0.3150$

31.5% of the dose is retained.

59. $P = 375(1.3)^{t}$

$= 375(1.3)^{3}$

$= 823.875$

The population will be 823.

61.
$$C = C_0(0.7)^{t}$$
$$2.471 \times 10^{-5} = C_0(0.7)^{7}$$
$$2.471 \times 10^{-5} = C_0(0.0823543)$$
$$\dfrac{2.471 \times 10^{-5}}{0.0823543} = C_0$$
$$3 \times 10^{-4} = C_0$$

63. $A = A_0\left(1 + \frac{r}{k}\right)^{kt}$

$\quad = 1500\left(1 + \frac{0.21}{12}\right)^{12 \cdot 1}$

$\quad = 1500(1.0175)^{12}$

$\quad = 1847.16$

She will pay \$1847.16.

65. $A = A_0(0.92)^t$

$\quad = 10{,}500(0.92)^9$

$\quad = 4957.69$

The boat will be worth \$4957.69.

67. $A = A_0\left(1 + \frac{r}{k}\right)^{kt}$

$\quad = 4400\left(1 + \frac{0.075}{12}\right)^{12 \cdot 5/12}$

$\quad = 4400(1.00625)^5$

$\quad = 4539.23$

He will pay \$4539.23.

69. $P = 14.7\left(2^{-0.000056a}\right)$

$\quad = 14.7(2^{-1.4784}) \quad (1 \text{ mi} = 5280 \text{ ft})$

$\quad = 5.2756$

The pressure will be 5.2756 lb/in^2.

71. $A = A_0\left(1 + \frac{r}{k}\right)^{kt}$

$\quad = (1.00666667)^{12t}$

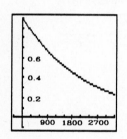

There will be \$119.59.

73. $A = 2^{-0.000635t}$

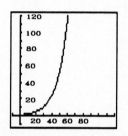

0.751 grams after 660 years
0.5 grams after 1600 years
0.273 grams after 3000 years

75 – 79. answers will vary

Exercise 6.2 (page 304)

1. $y = -e^x$

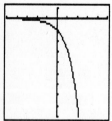

3. $y = e^{-0.5x}$

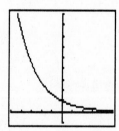

5. $y = 2e^{-x}$

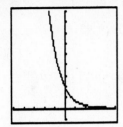

7. $y = e^x + 1$

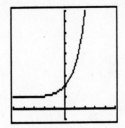

9. yes

11. no $[e^0 = 1]$

13. no $[e^2 \neq 2e]$

15. no $[e^{-2} \neq e]$

17. $A = A_0 e^{rt}$
$= 5000 e^{0.082 \cdot 12}$
$= 5000 e^{.984}$
$= 5000 \cdot 2.675135411$
$= 13375.68$

It will be worth \$13,375.68.

19. $A = A_0 e^{rt}$
$11,180 = A_0 e^{0.07 \cdot 7}$
$11,180 = A_0 e^{0.49}$
$11,180 = A_0 \cdot 1.63231622$
$\dfrac{11,180}{1.63231622} = A_0$
$6849.16 = A_0$

The original investment was \$6849.16.

21. $A = A_0 e^{rt}$
$= 5000 \cdot e^{0.085 \cdot 5}$
$= 5000 \cdot e^{0.425}$
$= 5000 \cdot 1.52959042$
$= 7647.95$ (continuous)

$A = A_0\left(1 + \dfrac{r}{k}\right)^{kt}$
$= 5000\left(1 + \dfrac{0.085}{1}\right)^{1 \cdot 5}$
$= 5000(1.085)^5$
$= 7518.28$ (annual)

23. $P = 173 e^{0.03t}$
$= 173 e^{0.03 \cdot 20}$
$= 173 e^{0.6}$
$= 315.23$

There will be 315 people.

25. $P = P_0 e^{0.27t}$
$= 2e^{0.27 \cdot 12}$
$= 2e^{3.24}$
$= 51.07$

There will be 51 infected cattle.

27. $P = P_0 e^{kt}$
$= 5.2 e^{0.019 \cdot 30}$
≈ 9.2

The population will be about 9.2 billion.

29. $P = P_0 e^{kt}$
$= 5.2 e^{0.019 \cdot 50}$
≈ 13.4457
$13.4457/5.2 \approx 2.59$

The population will increase by a factor of about 2.59.

31.
$$P = P_0 e^{kt}$$
$$2P_0 = P_0 e^{k \cdot 20}$$
$$2 = (e^k)^{20}$$
$$2^{1/20} = e^k$$
$$P = P_0(2^{1/20})^t$$
$$= 2 \times 10^5 \cdot (2^{1/20})^{35}$$
$$= 2 \times 10^5 \cdot 3.363585679$$
$$= 672,717$$
The population will be 672,717.

33.
$$x = 0.08(1 - e^{-0.1t})$$
$$= 0.08(1 - e^{-0.1 \cdot 30})$$
$$= 0.08 \cdot 0.9502129316$$
$$= 0.076$$
The concentration will be 0.076.

35. $P = \dfrac{1,200,000}{1 + (1200 - 1)e^{-0.4t}} = \dfrac{1,200,000}{1 + (1199)e^{-0.4 \cdot 5}} = \dfrac{1,200,000}{163.267} = 7350$

37.
$$A = A_0 e^{kt}$$
$$0.5A_0 = A_0 e^{k \cdot 2.3}$$
$$0.5 = e^{k \cdot 2.3}$$
$$(0.5)^{1/2.3} = e^k$$
$$A = A_0 \cdot \left[(0.5)^{1/2.3}\right]^t$$
$$= A_0 \cdot (0.739805)^{24}$$
$$= 0.000722 A_0$$
About 0.07% will remain.

39.
$$v = 50(1 - e^{-0.2t})$$
$$= 50(1 - e^{-0.2 \cdot 20})$$
$$= 50(0.9816843611)$$
$$= 49.084$$
The velocity will be about 49 meters per second.

41.
$$e^{t+3} = ke^t$$
$$e^t e^3 = ke^t$$
$$e^3 = k$$

43.

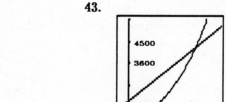

$y = 1000e^{0.02x}$ and $y = 31x + 2000$
The food will be adequate for about 72.2 years.

45. $1 + 1 + \dfrac{1}{2} + \dfrac{1}{2 \cdot 3} + \dfrac{1}{2 \cdot 3 \cdot 4} + \dfrac{1}{2 \cdot 3 \cdot 4 \cdot 5} + \dfrac{1}{2 \cdot 3 \cdot 4 \cdot 5 \cdot 6} + \dfrac{1}{2 \cdot 3 \cdot 4 \cdot 5 \cdot 6 \cdot 7}$
$$= 2.718253968 \quad - \quad \text{accurate to 4 decimal places}$$

47. $y = \dfrac{1{,}200{,}000}{1 + (1199)e^{-0.4t}}$

Exercise 6.3 (page 312)

1. $3^4 = 81$ **3.** $\left(\frac{1}{2}\right)^3 = \frac{1}{8}$ **5.** $4^{-3} = \frac{1}{64}$ **7.** $x^z = y$

9. $\log_8 64 = 2$ **11.** $\log_4 \frac{1}{16} = -2$ **13.** $\log_{1/2} 32 = -5$ **15.** $\log_x z = y$

17. $\log_2 8 = x$ **19.** $\log_4 64 = x$ **21.** $\log_{1/2} \frac{1}{8} = x$ **23.** $\log_9 3 = x$

$\qquad 2^x = 8 \qquad\qquad\qquad 4^x = 64 \qquad\qquad \left(\frac{1}{2}\right)^x = \frac{1}{8} \qquad\qquad 9^x = 3$

$\qquad x = 3 \qquad\qquad\qquad\quad x = 3 \qquad\qquad\qquad x = 3 \qquad\qquad\qquad x = \frac{1}{2}$

25. $\log_{1/2} 8 = x$ **27.** $\log_8 x = 2$ **29.** $\log_7 x = 1$ **31.** $\log_{25} x = \frac{1}{2}$

$\quad \left(\frac{1}{2}\right)^x = 8 \qquad\qquad 8^2 = x \qquad\qquad\quad 7^1 = x \qquad\qquad 25^{1/2} = x$

$\qquad x = -3 \qquad\qquad 64 = x \qquad\qquad\quad 7 = x \qquad\qquad\quad 5 = x$

33. $\log_5 x = -2$ **35.** $\log_{36} x = -\frac{1}{2}$

$\qquad 5^{-2} = x \qquad\qquad\qquad\qquad\qquad 36^{-1/2} = x$

$\qquad \frac{1}{25} = x \qquad\qquad\qquad\qquad\qquad\quad \frac{1}{6} = x$

37. $\log_{100} \frac{1}{1000} = x$ **39.** $\log_{27} 9 = x$

$\qquad\quad 100^x = \frac{1}{1000} \qquad\qquad\qquad 27^x = 9$

$\qquad\quad (10^2)^x = 10^{-3} \qquad\qquad\quad (3^3)^x = 3^2$

$\qquad\qquad 2x = -3 \qquad\qquad\qquad\quad 3x = 2$

$\qquad\qquad\quad x = -\frac{3}{2} \qquad\qquad\qquad\quad x = \frac{2}{3}$

41. $\log_x 5^3 = 3$ **43.** $\log_x \frac{9}{4} = 2$

$\qquad\quad x^3 = 5^3 \qquad\qquad\qquad\qquad\qquad x^2 = \frac{9}{4}$

$\qquad\quad x = 5$

$\qquad\qquad\qquad\qquad\qquad\qquad\qquad x = \frac{3}{2} \quad \left(-\frac{3}{2} \text{ is not valid.}\right)$

45. $\log_x \dfrac{1}{64} = -3$

$\qquad x^{-3} = \dfrac{1}{64}$

$\qquad x = 4$

47. $\log_{2\sqrt{2}} x = 2$

$\qquad \left(2\sqrt{2}\right)^2 = x$

$\qquad 8 = x$

49. $2^{\log_2 5} = x$

$\qquad 5 = x$

51. $x^{\log_4 6} = 6$

$\qquad x = 4$

53. $\log_{10} 10^3 = x$

$\qquad x = 3$

55. $10^{\log_{10} x} = 100$

$\qquad x = 100$

57. $\log 3.25 = 0.5119$

59. $\log 0.00467 = -2.3307$

61. $\ln 0.93 = -0.0726$

63. $\ln 37.896 = 3.6348$

65. $\log (\ln 1.7) = \log (0.5306282511)$

$\qquad = -0.2752$

67. $\ln (\log 0.1) = \ln (-1)$

$\qquad = \text{undefined}$

69. $\log y = 1.4023$

$\qquad y = 25.25$

71. $\ln y = 4.24$

$\qquad y = 69.41$

73. $\log y = -3.71$

$\qquad y = 0.00$

75. $\log y = \ln 8$

$\qquad \log y = 2.079441542$

$\qquad y = 120.07$

77. $y = \log_3 x$

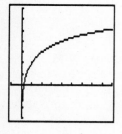

79. $y = \log_{1/3} x$

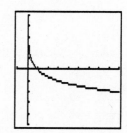

81. $y = 2 + \log_2 x$

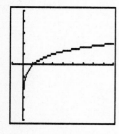

83. $y = \log_3 (x + 2)$

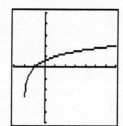

85. $y = 1 + \ln x$

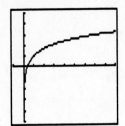

87. $y = 2 \ln x$

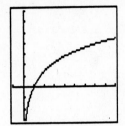

89.

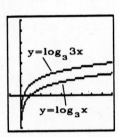

91.

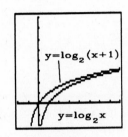

93.

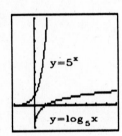

95.
$y = \log_b x$
$2 = \log_b 9$
$b^2 = 9$
$b = 3 \quad (-3 \text{ is not valid})$

97.
$y = \log_b x$
$1 = \log_b \frac{1}{2}$

$b^1 = \frac{1}{2}$

99.
$y = \ln x$
yes $(\ln 1 = 0)$

101. no $(\ln 0 \neq 1)$

103 – 105. answers will vary

Exercise 6.4 (page 319)

1. $\log_4 1 = 0$ **3.** $\log_4 4^7 = 7$ **5.** $5^{\log_5 10} = 10$ **7.** $\log_5 5 = 1$

9 – 13. check with a calculator

15. $\log_b 2xy = \log_b 2 + \log_b x + \log_b y$

17. $\log_b \frac{2x}{y} = \log_b 2x - \log_b y$
$= \log_b 2 + \log_b x - \log_b y$

19. $\log_b x^2 y^3 = \log_b x^2 + \log_b y^3$
$= 2\log_b x + 3\log_b y$

191

21. $\log_b(xy)^{1/3} = \frac{1}{3}\log_b xy = \frac{1}{3}(\log_b x + \log_b y) = \frac{1}{3}\log_b x + \frac{1}{3}\log_b y$

23. $\log_b x\sqrt{z} = \log_b xz^{1/2}$
$= \log_b x + \log_b z^{1/2}$
$= \log_b x + \frac{1}{2}\log_b z$

25. $\log_b \dfrac{\sqrt[3]{x}}{\sqrt[3]{yz}} = \log_b \dfrac{x^{1/3}}{(yz)^{1/3}}$
$= \log_b x^{1/3} - \log_b (yz)^{1/3}$
$= \frac{1}{3}\log_b x - \frac{1}{3}(\log_b yz)$
$= \frac{1}{3}\log_b x - \frac{1}{3}(\log_b y + \log_b z)$
$= \frac{1}{3}\log_b x - \frac{1}{3}\log_b y - \frac{1}{3}\log_b z$

27. $\log_b(x+1) - \log_b x = \log_b \dfrac{x+1}{x}$

29. $2\log_b x + \frac{1}{3}\log_b y = \log_b x^2 + \log_b y^{1/3}$
$= \log_b (x^2 \, y^{1/3})$

31. $-3\log_b x - 2\log_b y + \frac{1}{2}\log_b z = \log_b x^{-3} + \log_b y^{-2} + \log_b z^{1/2}$
$= \log_b x^{-3}y^{-2}z^{1/2}$
$= \log_b \frac{1}{x^3}\cdot\frac{1}{y^2}\cdot\sqrt{z}$
$= \log_b \dfrac{\sqrt{z}}{x^3 y^2}$

33. $\log_b\left(\dfrac{x}{z}+x\right) - \log_b\left(\dfrac{y}{z}+y\right) = \log_b \dfrac{\frac{x}{z}+x}{\frac{y}{z}+y} = \log_b \dfrac{\frac{x}{z}+x}{\frac{y}{z}+y}\cdot\dfrac{\frac{z}{1}}{\frac{z}{1}} = \log_b \dfrac{x+xz}{y+yz}$
$= \log_b \dfrac{x(1+z)}{y(1+z)}$
$= \log_b \dfrac{x}{y}$

35. $\log_b ab = \log_b a + \log_b b$
$= \log_b a + 1 \;\Rightarrow\; \text{TRUE}$

37. $\log_b 0 \neq 1$ [it is undefined] $\Rightarrow$ FALSE

39. TRUE $[\log_b xy = \log_b x + \log_b y]$

41. FALSE [only true if $a = b$]

43. TRUE

45. FALSE $[\log_b x^{-1} = -\log_b x]$

47. FALSE $[\log_b \frac{A}{B} = \log_b A - \log_b B]$

49. $\log_b \frac{1}{5} = \log_b 5^{-1} = -\log_b 5 \Rightarrow$ TRUE

51. $\frac{1}{3}\log_b a^3 = 3\cdot\frac{1}{3}\log_b a = \log_b a$
 TRUE

53. TRUE $[x = \log_b y \Rightarrow \log_{1/b} y = -x]$

55. $\log_{10}28 = \log_{10}4 \cdot 7 = \log_{10}4 + \log_{10}7 = 0.6021 + 0.8451 = 1.4472$

57. $\log_{10}2.25 = \log_{10}\frac{9}{4} = \log_{10}9 - \log_{10}4 = 0.9542 - 0.6021 = 0.3521$

59. $\log_{10}\frac{63}{4} = \log_{10}63 - \log_{10}4 = \log_{10}7 \cdot 9 - \log_{10}4 = \log_{10}7 + \log_{10}9 - \log_{10}4$
$$= 0.8451 + 0.9542 - 0.6021$$
$$= 1.1972$$

61. $\log_{10}252 = \log_{10}4 \cdot 7 \cdot 9 = \log_{10}4 + \log_{10}7 + \log_{10}9 = 0.6021 + 0.8451 + 0.9542$
$$= 2.4014$$

63. $\log_{10}112 = \log_{10}4^2 \cdot 7 = \log_{10}4^2 + \log_{10}7 = 2\log_{10}4 + \log_{10}7$
$$= 2(0.6021) + 0.8451 = 2.0493$$

65. $\log_{10}\frac{144}{49} = \log_{10}\frac{4^2 \cdot 9}{7^2} = \log_{10}4^2 + \log_{10}9 - \log_{10}7^2$
$$= 2\log_{10}4 + \log_{10}9 - 2\log_{10}7$$
$$= 2(0.6021) + 0.9542 - 2(0.8451) = 0.4682$$

67. $\log_3 7 = \frac{\log_{10}7}{\log_{10}3} = 1.7712$ 69. $\log_\pi 3 = \frac{\log_{10}3}{\log_{10}\pi} = 0.9597$

71. $\log_3 8 = \frac{\log_{10}8}{\log_{10}3} = 1.8928$ 73. $\log_{\sqrt{2}}\sqrt{5} = \frac{\log_{10}\sqrt{5}}{\log_{10}\sqrt{2}} = 2.3219$

75. Let $x = \log_b M$ and $y = \log_b N$.   77. $e^{x \ln a} = (e^{\ln a})^x = a^x$
Then $M = b^x$ and $N = b^y$, and

$\frac{M}{N} = \frac{b^x}{b^y} = b^{x-y}$.

So $b^{x-y} = \frac{M}{N}$, and

$\log_b \frac{M}{N} = x - y = \log_b M - \log_b N$.

79. $\ln(e^x) = x \ln e = x$ 81. $\ln(\log 0.9) = \ln(-0.0458)$
The logarithm of any negative
number is undefined.

Exercise 6.5   (page 325)

1. $\text{pH} = -\log[\text{H}^+] = -\log(1.7 \times 10^{-5}) = -(-4.77) = 4.77$

3.

$$\text{pH} = -\log[\text{H}^+]$$
$$2.9 = -\log[\text{H}^+]$$
$$-2.9 = \log[\text{H}^+]$$
$$1.26 \times 10^{-3} = [\text{H}^+]$$

$$3.3 = -\log[\text{H}^+]$$
$$-3.3 = \log[\text{H}^+]$$
$$5.01 \times 10^{-4} = [\text{H}^+]$$

5.

$$\text{db gain} = 20 \log \frac{E_O}{E_I}$$
$$29 = 20 \log \frac{20}{E_I}$$
$$\frac{29}{20} = \log \frac{20}{E_I}$$
$$28.1838 = \frac{20}{E_I}$$
$$28.1838 E_I = 20$$
$$E_I = 0.7096 \text{ volts}$$

7.

$$P = kE^2$$
$$\log P = \log kE^2$$
$$\log P = \log k + 2 \log E$$
$$\log P - \log k = 2 \log E$$
$$\tfrac{1}{2}(\log P - \log k) = \log E$$
$$\tfrac{1}{2} \log \frac{P}{k} = \log E$$

$$\text{db gain} = 20 \log \frac{E_O}{E_I}$$
$$= 20 (\log E_O - \log E_I)$$
$$= 20 \left(\tfrac{1}{2} \log \frac{P_O}{k} - \tfrac{1}{2} \log \frac{P_I}{k} \right)$$
$$= 10 \left(\log \frac{P_O}{k} - \log \frac{P_I}{k} \right)$$
$$= 10 \log \frac{\frac{P_O}{k}}{\frac{P_I}{k}}$$
$$= 10 \log \frac{P_O}{P_I}$$

9. $R = \log \frac{A}{P} = \log \frac{5000}{0.2} = 4.40$

11.

$$R = \log \frac{A}{P}$$
$$4 = \log \frac{A}{0.25}$$
$$10,000 = \frac{A}{0.25}$$
$$2500 = A \text{ (in micrometers)}$$

13.
$$C = M(1 - e^{-kt})$$
$$0.5M = M(1 - e^{-k \cdot 6})$$
$$0.5 = 1 - e^{-6k}$$
$$e^{-6k} = 0.5$$
$$e^{-k} = (0.5)^{1/6}$$
$$C = M(1 - e^{-kt})$$
$$0.90M = M[1 - (e^{-k})^t]$$
$$0.90 = 1 - [(0.5)^{1/6}]^t$$
$$[(0.5)^{1/6}]^t = 0.10$$
$$\log [(0.5)^{1/6}]^t = \log 0.10$$
$$t \log (0.5)^{1/6} = \log 0.1$$
$$t = \frac{\log 0.1}{\log (0.5)^{1/6}}$$
$$t = 19.93 \text{ hours}$$

15.
$$L_{old} = k \ln I$$
$$L_{new} = k \ln (2I)$$
$$= k (\ln 2 + \ln I)$$
$$= k \ln 2 + k \ln I$$
$$= k \ln 2 + L_{old}$$
$$L_{new} = L_{old} + k \ln 2$$

17. $t = \frac{\ln 2}{r} = \frac{\ln 2}{0.12} = 5.78$ years

19.
$$E = RT \ln \frac{V_f}{V_i}$$
$$= 8.314 \cdot 400 \cdot \ln \frac{3V_i}{V_i}$$
$$= 8.314(400)(\ln 3)$$
$$= 3653.5 \text{ joules}$$

21. $n = \dfrac{\log V - \log C}{\log \left(1 - \frac{2}{N}\right)} = \dfrac{\log 8000 - \log 37{,}000}{\log \left(1 - \frac{2}{5}\right)} = \dfrac{-0.6651117371}{-0.2218487496} = 3.0$ years old

23. $n = \dfrac{\log\left(\frac{Ar}{P} + 1\right)}{\log (1 + r)} = \dfrac{\log \left(\dfrac{20{,}000(0.12)}{1000} + 1\right)}{\log (1 + 0.12)} = \dfrac{\log 3.4}{\log 1.12} = 10.80$ years

25. $V = ER_1 \ln \dfrac{R_2}{R_1} = 400{,}000(0.25) \ln \dfrac{2}{0.25} = 100{,}000 \ln 8 = 207{,}944$ volts

27.
$$P = P_0 e^{rt}$$
$$= P_0 e^{r \cdot \frac{\ln 2}{r}}$$
$$= P_0 e^{\ln 2}$$
$$= P_0 \cdot 2$$
So the population will be twice the original population when $t = (\ln 2)/r$.

29.
$$y = \frac{1}{1 + e^{-2x}} \qquad y\text{-int: set } x = 0$$
$$= \frac{1}{1 + e^0}$$
$$= \frac{1}{2}$$
$\left(0, \frac{1}{2}\right)$ is the y-intercept.

Exercise 6.6 (page 332)

1.
$$4^x = 5$$
$$\log 4^x = \log 5$$
$$x \log 4 = \log 5$$
$$x = \frac{\log 5}{\log 4} \approx 1.1610$$

3.
$$13^{x-1} = 2$$
$$\log 13^{x-1} = \log 2$$
$$(x-1) \log 13 = \log 2$$
$$x - 1 = \frac{\log 2}{\log 13}$$
$$x = \frac{\log 2}{\log 13} + 1 \approx 1.2702$$

5.
$$2^{x+1} = 3^x$$
$$\log 2^{x+1} = \log 3^x$$
$$(x+1) \log 2 = x \log 3$$
$$x \log 2 + \log 2 = x \log 3$$
$$x \log 2 - x \log 3 = -\log 2$$
$$x(\log 2 - \log 3) = -\log 2$$
$$x = \frac{-\log 2}{\log 2 - \log 3}$$
$$\approx 1.7095$$

7.
$$2^x = 3^x$$
$$\log 2^x = \log 3^x$$
$$x \log 2 = x \log 3$$
$$x \log 2 - x \log 3 = 0$$
$$x(\log 2 - \log 3) = 0$$
$$x = \frac{0}{\log 2 - \log 3}$$
$$= 0$$

9.
$$7^{x^2} = 10$$
$$\log 7^{x^2} = \log 10$$
$$x^2 \log 7 = \log 10$$
$$x^2 = \frac{\log 10}{\log 7}$$
$$x = \pm\sqrt{\frac{\log 10}{\log 7}}$$
$$x \approx \pm 1.0878$$

11.
$$8^{x^2} = 9^x$$
$$\log 8^{x^2} = \log 9^x$$
$$x^2 \log 8 = x \log 9$$
$$x^2 \log 8 - x \log 9 = 0$$
$$x(x \log 8 - \log 9) = 0$$
$$x = 0 \quad \text{or} \quad x \log 8 - \log 9 = 0$$
$$x \log 8 = \log 9$$
$$x = \frac{\log 9}{\log 8}$$
$$\approx 1.0566$$

$$x = 0 \text{ or } x \approx 1.0566$$

13.
$$2^{x^2 - 2x} = 8$$
$$2^{x^2 - 2x} = 2^3$$
$$x^2 - 2x = 3$$
$$x^2 - 2x - 3 = 0$$
$$(x-3)(x+1) = 0$$
$$x = 3 \text{ or } x = -1$$

15.
$$3^{x^2 + 4x} = \frac{1}{81}$$
$$3^{x^2 + 4x} = 3^{-4}$$
$$x^2 + 4x = -4$$
$$x^2 + 4x + 4 = 0$$
$$(x+2)(x+2) = 0$$
$$x = -2$$

196

17.
$$4^{x+2} - 4^x = 15$$
$$4^x \cdot 4^2 - 4^x = 15$$
$$16 \cdot 4^x - 4^x = 15$$
$$15 \cdot 4^x = 15$$
$$4^x = 1$$
$$x = 0$$

19.
$$2(3^x) = 6^{2x}$$
$$\log 2(3^x) = \log 6^{2x}$$
$$\log 2 + \log 3^x = 2x \log 6$$
$$\log 2 + x \log 3 = 2x \log 6$$
$$\log 2 = 2x \log 6 - x \log 3$$
$$\log 2 = x(2 \log 6 - \log 3)$$
$$\frac{\log 2}{2 \log 6 - \log 3} = x$$
$$0.2789 \approx x$$

21.
$$2^{2x} - 10(2^x) + 16 = 0$$
$$y^2 - 10y + 16 = 0$$
$$(y-2)(y-8) = 0$$
$$y = 2 \quad \text{or} \quad y = 8$$
$$2^x = 2 \quad \bigg| \quad 2^x = 8$$
$$x = 1 \quad \bigg| \quad x = 3$$

23.
$$2^{2x+1} - 2^x = 1$$
$$2^{2x} \cdot 2^1 - 2^x - 1 = 0$$
$$2y^2 - y - 1 = 0$$
$$(2y+1)(y-1) = 0$$
$$y = -\frac{1}{2} \quad \text{or} \quad y = 1$$
$$2^x = -\frac{1}{2} \quad \bigg| \quad 2^x = 1$$
$$\text{(no solution)} \quad \bigg| \quad x = 0$$
The solution set is $\{0\}$.

25.
$$\log (2x - 3) = \log (x + 4)$$
$$2x - 3 = x + 4$$
$$x = 7$$

27.
$$\log \frac{4x+1}{2x+9} = 0$$
$$\frac{4x+1}{2x+9} = 1$$
$$4x + 1 = 2x + 9$$
$$2x = 8$$
$$x = 4$$

29.
$$\log x^2 = 2$$
$$x^2 = 100$$
$$x = \pm \sqrt{100}$$
$$x = \pm 10$$

31.
$$\log x + \log (x - 48) = 2$$
$$\log x(x - 48) = 2$$
$$x^2 - 48x = 100$$
$$x^2 - 48x - 100 = 0$$
$$(x - 50)(x + 2) = 0$$
$$x = 50 \quad [x = -2 \text{ is not valid}]$$

33.
$$\log x + \log (x - 15) = 2$$
$$\log x(x - 15) = 2$$
$$x^2 - 15x = 100$$
$$x^2 - 15x - 100 = 0$$
$$(x - 20)(x + 5) = 0$$
$$x = 20 \quad [x = -5 \text{ is not valid}]$$

35.
$$\log (x + 90) = 3 - \log x$$
$$\log (x + 90) + \log x = 3$$
$$\log x(x + 90) = 3$$
$$x^2 + 90x = 1000$$
$$x^2 + 90x - 1000 = 0$$
$$(x + 100)(x - 10) = 0$$
$$x = 10 \quad [x = -100 \text{ is not valid}]$$

37.
$$\log(x-1) - \log 6 = \log(x-2) - \log x$$
$$\log(x-1) + \log x - \log(x-2) = \log 6$$
$$\log \frac{x(x-1)}{x-2} = \log 6$$
$$\frac{x^2 - x}{x-2} = 6$$
$$x^2 - x = 6x - 12$$
$$x^2 - 7x + 12 = 0$$
$$(x-3)(x-4) = 0$$
$$x = 3 \text{ or } x = 4$$

39.
$$\log x^2 = (\log x)^2$$
$$2 \log x = (\log x)^2$$
$$0 = (\log x)^2 - 2 \log x$$
$$0 = \log x \, (\log x - 2)$$
$$\log x = 0 \quad \text{or} \quad \log x - 2 = 0$$
$$x = 1 \qquad \qquad \log x = 2$$
$$x = 100$$

41.
$$\frac{\log(3x+4)}{\log x} = 2$$
$$\log(3x+4) = 2 \log x$$
$$\log(3x+4) = \log x^2$$
$$3x + 4 = x^2$$
$$0 = x^2 - 3x - 4$$
$$0 = (x-4)(x+1)$$
$$x = 4 \; [x = -1 \text{ is not valid}]$$

43.
$$\frac{\log(5x+6)}{2} = \log x$$
$$\log(5x+6) = 2 \log x$$
$$\log(5x+6) = \log x^2$$
$$5x + 6 = x^2$$
$$0 = x^2 - 5x - 6$$
$$0 = (x-6)(x+1)$$
$$x = 6 \; [x = -1 \text{ is not valid}]$$

45.
$$\log_3 x = \log_3 \left(\frac{1}{x}\right) + 4$$
$$\log_3 x = \log_3 x^{-1} + 4$$
$$\log_3 x = -\log_3 x + 4$$
$$2 \log_3 x = 4$$
$$\log_3 x = 2$$
$$x = 9$$

47.
$$2 \log_2 x = 3 + \log_2(x-2)$$
$$\log_2 x^2 - \log_2(x-2) = 3$$
$$\log_2 \frac{x^2}{x-2} = 3$$
$$\frac{x^2}{x-2} = 8$$
$$x^2 = 8x - 16$$
$$x^2 - 8x + 16 = 0$$
$$(x-4)(x-4) = 0 \qquad x = 4$$

49.

$$\log(7y+1) = 2\log(y+3) - \log 2$$
$$\log(7y+1) = \log(y+3)^2 - \log 2$$
$$\log(7y+1) = \log\frac{(y+3)^2}{2}$$
$$7y+1 = \frac{y^2+6y+9}{2}$$
$$14y+2 = y^2+6y+9$$
$$0 = y^2-8y+7$$
$$0 = (y-7)(y-1) \quad y=7 \text{ or } y=1$$

51.

$$A = A_0 2^{-t/h}$$
$$0.75A_0 = A_0 2^{-t/(12.4)}$$
$$\log 0.75 = \log 2^{-t/(12.4)}$$
$$\log 0.75 = -\frac{t}{12.4}\log 2$$
$$-12.4\log 0.75 = t\log 2$$
$$-\frac{12.4\log 0.75}{\log 2} = t$$
$$5.15 \approx x$$

It will take about 5.2 years.

53.

$$A = A_0 2^{-t/h}$$
$$0.20A_0 = A_0 2^{-t/(18.4)}$$
$$\log 0.20 = \log 2^{-t/(18.4)}$$
$$\log 0.20 = -\frac{t}{18.4}\log 2$$
$$-18.4\log 0.20 = t\log 2$$
$$-\frac{18.4\log 0.20}{\log 2} = t$$
$$42.72 \approx x$$

It will take about 42.7 days.

55.

$$A = A_0 2^{-t/h}$$
$$0.60A_0 = A_0 2^{-t/5700}$$
$$\log 0.60 = \log 2^{-t/5700}$$
$$\log 0.60 = -\frac{t}{5700}\log 2$$
$$-5700\log 0.60 = t\log 2$$
$$-\frac{5700\log 0.60}{\log 2} = t$$
$$4200 \approx t$$

It is about 4200 years old.

57.

$$A = A_0\left(1+\frac{r}{k}\right)^{kt}$$
$$800 = 500\left(1+\frac{0.085}{2}\right)^{2t}$$
$$\frac{800}{500} = (1.0425)^{2t}$$
$$\log 1.6 = \log(1.0425)^{2t}$$
$$\log 1.6 = 2t\log 1.0425$$
$$\frac{\log 1.6}{2\log 1.0425} = t$$
$$5.646 \approx t$$

It will take about 5.6 years.

59.

$$A = A_0\left(1+\frac{r}{k}\right)^{kt}$$
$$2100 = 1300\left(1+\frac{0.09}{4}\right)^{4t}$$
$$\frac{2100}{1300} = (1.0225)^{4t}$$
$$\log\frac{21}{13} = 4t\log 1.0225$$
$$\frac{\log\frac{21}{13}}{4\log 1.0225} = t, \text{ or } t \approx 5.4 \text{ years}$$

61.

$$A = A_0 e^{rt}$$
$$2A_0 = A_0 e^{rt}$$
$$2 = e^{rt}$$
$$\ln 2 = rt\ln e$$
$$\ln 2 = rt$$
$$\frac{\ln 2}{r} = t$$
$$\frac{0.69}{r} \approx t, \text{ or } t \approx \frac{70}{100\cdot r}$$

63.

$$I = I_0 e^{kx}$$

$$0.70I_0 = I_0 e^{k \cdot 6}$$

$$0.70 = e^{6k}$$

$$\ln 0.70 = \ln e^{6k}$$

$$\ln 0.70 = 6k$$

$$\frac{\ln 0.70}{6} = k$$

$$I = I_0 e^{\frac{\ln 0.70}{6} x}$$

$$0.20I_0 = I_0 e^{\frac{\ln 0.70}{6} x}$$

$$0.20 = e^{\frac{\ln 0.70}{6} x}$$

$$\ln 0.20 = \ln e^{\frac{\ln 0.70}{6} x}$$

$$\ln 0.20 = \frac{\ln 0.70}{6} x$$

$$\frac{6 \ln 0.20}{\ln 0.70} = x \quad \text{or} \quad x \approx 27 \text{ meters}$$

65.

$$T = 60 + 40e^{kt}$$

$$90 = 60 + 40e^{k \cdot 3}$$

$$30 = 40e^{3k}$$

$$\frac{30}{40} = e^{3k}$$

$$0.75 = e^{3k}$$

$$\ln 0.75 = \ln e^{3k}$$

$$\ln 0.75 = 3k$$

$$\frac{\ln 0.75}{3} = k$$

$$-0.0959 \approx k$$

67.

$$T = 300 - 300e^{kt}$$

$$100 = 300 - 300e^{k \cdot 5}$$

$$-200 = -300e^{5k}$$

$$\frac{2}{3} = e^{5k}$$

$$\ln \frac{2}{3} = \ln e^{5k}$$

$$\ln \frac{2}{3} = 5k$$

$$\frac{\ln \frac{2}{3}}{5} = k \quad \text{or} \quad k \approx -0.0811$$

69. Graph $y = \log x + \log (x - 15)$ and find where $y = 2$. Solution: $x = 20$

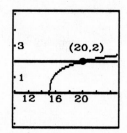

71. Graph $y = 2^{x+1}$ and find where $y = 7$.
Solution: $x \approx 1.8$

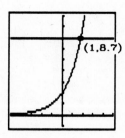

Chapter 6 Review Exercises (page 334)

1. $y = \left(\frac{6}{5}\right)^x$

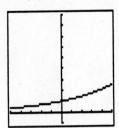

3. $y = \log x$

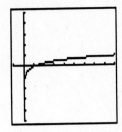

5.

7.

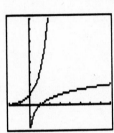

9. $\log_3 9 = 2$

11. $\log_\pi 1 = 0$

13. $\log_a \sqrt{a} = \log_a a^{1/2} = \frac{1}{2}$

15. $\ln e^4 = 4$

17. $10^{\log_{10} 7} = 7$

19. $\log_b b^4 = 4$

21. $\log_2 x = 3$
$2^3 = x$
$8 = x$

23. $\log_x 9 = 2$
$x^2 = 9$
$x = 3$
$[x = -3 \text{ not valid}]$

25. $\log_7 7 = x$
$7^x = 7$
$x = 1$

27. $\log_8 \sqrt{2} = x$
$8^x = \sqrt{2}$
$(2^3)^x = 2^{1/2}$
$2^{3x} = 2^{1/2}$
$3x = \frac{1}{2}$
$\frac{1}{3} \cdot 3x = \frac{1}{3} \cdot \frac{1}{2}$
$x = \frac{1}{6}$

29. $\log_{1/3} 9 = x$
$\left(\frac{1}{3}\right)^x = 9$
$(3^{-1})^x = 3^2$
$3^{-x} = 3^2$
$-x = 2$
$x = -2$

201

31.
$$\log_x 3 = \tfrac{1}{3}$$
$$x^{1/3} = 3$$
$$x = 27$$

33.
$$\log_2 x = 5$$
$$2^5 = x$$
$$32 = x$$

35.
$$\log_{\sqrt{3}} x = 6$$
$$(\sqrt{3})^6 = x$$
$$(3^{1/2})^6 = x$$
$$27 = x$$

37.
$$\log_x 2 = -\tfrac{1}{3}$$
$$x^{-1/3} = 2$$
$$(x^{-1/3})^{-3} = 2^{-3}$$
$$x = \tfrac{1}{8}$$

39.
$$\log_{0.25} x = -1$$
$$\left(\tfrac{1}{4}\right)^{-1} = x$$
$$4 = x$$

41.
$$\log_{\sqrt{2}} 32 = x$$
$$\left(2^{1/2}\right)^x = 2^5$$
$$\tfrac{1}{2}x = 5$$
$$x = 10$$

43.
$$\log_{\sqrt{3}} 9\sqrt{3} = x$$
$$\left(3^{1/2}\right)^x = 3^2 \cdot 3^{1/2}$$
$$3^{1/2 \cdot x} = 3^{5/2}$$
$$\tfrac{1}{2}x = \tfrac{5}{2}$$
$$x = 5$$

45.
$$\log_b \frac{x^2 y^3}{z^4} = \log_b x^2 y^3 - \log_b z^4 = \log_b x^2 + \log_b y^3 - \log_b z^4$$
$$= 2\log_b x + 3\log_b y - 4\log_b z$$

47.
$$3\log_b x - 5\log_b y + 7\log_b z = \log_b x^3 - \log_b y^5 + \log_b z^7 = \log_b \frac{x^3}{y^5} + \log_b z^7$$
$$= \log_b \frac{x^3 z^7}{y^5}$$

49. $\log abc = \log a + \log b + \log c = 0.60 + 0.36 + 2.40 = 3.36$

51. $\log \frac{ac}{b} = \log a + \log c - \log b = 0.60 + 2.40 - 0.36 = 2.64$

53.
$$3^x = 7$$
$$\log 3^x = \log 7$$
$$x \log 3 = \log 7$$
$$x = \frac{\log 7}{\log 3}$$

55.
$$2^x = 3^{x-1}$$
$$\log 2^x = \log 3^{x-1}$$
$$x \log 2 = (x-1) \log 3$$
$$x \log 2 = x \log 3 - \log 3$$
$$x \log 2 - x \log 3 = -\log 3$$
$$x(\log 2 - \log 3) = -\log 3$$
$$x = \frac{\log 3}{\log 3 - \log 2}$$

57.
$$\log x + \log (29 - x) = 2$$
$$\log x(29 - x) = 2$$
$$29x - x^2 = 100$$
$$x^2 - 29x + 100 = 0$$
$$(x - 25)(x - 4) = 0$$
$$x = 25 \ \text{ or } \ x = 4$$

59.
$$\log_2 (x + 2) + \log_2 (x - 1) = 2$$
$$\log_2 (x + 2)(x - 1) = 2$$
$$x^2 + x - 2 = 4$$
$$x^2 + x - 6 = 0$$
$$(x + 3)(x - 2) = 0$$
$$x = 2 \ \ [x = -3 \text{ is not valid}]$$

61.
$$\log x + \log (x - 5) = \log 6$$
$$\log x(x - 5) = \log 6$$
$$x^2 - 5x = 6$$
$$x^2 - 5x - 6 = 0$$
$$(x - 6)(x + 1) = 0$$
$$x = 6 \ \ [x = -1 \text{ is not valid}]$$

63.
$$e^{x \ln 2} = 9$$
$$\left(e^{\ln 2}\right)^x = 9$$
$$2^x = 9$$
$$\log 2^x = \log 9$$
$$x \log 2 = \log 9$$
$$x = \frac{\log 9}{\log 2}$$

65.
$$\ln x = \ln (x - 1) + 1$$
$$\ln x - \ln (x - 1) = 1$$
$$\ln \frac{x}{x - 1} = 1$$
$$\frac{x}{x - 1} = e$$
$$x = xe - e$$
$$x = \frac{e}{e - 1}$$

67.
$$A = A_0 2^{-t/h}$$
$$\tfrac{2}{3} A_0 = A_0 \, 2^{-t/5700}$$
$$\tfrac{2}{3} = 2^{-t/5700}$$
$$\log \tfrac{2}{3} = \log 2^{-t/5700}$$
$$\log 2 - \log 3 = -\frac{t}{5700} \log 2$$
$$t = 5700 \cdot \frac{\log 3 - \log 2}{\log 2}$$
$$t \approx 3334 \text{ years old}$$

69.

$$A = A_0 2^{-t/h}$$

$$\tfrac{2}{3}A_0 = A_0\, 2^{-20/h}$$

$$\tfrac{2}{3} = 2^{-20/h}$$

$$\log \tfrac{2}{3} = \log 2^{-20/h}$$

$$\log 2 - \log 3 = -\tfrac{20}{h}\log 2$$

$$h(\log 2 - \log 3) = -20\log 2$$

$$h = -\frac{20\log 2}{\log 2 - \log 3}$$

$$h \approx 34 \text{ years}$$

Chapter 6 Test (page 336)

1. $y = 2^x + 1$

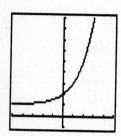

5. $y = e^x$

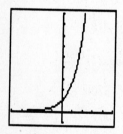

3.
$$A = A_0(2)^{-t}$$
$$= 3 \cdot 2^{-6}$$

$$= 3 \cdot \frac{1}{2^6} = \frac{3}{64} \text{ gram left}$$

7.
$$A = A_0 e^{rt}$$
$$= 2000 \cdot e^{0.08(10)}$$
$$= 2000 \cdot e^{0.8}$$
$$\approx \$4451.08 \text{ after 10 years}$$

9. $\log_3 \frac{1}{27} = \log_3 3^{-3} = -3$

11. $\log_{3/2} \frac{9}{4} = \log_{3/2}\left(\frac{3}{2}\right)^2 = 2$

13. $y = \log(x - 1)$

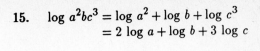

15. $\log a^2 bc^3 = \log a^2 + \log b + \log c^3$
$$= 2\log a + \log b + 3\log c$$

17. $\frac{1}{2}\log(a+2) + \log b - 2\log c = \log(a+2)^{1/2} + \log b - \log c^2 = \log \dfrac{b\sqrt{a+2}}{c^2}$

19. $\log 24 = \log 3 \cdot 8 = \log 3 + \log 8 = \log 3 + \log 2^3 = \log 3 + 3\log 2 = 1.3801$

21. $\log_7 3 = \dfrac{\log 3}{\log 7}$

23. $\log_a ab = \log_a a + \log_a b = 1 + \log_a b$
   TRUE

25. $\log a^{-3} = -3\log a$ FALSE

27. $\text{pH} = -\log[\text{H}^+] = -\log[3.7 \times 10^{-7}] = -(\log 3.7 + \log 10^{-7}) \approx -(0.57 - 7)$
$$\approx 6.43$$

29.
$$3^{x-1} = 100^x$$
$$\log 3^{x-1} = \log 100^x$$
$$(x-1)\log 3 = x\log 100$$
$$x\log 3 - \log 3 = 2x$$
$$x\log 3 - 2x = \log 3$$
$$x(\log 3 - 2) = \log 3$$
$$x = \frac{\log 3}{\log 3 - 2}$$

31. $\log(5x+2) = \log(2x+5)$
$$5x + 2 = 2x + 5$$
$$x = 1$$

Cumulative Review Exercises (page 337)

1. function

3. function

5. domain = $\{x \mid x \in \mathbb{R}\}$
range = $\{y \mid y \in \mathbb{R}, y \geq 5\}$

7. domain = $\{x \mid x \in \mathbb{R}, x \geq 2\}$
range = $\{y \mid y \in \mathbb{R}, y \leq 0\}$

9. $y = x^2 + 5x - 6$
$a = 1, b = 5, c = -6$

$h = -\dfrac{b}{2a} = -\dfrac{5}{2(1)} = -\dfrac{5}{2}$

$k = c - \dfrac{b^2}{4a} = -6 - \dfrac{5^2}{4(1)} = -\dfrac{49}{4}$

vertex: $\left(-\dfrac{5}{2}, -\dfrac{49}{4}\right)$

11. $y = x^2 - 4$

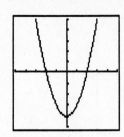

13. $y = x^3 + x$

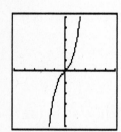

15. $y = \dfrac{x}{x - 3}$

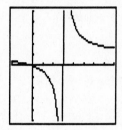

17. $(f + g)(x) = f(x) + g(x) = 3x - 4 + x^2 + 1 = x^2 + 3x - 3$
domain $= \{x \mid x \in \mathbb{R}\}$

19. $(f \cdot g)(x) = f(x) \cdot g(x) = (3x - 4)(x^2 + 1) = 3x^3 - 4x^2 + 3x - 4$
domain $= \{x \mid x \in \mathbb{R}\}$

21. $(f \circ g)(2) = f(g(2)) = f(2^2 + 1) = f(5) = 3(5) - 4 = 11$

23. $(f \circ g)(x) = f(g(x)) = f(x^2 + 1) = 3(x^2 + 1) - 4 = 3x^2 - 1$

25.
$$y = 3x + 2$$
$$x = 3y + 2$$
$$x - 2 = 3y$$
$$\frac{x - 2}{3} = y$$

27.
$$y = x^2 + 5$$
$$x = y^2 + 5$$
$$x - 5 = y^2$$
$$\pm\sqrt{x - 5} = y$$

This is not a function, since the original function was not one-to-one. You would need to restrict the domain of the original function so that it becomes one-to-one, and then pick the sign for the inverse accordingly.

29. $y = kwz$

31.
$$\begin{array}{c|cccc}
1 & 4 & 0 & 3 & 2 \\
  &   & 4 & 4 & 7 \\
\hline
  & 4 & 4 & 7 & \big|\ 9 \\
\end{array}$$

33.
$$\begin{array}{c|cccc}
\frac{1}{2} & 4 & 0 & 3 & 2 \\
  &   & 2 & 1 & 2 \\
\hline
  & 4 & 2 & 4 & \big|\ 4 \\
\end{array}$$

35.
$$\begin{array}{c|cccc}
-1 & 1 & 2 & -1 & -2 \\
   &   & -1 & -1 & 2 \\
\hline
   & 1 & 1 & -2 & \big|\ 0 \\
\end{array}$$
Since the remainder is 0, it is a factor.

37.
$$\begin{array}{c|cccc}
1 & 1 & 2 & -1 & -2 \\
  &   & 1 & 3 & 2 \\
\hline
  & 1 & 3 & 2 & \big|\ 0 \\
\end{array}$$
Since the remainder is 0, it is a factor.

39. 12 roots

41. $P(x) = x^4 + 2x^3 - 3x^2 + x + 2 \Rightarrow$ 2 sign changes $\Rightarrow$ at most 2 + rational roots
$P(-x) = x^4 - 2x^3 - 3x^2 - x + 2 \Rightarrow$ 2 sign changes $\Rightarrow$ at most 2 − rational roots

43. $x^3 + x^2 - 9x - 9 = 0$
$$\begin{array}{c|cccc}
-1 & 1 & 1 & -9 & -9 \\
   &   & -1 & 0 & 9 \\
\hline
   & 1 & 0 & -9 & \big|\ 0 \\
\end{array}$$
$x^3 + x^2 - 9x - 9 = (x + 1)(x^2 - 9)$
$\qquad\qquad\qquad\quad = (x + 1)(x + 3)(x - 3)$
ROOTS: $-1, -3, 3$

45. $y = 3^x - 2$

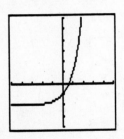

47. $y = \log_3 x$

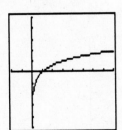

49. $\log_2 64 = \log_2 2^6 = 6$

51. $\ln e^3 = 3$

53. $\log abc = \log a + \log b + \log c$

55. $\log \sqrt{\dfrac{ab}{c^2}} = \log \left(\dfrac{ab}{c^2}\right)^{1/2}$

$$= \tfrac{1}{2} \log \dfrac{ab}{c^2}$$

$$= \tfrac{1}{2}(\log ab - \log c^2)$$

$$= \tfrac{1}{2}(\log a + \log b - 2 \log c)$$

$$= \tfrac{1}{2} \log a + \tfrac{1}{2} \log b - \log c$$

57. $3 \log a - 3 \log b = 3(\log a - \log b) = 3 \log \dfrac{a}{b}$

59.
$$3^{x+1} = 8$$
$$\log 3^{x+1} = \log 8$$
$$(x+1) \log 3 = \log 8$$
$$x \log 3 + \log 3 = \log 8$$
$$x \log 3 = \log 8 - \log 3$$
$$x = \dfrac{\log 8 - \log 3}{\log 3}$$

61.
$$\log x + \log 2 = 3$$
$$\log 2x = 3$$
$$2x = 1000$$
$$x = 500$$

Exercise 7.1 (page 348)

1.

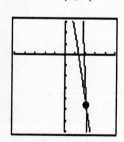

solution: $(1, 2)$

3.

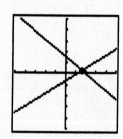

solution: $(-2, 4)$

5.

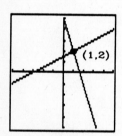

solution: $(2.2, -4.7)$

7.

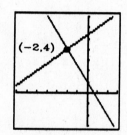

solution: $(1.7, 0.3)$

9. $y = x - 1$

$\boxed{y = 2x}$ {substitution}

$y = x - 1$

$2x = x - 1$

$x = -1$

Substitute $x = -1$.

$y = 2x$

$= 2(-1) = -2$

The solution is $(-1, -2)$.

11. $2x + 3y = 0$

$\boxed{y = 3x - 11}$ {substitution}

$2x + 3y = 0$

$2x + 3(3x - 11) = 0$

$2x + 9x - 33 = 0$

$11x = 33$

$x = 3$

Substitute $x = 3$.

$y = 3x - 11$

$= 3(3) - 11 = -2$

The solution is $(3, -2)$.

13. $4x + 3y = 3$

$2x - 6y = -1$

$$\boxed{x = 3y - \frac{1}{2}}$$

$4x + 3y = 3$

$4\left(3y - \frac{1}{2}\right) + 3y = 3$

$15y = 5$

$y = \frac{1}{3}$

$x = 3y - \frac{1}{2}$

$= 3\left(\frac{1}{3}\right) - \frac{1}{2}$

$= 1 - \frac{1}{2} = \frac{1}{2}$

The solution is $\left(\frac{1}{2}, \frac{1}{3}\right)$.

15. $x + 3y = 1 \Rightarrow \boxed{x = 1 - 3y}$

$2x + 6y = 3$

$2x + 6y = 3$

$2(1 - 3y) + 6y = 3$

$2 - 6y + 6y = 3$

$2 = 3$ **FALSE**

There is no solution.

The system is inconsistent.

17. $5x - 3y = 12 \Rightarrow$

$2x - 3y = 3 \Rightarrow (\times -1)$

$5x - 3y = 12$

$-2x + 3y = -3$

$3x = 9 \Rightarrow x = 3$

Substitute $x = 3$ into one of the equations and solve for y.

$2x - 3y = 3$

$2(3) - 3y = 3$

$-3y = -3 \Rightarrow y = 1$

The solution is $(3, 1)$.

19. $x - 7y = -11 \Rightarrow (\times -8)$

$8x + 2y = 28 \Rightarrow$

$-8x + 56y = 88$

$8x + 2y = 28$

$58y = 116 \Rightarrow y = 2$

Substitute $y = 2$ into one of the equations and solve for x.

$x - 7y = -11$

$x - 7(2) = -11$

$x - 14 = -11 \Rightarrow x = 3$

The solution is $(3, 2)$.

21.

$$3(x-y) = y-9$$
$$3x-3y = y-9$$

$$3x-4y = -9 \quad (1)$$

$$5(x+y) = -15$$
$$5x+5y = -15 \quad (2)$$

$$5 \cdot (1): \quad 15x-20y = -45$$
$$4 \cdot (2): \quad \underline{20x+20y = -60}$$
$$35x = -105$$
$$x = -3$$

Substitute $x = -3$ into one of the equations and solve for y.
$$3x-4y = -9$$
$$3(-3)-4y = -9$$
$$-9-4y = -9 \implies y = 0 \qquad \text{The solution is } (-3, 0).$$

23.

$$2 = \frac{1}{x+y}$$
$$2(x+y) = 1$$
$$2x+2y = 1 \quad (1)$$

$$2 = \frac{3}{x-y}$$
$$2(x-y) = 3$$
$$2x-2y = 3 \quad (2)$$

$$(1) \quad 2x+2y = 1$$
$$(2) \quad \underline{2x-2y = 3}$$
$$4x = 4$$
$$x = 1$$

Substitute $x = 1$ into one of the equations and solve for y.
$$2x+2y = 1$$
$$2(1)+2y = 1$$
$$2+2y = 1 \implies y = -\frac{1}{2} \qquad \text{The solution is } \left(1, -\frac{1}{2}\right).$$

25.

$$0.5x = 0.1y \qquad \Rightarrow$$
$$y-0.5x = 1.5 \qquad \Rightarrow$$

$$5x-y = 0$$
$$\underline{-5x+10y = 15}$$
$$9y = 15 \implies y = \frac{5}{3}$$

Substitute $y = \frac{5}{3}$ into one of the equations and solve for x.
$$5x-y = 0$$
$$5x-\frac{5}{3} = 0$$
$$5x = \frac{5}{3} \implies x = \frac{1}{3}. \qquad \text{The solution is } \left(\frac{1}{3}, \frac{5}{3}\right).$$

27.

$$x+2(x-y) = 2$$
$$x+2x-2y = 2$$

$$3x-2y = 2 \quad (1)$$

$$3(y-x)-y = 5$$
$$3y-3x-y = 5$$

$$-3x+2y = 5 \quad (2)$$

$$(1) \quad 3x-2y = 2$$
$$(2) \quad \underline{-3x+2y = 5}$$
$$0 = 7$$
$$\text{FALSE}$$

The system is inconsistent, and there is no solution.

29.

$$\frac{3}{2}x+\frac{1}{3}y = 2$$
$$9x+2y = 12 \quad (1)$$

$$\frac{2}{3}x+\frac{1}{9}y = 1$$
$$6x+y = 9 \quad (2)$$

$$(1) \qquad 9x+2y = 12$$
$$-2 \cdot (2) \quad \underline{-12x-2y = -18}$$
$$-3x = -6 \implies x = 2$$

Substitute $x = 2$ into one of the equations and solve for y.
$$6x+y = 9$$
$$6(2)+y = 9$$
$$12+y = 9 \implies y = -3 \qquad \text{The solution is } (2, -3).$$

31.

$$\frac{x-y}{5}+\frac{x+y}{2}=6$$
$$2(x-y)+5(x+y)=60$$
$$2x-2y+5x+5y=60$$
$$7x+3y=60 \quad (1)$$

$$\frac{x-y}{2}-\frac{x+y}{4}=3$$
$$4(x-y)-2(x+y)=24$$
$$4x-4y-2x-2y=24$$
$$2x-6y=24 \quad (2)$$

$2\cdot(1)$ $\qquad 14x+6y=120$
(2) $\qquad \underline{2x-6y=24}$

$$16x=144 \;\Rightarrow\; x=9$$

Substitute $x=9$ into one of the equations and solve for y.

$$7x+3y=60$$
$$7(9)+3y=60$$
$$63+3y=60 \;\Rightarrow\; y=-1 \qquad \text{The solution is } (9,-1).$$

33.

$$x + y + z = 3 \quad (1)$$
$$2x + y + z = 4 \quad (2)$$
$$3x + y - z = 5 \quad (3)$$

Add (1) and (3):

$$x + y + z = 3$$
$$3x + y - z = 5$$

$$\overline{4x + 2y \qquad = 8} \quad (4)$$

Add (2) and (3):

$$2x + y + z = 4$$
$$3x + y - z = 5$$

$$\overline{5x + 2y \qquad = 9} \quad (5)$$

Add $-$ **(4)** and **(5)**:

$$-4x - 2y = -8$$
$$5x + 2y = 9$$

$$\overline{x \qquad = 1}$$

Substitute $x=1$ into
(4) and solve for y:

$$4x+2y=8$$
$$4(1)+2y=8$$
$$y=2$$

The solution is $x=1$, $y=2$ and $z=0$.

Substitute $x=1$ and $y=2$
into **(1)** and solve for z:

$$x+y+z=3$$
$$1+2+z=3$$
$$z=0$$

35.

$$x - y + z = 0 \quad \textbf{(1)}$$
$$x + y + 2z = -1 \quad \textbf{(2)}$$
$$-x - y + z = 0 \quad \textbf{(3)}$$

Add **(1)** and **(3)**:

$$x - y + z = 0$$
$$-x - y + z = 0$$

$$\overline{\quad -2y + 2z = 0 \quad \textbf{(4)}}$$

Add **(2)** and **(3)**:

$$x + y + 2z = -1$$
$$-x - y + z = 0$$

$$\overline{\quad 3z = -1 \quad \textbf{(5)}}$$

From **(5)**, $z = -\frac{1}{3}$. Substituting this into **(4)** yields

$$-2y + 2z = 0$$
$$-2y + 2\left(-\frac{1}{3}\right) = 0$$
$$-2y = \frac{2}{3}$$
$$y = -\frac{1}{3}$$

Substitute $y = z = -\frac{1}{3}$ into **(1)** and solve for x:
$$x - y + z = 0$$
$$x - -\frac{1}{3} + -\frac{1}{3} = 0$$
$$x = 0$$

The solution is $x = 0$, $y = z = -\frac{1}{3}$.

37.

$$2x + y \qquad = 4 \quad \textbf{(1)}$$
$$x \qquad - z = 2 \quad \textbf{(2)}$$
$$y + z = 1 \quad \textbf{(3)}$$

Add **(2)** and **(3)**:

$$x \quad - z = 2$$
$$y + z = 1$$

$$\overline{\quad x + y \qquad = 3 \quad \textbf{(4)}}$$

Add **(1)** and $-$ **(4)**:

$$2x + y \qquad = 4$$
$$-x - y \qquad = -3$$

$$\overline{\quad x \qquad = 1 \quad \textbf{(5)}}$$

Substitute $x = 1$ into
(4) and solve for y:
$$x + y = 3$$
$$1 + y = 3$$
$$y = 2$$

Substitute $x = 1$
into **(2)** and solve for z:
$$x - z = 2$$
$$1 - z = 2$$
$$z = -1$$

The solution is $x = 1$, $y = 2$ and $z = -1$.

39.

$$
\begin{aligned}
x + y + z &= 6 \quad \textbf{(1)} \\
2x + y + 3z &= 17 \quad \textbf{(2)} \\
x + y + 2z &= 11 \quad \textbf{(3)}
\end{aligned}
$$

Add **(1)** and $-$ **(2)**:

$$
\begin{aligned}
x + y + z &= 6 \\
-2x - y - 3z &= -17 \\
\hline
-x \quad\quad - 2z &= -11 \quad \textbf{(4)}
\end{aligned}
$$

Add **(1)** and $-$ **(3)**:

$$
\begin{aligned}
x + y + z &= 6 \\
-x - y - 2z &= -11 \\
\hline
- z &= -5 \quad \textbf{(5)}
\end{aligned}
$$

From **(5)**, $z = 5$. Substituting this into **(4)** yields

$$
\begin{aligned}
-x - 2z &= -11 \\
-x - 2(5) &= -11 \\
-x &= -1 \\
x &= 1
\end{aligned}
$$

Substitute $x = 1$ and $z = 5$ into **(1)** and solve for x:

$$
\begin{aligned}
x + y + z &= 6 \\
1 + y + 5 = 6 &\Rightarrow y = 0
\end{aligned}
$$

The solution is $x = 1$, $y = 0$ and $z = 5$.

41.

$$
\begin{aligned}
x + y + z &= 3 \quad \textbf{(1)} \\
x \quad\quad + z &= 2 \quad \textbf{(2)} \\
2x + 2y + 2z &= 3 \quad \textbf{(3)}
\end{aligned}
$$

Add $-2 \cdot$ **(1)** and **(3)**:

$$
\begin{aligned}
-2x - 2y - 2z &= -6 \\
2x + 2y + 2z &= 3 \\
\hline
0 &= -3 \quad \text{FALSE} \quad \text{The system is inconsistent} \Rightarrow \text{no solution.}
\end{aligned}
$$

43.

$$
\begin{aligned}
x + 2y - z &= 2 \quad \textbf{(1)} \\
2x - y \quad &= -1 \quad \textbf{(2)} \\
3x + y + z &= 1 \quad \textbf{(3)}
\end{aligned}
$$

Add **(1)** and $2 \cdot$ **(2)**:

$$
\begin{aligned}
x + 2y - z &= 2 \\
4x - 2y \quad &= -2 \\
\hline
5x \quad\quad - z &= 0 \quad \textbf{(4)}
\end{aligned}
$$

Add **(1)** and $-2 \cdot$ **(3)**:

$$
\begin{aligned}
x + 2y - z &= 2 \\
-6x - 2y - 2z &= -2 \\
\hline
-5x \quad\quad - 3z &= 0 \quad \textbf{(5)}
\end{aligned}
$$

Add **(4)** and **(5)**:

$$
\begin{aligned}
5x - z &= 0 \\
-5x - 3z &= 0 \\
\hline
-4z &= 0 \quad \Rightarrow z = 0
\end{aligned}
$$

Substitute $z = 0$ into **(4)** and solve for x:

$$
\begin{aligned}
5x - z &= 0 \\
5x - 0 &= 0 \\
x &= 0
\end{aligned}
$$

Substitute $x = 0$ and $z = 0$ into **(1)** and solve for y:

$$
\begin{aligned}
x + 2y - z &= 2 \\
0 + 2y - 0 &= 2 \\
y &= 1
\end{aligned}
$$

The solution is $x = 0$, $y = 1$ and $z = 0$.

45.

$$3x + 4y + 2z = 4 \quad (1)$$
$$6x - 2y + z = 4 \quad (2)$$
$$3x - 8y - 6z = -3 \quad (3)$$

Add $-2 \cdot (1)$ and (2):

$$-6x - 8y - 4z = -8$$
$$6x - 2y + z = 4$$
$$\overline{\qquad\qquad\qquad}$$
$$-10y - 3z = -4 \quad (4)$$

Add $-2 \cdot (3)$ and (2):

$$-6x + 16y + 12z = 6$$
$$6x - 2y + z = 4$$
$$\overline{\qquad\qquad\qquad}$$
$$14y + 13z = 10 \quad (5)$$

Add $7 \cdot (4)$ and $5 \cdot (5)$:

$$-70y - 21z = -28$$
$$70y + 65z = 50$$
$$\overline{\qquad\qquad\qquad}$$
$$44z = 22 \Rightarrow z = \tfrac{1}{2}$$

Substitute $z = \tfrac{1}{2}$ into (4):

$$-10y - 3z = -4$$
$$-10y - 3 \cdot \tfrac{1}{2} = -4$$
$$-10y = -\tfrac{5}{2}$$
$$y = \tfrac{1}{4}$$

Substitute $z = \tfrac{1}{2}$ and $y = \tfrac{1}{4}$ into (1):

$$3x + 4y + 2z = 4$$
$$3x + 4 \cdot \tfrac{1}{4} + 2 \cdot \tfrac{1}{2} = 4$$
$$3x = 2$$
$$x = \tfrac{2}{3}$$

The solution is $x = \tfrac{2}{3}$, $y = \tfrac{1}{4}$ and $z = \tfrac{1}{2}$.

47.

$$2x - y - z = 0 \quad (1)$$
$$x - 2y - z = -1 \quad (2)$$
$$x - y - 2z = -1 \quad (3)$$

Add (1) and $-2 \cdot (3)$:

$$2x - y - z = 0$$
$$-2x + 2y + 4z = 2$$
$$\overline{\qquad\qquad\qquad}$$
$$y + 3z = 2 \quad (4)$$

Add (1) and $-2 \cdot (2)$:

$$2x - y - z = 0$$
$$-2x + 4y + 2z = 2$$
$$\overline{\qquad\qquad\qquad}$$
$$3y + z = 2 \quad (5)$$

Add $-3 \cdot (4)$ and (5):

$$-3y - 9z = -6$$
$$3y + z = 2$$
$$\overline{\qquad\qquad\qquad}$$
$$-8z = -4 \Rightarrow z = \tfrac{1}{2}$$

Substitute $z = \tfrac{1}{2}$ into (4):

$$y + 3z = 2$$
$$y + 3 \cdot \tfrac{1}{2} = 2$$
$$y = \tfrac{1}{2}$$

Substitute $z = \tfrac{1}{2}$ and $y = \tfrac{1}{2}$ into (1):

$$2x - y - z = 0$$
$$2x - \tfrac{1}{2} - \tfrac{1}{2} = 0$$
$$2x = 1 \Rightarrow x = \tfrac{1}{2}$$

The solution is $x = \tfrac{1}{2}$, $y = \tfrac{1}{2}$ and $z = \tfrac{1}{2}$.

49. Simplifying the second equation yields $0 = 0$. This means the equations are dependent, and there are infinitely many solutions.

51. Let $x =$ the number of acres of corn planted and $y =$ the acres of beans.

$$x + y = 350$$
$$\boxed{x = y + 100} \quad \{\text{substitute}\}$$

$$x + y = 350$$
$$y + 100 + y = 350$$
$$2y = 250$$
$$y = 125$$

$x = y + 100 = 125 + 100 = 225$

He plants 125 acres of beans and 225 acres of corn.

53. Let $x =$ the initiation fee and $y =$ the amount of monthly dues.

$$x + 7y = 3025 \quad \Rightarrow$$
$$x + 18y = 3850 \quad \Rightarrow$$

$$-x - 7y = -3025$$
$$\underline{x + 18y = 3850}$$
$$11y = 825 \quad \Rightarrow \quad y = 75$$

$$x + 7y = 3025$$
$$x + 7(75) = 3025$$
$$x = 2500 \qquad \text{The initiation fee is \$2500, with monthly dues of \$75.}$$

55. Let $r =$ the speed in still water and $c =$ the speed of the current. Note that the distance down is 30 km and the distance back is 30 km.

$$3(r + c) = 30 \Rightarrow \quad 3r + 3c = 30 \Rightarrow \quad 15r + 15c = 150$$
$$5(r - c) = 30 \Rightarrow \quad 5r - 5c = 30 \Rightarrow \quad \underline{15r - 15c = 90}$$
$$30r = 240 \Rightarrow \quad r = 8$$

$$3r + 3c = 30$$
$$3(8) + 3c = 30 \Rightarrow \quad c = 2 \qquad \text{The boat's speed in still water is 8 km per hour.}$$

57. Let $x =$ the weight of the first and $y =$ the weight of the second.

$$2x = 3y \Rightarrow \quad 2x - 3y = 0 \quad \Rightarrow \quad 6x - 9y = 0$$
$$2(y + 5) = 3x \Rightarrow \quad -3x + 2y = -10 \quad \Rightarrow \quad \underline{-6x + 4y = -20}$$
$$-5y = -20 \Rightarrow y = 4$$

$$2x = 3y$$
$$2x = 3(4) \Rightarrow \quad x = 6 \qquad \text{The weights are 4 pounds and 6 pounds.}$$

59. Let $r =$ the speed of the plane with no wind and $w =$ the speed of the wind.

$$r + c = 300$$
$$\underline{r - c = 220}$$
$$2r = 520 \quad \Rightarrow \quad r = 260 \qquad \text{The range in 5 hours} = 5 \cdot 260 = 1300 \text{ miles.}$$

61. $E = 742.72 + 43.53x$
$R = 89.95x$

$$E = R$$
$$742.72 + 43.53x = 89.95x$$
$$742.72 = 46.42x$$
$$16 = x \qquad \text{The break-even point is 16 pairs.}$$

63. Let $x =$ hours at restaurant, $y =$ hours at gas station and $z =$ hours of janitorial.

$$\boxed{x = 15} \quad \text{\{substitute\}}$$

$$
\begin{aligned}
x + y + z &= 30 \\
2.85x + 3.15y + 5z &= 99.25
\end{aligned}
\quad
\begin{aligned}
&\Rightarrow \\
&\Rightarrow
\end{aligned}
\quad
\begin{aligned}
15 + y + z &= 30 \\
42.75 + 3.15y + 5z &= 99.25
\end{aligned}
$$

$$
\begin{aligned}
y + z &= 15 \\
3.15y + 5z &= 56.50
\end{aligned}
\quad
\begin{aligned}
&\Rightarrow \\
&\Rightarrow
\end{aligned}
\quad
\begin{aligned}
-5y - 5z &= -75 \\
\underline{3.15y + 5z} &= \underline{56.50}
\end{aligned}
$$

$$-1.85y = -18.50 \Rightarrow y = 10$$

$$
\begin{aligned}
y + z &= 15 \\
10 + z &= 15 \Rightarrow z = 5
\end{aligned}
$$
He spends 15 hours cooking, 10 hours working at the gas station and 5 hours doing janitorial work.

65. Let $x =$ population from $0 - 14$, $y =$ population from $15 - 49$ and $z =$ population 50 and over.

$$
\begin{aligned}
x + y + z &= 3 \quad \textbf{(1)}\\
x + y &= 2.61 \quad \textbf{(2)}\\
y + z &= 1.95 \quad \textbf{(3)}
\end{aligned}
$$

Add **(1)** and $-$**(2)**:

$$
\begin{aligned}
x + y + z &= 3 \\
-x - y &= -2.61 \\
\hline
z &= 0.39 \quad \textbf{(4)}
\end{aligned}
$$

Add **(1)** and $-$**(3)**:

$$
\begin{aligned}
x + y + z &= 3 \\
- y - z &= -1.95 \\
\hline
x &= 1.05 \quad \textbf{(5)}
\end{aligned}
$$

$$
\begin{aligned}
x + y &= 2.61 \\
1.05 + y &= 2.61 \Rightarrow y = 1.56
\end{aligned}
$$
There are 1.05 million from $0 - 14$, 1.56 million from $15 - 49$, and 0.39 million 50 and over.

67. Let $x =$ smallest, $y =$ middle and $z =$ largest angle.

$$
\begin{aligned}
x + y + z &= 180 \\
z &= x + y + 20 \\
z &= 3x + 10
\end{aligned}
$$

$$
\begin{aligned}
x + y + z &= 180 \quad \textbf{(1)}\\
-x - y + z &= 20 \quad \textbf{(2)}\\
-3x + z &= 10 \quad \textbf{(3)}
\end{aligned}
$$

Add **(1)** and **(2)**:

$$
\begin{aligned}
x + y + z &= 180 \\
-x - y + z &= 20 \\
\hline
2z &= 200 \Rightarrow z = 100
\end{aligned}
$$

Substitute $z = 100$ into **(3)**:

$$
\begin{aligned}
-3x + z &= 10 \\
-3x + 100 &= 10 \\
-3x &= -90 \\
x &= 30
\end{aligned}
$$

Substitute $x = 30$ and $z = 100$ into **(1)**:

$$
\begin{aligned}
x + y + z &= 180 \\
30 + y + 100 &= 180 \\
y &= 50
\end{aligned}
$$

The angles measure 30°, 50° and 100°.

69. Let $x =$ dictionary width, $y =$ atlas width and $z =$ thesaurus width.

$$\begin{aligned} 3x + 5y + z &= 35 \quad \textbf{(1)} \\ 6x \qquad\; + 2z &= 35 \quad \textbf{(2)} \\ 2x + 4y + 3z &= 35 \quad \textbf{(3)} \end{aligned}$$

Add $-2\cdot\textbf{(1)}$ and $\textbf{(2)}$:

$$\begin{aligned} -6x - 10y - 2z &= -70 \\ 6x \qquad\quad + 2z &= 35 \end{aligned}$$

$$-10y \qquad = -35 \;\Rightarrow\; y = 3.5$$

Add $-3\cdot\textbf{(3)}$ and $\textbf{(2)}$:

$$\begin{aligned} -6x - 12y - 9z &= -105 \\ 6x \qquad\quad + 2z &= 35 \end{aligned}$$

$$-12y - 7z = -70 \quad \textbf{(4)}$$

Substitute $y = 3.5$ into $\textbf{(4)}$:

$$\begin{aligned} -12y - 7z &= -70 \\ -12(3.5) - 7z &= -70 \\ -7z &= -28 \\ z &= 4 \end{aligned}$$

Substitute $z = 4$ into $\textbf{(2)}$:

$$\begin{aligned} 6x + 2z &= 35 \\ 6x + 2(4) &= 35 \\ 6x &= 27 \\ x &= 4.5 \end{aligned}$$

The dictionary width is 4.5 inches, the atlas width is 3.5 inches and the thesaurus width is 4 inches.

Exercise 7.2 (page 359)

NOTE: In these exercises, a notation like R_1 refers to the first row in the immediately preceding matrix or system. $2R_1 + R_2$ indicates the row operation of multiplying the first row by 2 and adding the result to the second row.

1.
$$\begin{aligned} x + y &= 7 \\ x - 2y &= -1 \end{aligned}$$

$$\begin{aligned} x + y &= 7 \quad (R_1) \\ 3y &= 8 \quad (-R_2 + R_1) \end{aligned}$$

$$\begin{aligned} x + y &= 7 \quad (R_1) \\ y &= \tfrac{8}{3} \quad (\tfrac{1}{3}R_2) \end{aligned}$$

$$\begin{aligned} x + y &= 7 \\ x + \tfrac{8}{3} &= 7 \\ x &= \tfrac{13}{3} \end{aligned}$$

$$x = \tfrac{13}{3}, \; y = \tfrac{8}{3}$$

3.
$$\begin{aligned} x - y &= 1 \\ 2x - y &= 8 \end{aligned}$$

$$\begin{aligned} x - y &= 1 \quad (R_1) \\ y &= 6 \quad (-2R_1 + R_2) \end{aligned}$$

$$\begin{aligned} x - y &= 1 \\ x - 6 &= 1 \\ x &= 7 \end{aligned}$$

$$x = 7, \; y = 6$$

5.
$$x + 2y - z = 2$$
$$x - 3y + 2z = 1$$
$$x + y - 3z = -6$$

$$
\begin{array}{ll}
x + 2y - z = & 2 \quad (R_1) \\
5y - 3z = & 1 \quad (-R_2 + R_1) \\
y + 2z = & 8 \quad (-R_3 + R_1)
\end{array}
$$

$$
\begin{array}{ll}
x + 2y - z = & 2 \quad (R_1) \\
5y - 3z = & 1 \quad (R_2) \\
- 13z = & -39 \quad (-5\,R_3 + R_2)
\end{array}
$$

$$
\begin{array}{ll}
x + 2y - z = & 2 \quad (R_1) \\
y - (3/5)z = & 1/5 \quad (1/5\,R_2) \\
z = & 3 \quad (1/13\,R_3)
\end{array}
$$

$$y - (3/5)z = 1/5$$
$$y - (3/5)(3) = 1/5$$
$$y = 10/5 \;\Rightarrow\; y = 2$$

$$x + 2y - z = 2$$
$$x + 2(2) - 3 = 2$$
$$x = 1$$

$$\boxed{x = 1,\, y = 2,\, z = 3}$$

7.
$$x - y - z = -3$$
$$5x + y = 6$$
$$y + z = 4$$

$$
\begin{array}{ll}
x - y - z = & -3 \quad (R_1) \\
6y + 5z = & 21 \quad (-5\,R_1 + R_2) \\
y + z = & 4 \quad (R_3)
\end{array}
$$

$$
\begin{array}{ll}
x - y - z = & -3 \quad (R_1) \\
6y + 5z = & 21 \quad (R_2) \\
- z = & -3 \quad (-6\,R_3 + R_2)
\end{array}
$$

$$
\begin{array}{ll}
x - y - z = & -3 \quad (R_1) \\
y + (5/6)z = & 7/2 \quad (1/6\,R_2) \\
z = & 3 \quad (-R_3)
\end{array}
$$

$$y + (5/6)z = 7/2$$
$$y + (5/6)(3) = 7/2$$
$$y = 2/2 \;\Rightarrow\; y = 1$$

$$x - y - z = -3$$
$$x - 1 - 3 = -3$$
$$x = 1$$

$$\boxed{x = 1,\, y = 1,\, z = 3}$$

9. row echelon form

11. reduced row echelon form

13.
$$-2\,R_2 + R_1 \Rightarrow R_2$$
$$
\begin{bmatrix} 2 & 1 & | & 3 \\ 1 & -3 & | & 5 \end{bmatrix}
\Rightarrow
\begin{bmatrix} 2 & 1 & | & 3 \\ 0 & 7 & | & -7 \end{bmatrix}
$$

$$(1/2)\,R_1 \Rightarrow R_1, \; (1/7)\,R_2 \Rightarrow R_2$$
$$
\Rightarrow
\begin{bmatrix} 1 & 1/2 & | & 3/2 \\ 0 & 1 & | & -1 \end{bmatrix}
$$

From R_2 of the last matrix, $y = -1$. Substitute into R_1:

$$x + (1/2)y = 3/2$$
$$x - (1/2) = 3/2$$
$$x = 2$$

$$\boxed{x = 2,\, y = -1}$$

15.
$$-5\,R_1 + R_2 \Rightarrow R_2$$
$$
\begin{bmatrix} 1 & -7 & | & -2 \\ 5 & -2 & | & -10 \end{bmatrix}
\Rightarrow
\begin{bmatrix} 1 & -7 & | & -2 \\ 0 & 33 & | & 0 \end{bmatrix}
$$

$$(1/33)\,R_2 \Rightarrow R_2$$
$$
\Rightarrow
\begin{bmatrix} 1 & -7 & | & -2 \\ 0 & 1 & | & 0 \end{bmatrix}
$$

From R_2 of the last matrix, $y = 0$. Substitute into R_1:

$$x - 7y = -2$$
$$x - 7(0) = -2$$
$$x = -2$$

$$\boxed{x = -2,\, y = 0}$$

17.

$$-2\,R_2 + R_1 \Rightarrow R_2 \qquad\qquad (1/2)\,R_1 \Rightarrow R_1,\ -(1/7)\,R_2 \Rightarrow R_2$$

$$\begin{bmatrix} 2 & -1 & \bigm| & 5 \\ 1 & 3 & \bigm| & 6 \end{bmatrix} \Rightarrow \begin{bmatrix} 2 & -1 & \bigm| & 5 \\ 0 & -7 & \bigm| & -7 \end{bmatrix} \Rightarrow \begin{bmatrix} 1 & -1/2 & \bigm| & 5/2 \\ 0 & 1 & \bigm| & 1 \end{bmatrix}$$

From R_2 of the last matrix, $y = 1$. Substitute into R_1:
$$\begin{aligned} x - (1/2)y &= 5/2 \\ x - (1/2)(1) &= 5/2 \\ x &= 3 \end{aligned}$$

$$\boxed{x = 3,\ y = 1}$$

19.

$$2\,R_1 + R_2 \Rightarrow R_2 \qquad\qquad (1/12)\,R_2 \Rightarrow R_2$$

$$\begin{bmatrix} 1 & -2 & \bigm| & 3 \\ -2 & 4 & \bigm| & 6 \end{bmatrix} \Rightarrow \begin{bmatrix} 1 & -2 & \bigm| & 3 \\ 0 & 0 & \bigm| & 12 \end{bmatrix} \Rightarrow \begin{bmatrix} 1 & -2 & \bigm| & 3 \\ 0 & 0 & \bigm| & 1 \end{bmatrix}$$

R_2 of the last matrix indicates that $0x + 0y = 1$, but this is impossible.
NO SOLUTION – SYSTEM IS INCONSISTENT.

21. $\quad -2\,R_1 + R_2 \Rightarrow R_2,\ -R_1 + R_3 \Rightarrow R_3 \qquad\qquad -3\,R_2 + R_3 \Rightarrow R_3$

$$\begin{bmatrix} 1 & -1 & 1 & \bigm| & 3 \\ 2 & -1 & 1 & \bigm| & 4 \\ 1 & 2 & -1 & \bigm| & -1 \end{bmatrix} \Rightarrow \begin{bmatrix} 1 & -1 & 1 & \bigm| & 3 \\ 0 & 1 & -1 & \bigm| & -2 \\ 0 & 3 & -2 & \bigm| & -4 \end{bmatrix} \Rightarrow \begin{bmatrix} 1 & -1 & 1 & \bigm| & 3 \\ 0 & 1 & -1 & \bigm| & -2 \\ 0 & 0 & 1 & \bigm| & 2 \end{bmatrix}$$

From the last matrix, $z = 2$. Substitute back to find y and x:
$$\begin{array}{ll} y - z = -2 & x - y + z = 3 \\ y - 2 = -2 & x - 0 + 2 = 3 \\ \quad y = 0 & \qquad x = 1 \end{array} \qquad \boxed{x = 1,\ y = 0,\ z = 2}$$

23.

$$-3\,R_1 + R_2 \Rightarrow R_2 \qquad\qquad 2\,R_3 + R_2 \Rightarrow R_2$$

$$\begin{bmatrix} 1 & 1 & -1 & \bigm| & -1 \\ 3 & 1 & 0 & \bigm| & 4 \\ 0 & 1 & -2 & \bigm| & -4 \end{bmatrix} \Rightarrow \begin{bmatrix} 1 & 1 & -1 & \bigm| & -1 \\ 0 & -2 & 3 & \bigm| & 7 \\ 0 & 1 & -2 & \bigm| & -4 \end{bmatrix} \Rightarrow \begin{bmatrix} 1 & 1 & -1 & \bigm| & -1 \\ 0 & -2 & 3 & \bigm| & 7 \\ 0 & 0 & -1 & \bigm| & -1 \end{bmatrix}$$

From the last matrix, $z = 1$. Substitute back to find y and x:
$$\begin{array}{ll} -2y + 3z = 7 & x + y - z = -1 \\ -2y + 3 = 7 & x + -2 - 1 = -1 \\ \quad y = -2 & \qquad x = 2 \end{array} \qquad \boxed{x = 2,\ y = -2,\ z = 1}$$

25. $\quad -2\,R_1 + R_2 \Rightarrow R_2,\ -3\,R_1 + R_3 \Rightarrow R_3 \qquad\qquad -R_2 + R_3 \Rightarrow R_3$

$$\begin{bmatrix} 1 & -1 & 1 & \bigm| & 2 \\ 2 & 1 & 1 & \bigm| & 5 \\ 3 & 0 & -4 & \bigm| & -5 \end{bmatrix} \Rightarrow \begin{bmatrix} 1 & -1 & 1 & \bigm| & 2 \\ 0 & 3 & -1 & \bigm| & 1 \\ 0 & 3 & -7 & \bigm| & -11 \end{bmatrix} \Rightarrow \begin{bmatrix} 1 & -1 & 1 & \bigm| & 2 \\ 0 & 3 & -1 & \bigm| & 1 \\ 0 & 0 & -6 & \bigm| & -12 \end{bmatrix}$$

From the last matrix, $z = 2$. Substitute back to find y and x:
$$\begin{array}{ll} 3y - z = 1 & x - y + z = 2 \\ 3y - 2 = 1 & x - 1 + 2 = 2 \\ \quad y = 1 & \qquad x = 1 \end{array} \qquad \boxed{x = 1,\ y = 1,\ z = 2}$$

27. $R_1 + R_2 \Rightarrow R_2,\ -2\,R_1 + R_3 \Rightarrow R_3 \qquad\qquad -R_3 \Rightarrow R_2,\ -R_2 \Rightarrow R_3$

$$\left[\begin{array}{ccc|c} 1 & 1 & 2 & 4 \\ -1 & -1 & -3 & -5 \\ 2 & 1 & 1 & 2 \end{array}\right] \Rightarrow \left[\begin{array}{ccc|c} 1 & 1 & 2 & 4 \\ 0 & 0 & -1 & -1 \\ 0 & -1 & -3 & -6 \end{array}\right] \Rightarrow \left[\begin{array}{ccc|c} 1 & 1 & 2 & 4 \\ 0 & 1 & 3 & 6 \\ 0 & 0 & 1 & 1 \end{array}\right]$$

From the last matrix, $z = 1$. Substitute back to find y and x:

$$y + 3z = 6 \qquad\qquad\qquad x + y + 2z = 4$$
$$y + 3 = 6 \qquad\qquad\qquad x + 3 + 2 = 4$$
$$y = 3 \qquad\qquad\qquad x = -1 \qquad \boxed{x = -1,\ y = 3,\ z = 1}$$

29. $\qquad\qquad 2\,R_2 + R_1 \Rightarrow R_1$

$$\left[\begin{array}{cc|c} 1 & -2 & 7 \\ 0 & 1 & 3 \end{array}\right] \Rightarrow \left[\begin{array}{cc|c} 1 & 0 & 13 \\ 0 & 1 & 3 \end{array}\right] \Rightarrow \boxed{x = 13,\ y = 3}$$

31. $\qquad -2\,R_2 + R_1 \Rightarrow R_1 \qquad\qquad 7\,R_3 + R_1 \Rightarrow R_1,\ -3\,R_3 + R_2 \Rightarrow R_2$

$$\left[\begin{array}{ccc|c} 1 & 2 & -1 & 3 \\ 0 & 1 & 3 & 1 \\ 0 & 0 & 1 & -2 \end{array}\right] \Rightarrow \left[\begin{array}{ccc|c} 1 & 0 & -7 & 1 \\ 0 & 1 & 3 & 1 \\ 0 & 0 & 1 & -2 \end{array}\right] \Rightarrow \left[\begin{array}{ccc|c} 1 & 0 & 0 & -13 \\ 0 & 1 & 0 & 7 \\ 0 & 0 & 1 & -2 \end{array}\right]$$

$$\boxed{x = -13,\ y = 7,\ z = -2}$$

33. $\qquad -R_1 + R_2 \Rightarrow R_2 \qquad\quad 2\,R_1 + R_2 \Rightarrow R_1 \qquad (1/2)\,R_1 \Rightarrow R_1$
$$\qquad\qquad\qquad\qquad\qquad\qquad\qquad\qquad\qquad\qquad\qquad (1/2)\,R_3 \Rightarrow R_3$$

$$\left[\begin{array}{cc|c} 1 & -1 & 7 \\ 1 & 1 & 13 \end{array}\right] \Rightarrow \left[\begin{array}{cc|c} 1 & -1 & 7 \\ 0 & 2 & 6 \end{array}\right] \Rightarrow \left[\begin{array}{cc|c} 2 & 0 & 20 \\ 0 & 2 & 6 \end{array}\right] \Rightarrow \left[\begin{array}{cc|c} 1 & 0 & 10 \\ 0 & 1 & 3 \end{array}\right]$$

$$\boxed{x = 10,\ y = 3}$$

35. $\qquad -R_1 + R_2 \Rightarrow R_2 \qquad\quad 5\,R_1 + R_2 \Rightarrow R_1 \qquad (1/5)\,R_1 \Rightarrow R_1$
$$\qquad\qquad\qquad\qquad\qquad\qquad\qquad\qquad\qquad\qquad\qquad (2/5)\,R_2 \Rightarrow R_2$$

$$\left[\begin{array}{cc|c} 1 & -\frac{1}{2} & 0 \\ 1 & 2 & 0 \end{array}\right] \Rightarrow \left[\begin{array}{cc|c} 1 & -\frac{1}{2} & 0 \\ 0 & \frac{5}{2} & 0 \end{array}\right] \Rightarrow \left[\begin{array}{cc|c} 5 & 0 & 0 \\ 0 & \frac{5}{2} & 0 \end{array}\right] \Rightarrow \left[\begin{array}{cc|c} 1 & 0 & 0 \\ 0 & 1 & 0 \end{array}\right]$$

$$\boxed{x = 0,\ y = 0}$$

37.

$$-R_2 + R_1 \Rightarrow R_2$$
$$-R_3 + R_1 \Rightarrow R_3 \qquad\qquad R_2 \Leftrightarrow R_3$$

$$
\begin{bmatrix}
1 & 1 & 2 & | & 0 \\
1 & 1 & 1 & | & 2 \\
1 & 0 & 1 & | & 1
\end{bmatrix}
\Rightarrow
\begin{bmatrix}
1 & 1 & 2 & | & 0 \\
0 & 0 & 1 & | & -2 \\
0 & 1 & 1 & | & -1
\end{bmatrix}
\Rightarrow
\begin{bmatrix}
1 & 1 & 2 & | & 0 \\
0 & 1 & 1 & | & -1 \\
0 & 0 & 1 & | & -2
\end{bmatrix}
$$

$$-R_2 + R_1 \Rightarrow R_1 \qquad\qquad -R_3 + R_1 \Rightarrow R_1$$
$$\qquad\qquad\qquad\qquad\qquad -R_3 + R_2 \Rightarrow R_2$$

$$
\begin{bmatrix}
1 & 0 & 1 & | & 1 \\
0 & 1 & 1 & | & -1 \\
0 & 0 & 1 & | & -2
\end{bmatrix}
\Rightarrow
\begin{bmatrix}
1 & 0 & 0 & | & 3 \\
0 & 1 & 0 & | & 1 \\
0 & 0 & 1 & | & -2
\end{bmatrix}
\qquad
\boxed{x = 3,\; y = 1,\; z = -2}
$$

39.

$$2R_2 + R_1 \Rightarrow R_2 \qquad\qquad -3R_1 + R_2 \Rightarrow R_1$$
$$-2R_1 + R_3 \Rightarrow R_3 \qquad\qquad -3R_3 + R_2 \Rightarrow R_3$$

$$
\begin{bmatrix}
2 & 1 & -2 & | & 1 \\
-1 & 1 & -3 & | & 0 \\
4 & 3 & 0 & | & 4
\end{bmatrix}
\Rightarrow
\begin{bmatrix}
2 & 1 & -2 & | & 1 \\
0 & 3 & -8 & | & 1 \\
0 & 1 & 4 & | & 2
\end{bmatrix}
\Rightarrow
\begin{bmatrix}
-6 & 0 & -2 & | & -2 \\
0 & 3 & -8 & | & 1 \\
0 & 0 & -20 & | & -5
\end{bmatrix}
$$

$$-(1/20)R_3 \Rightarrow R_3 \qquad -R_3 + R_1 \Rightarrow R_1 \qquad (1/3)R_1 \Rightarrow R_1$$
$$-(1/2)R_1 \Rightarrow R_1 \qquad 8R_3 + R_2 \Rightarrow R_2 \qquad (1/3)R_2 \Rightarrow R_2$$

$$
\begin{bmatrix}
3 & 0 & 1 & | & 1 \\
0 & 3 & -8 & | & 1 \\
0 & 0 & 1 & | & \frac{1}{4}
\end{bmatrix}
\Rightarrow
\begin{bmatrix}
3 & 0 & 0 & | & \frac{3}{4} \\
0 & 3 & 0 & | & 3 \\
0 & 0 & 1 & | & \frac{1}{4}
\end{bmatrix}
\Rightarrow
\begin{bmatrix}
1 & 0 & 0 & | & \frac{1}{4} \\
0 & 1 & 0 & | & 1 \\
0 & 0 & 1 & | & \frac{1}{4}
\end{bmatrix}
$$

$$\boxed{x = \tfrac{1}{4},\; y = 1,\; z = \tfrac{1}{4}}$$

41.

$$R_1 + R_3 \Rightarrow R_1, \; -2\,R_2 + R_1 \Rightarrow R_2$$
$$2\,R_3 + R_1 \Rightarrow R_3, \; -R_4 + R_2 \Rightarrow R_4$$

$$
\begin{bmatrix}
2 & -2 & 3 & 1 & | & 2 \\
1 & 1 & 1 & 1 & | & 5 \\
-1 & 2 & -3 & 2 & | & 2 \\
1 & 1 & 2 & -1 & | & 4
\end{bmatrix}
\Rightarrow
\begin{bmatrix}
1 & 0 & 0 & 3 & | & 4 \\
0 & -4 & 1 & -1 & | & -8 \\
0 & 2 & -3 & 5 & | & 6 \\
0 & 0 & -1 & 2 & | & 1
\end{bmatrix}
$$

$$2\,R_3 + R_2 \Rightarrow R_3 \qquad\qquad 5\,R_2 + R_3 \Rightarrow R_2, \; -5\,R_4 + R_3 \Rightarrow R_4$$

$$
\begin{bmatrix}
1 & 0 & 0 & 3 & | & 4 \\
0 & -4 & 1 & -1 & | & -8 \\
0 & 0 & -5 & 9 & | & 4 \\
0 & 0 & -1 & 2 & | & 1
\end{bmatrix}
\Rightarrow
\begin{bmatrix}
1 & 0 & 0 & 3 & | & 4 \\
0 & -20 & 0 & 4 & | & -36 \\
0 & 0 & -5 & 9 & | & 4 \\
0 & 0 & 0 & -1 & | & -1
\end{bmatrix}
$$

$$3\,R_4 + R_1 \Rightarrow R_1 \qquad\qquad\qquad -R_4 \Rightarrow R_4$$
$$9\,R_4 + R_3 \Rightarrow R_3 \qquad\qquad\qquad -(1/20)\,R_2 \Rightarrow R_2$$
$$4\,R_4 + R_2 \Rightarrow R_2 \qquad\qquad\qquad -(1/5)\,R_3 \Rightarrow R_3$$

$$
\begin{bmatrix}
1 & 0 & 0 & 0 & | & 1 \\
0 & -20 & 0 & 0 & | & -40 \\
0 & 0 & -5 & 0 & | & -5 \\
0 & 0 & 0 & -1 & | & -1
\end{bmatrix}
\Rightarrow
\begin{bmatrix}
1 & 0 & 0 & 0 & | & 1 \\
0 & 1 & 0 & 0 & | & 2 \\
0 & 0 & 1 & 0 & | & 1 \\
0 & 0 & 0 & 1 & | & 1
\end{bmatrix}
$$

$$\boxed{x = 1, \; y = 2, \; z = 1, \; t = 1}$$

43.

$$-R_1 + R_2 \Rightarrow R_2, \ -2\,R_1 + R_3 \Rightarrow R_3$$
$$-R_1 + R_4 \Rightarrow R_4$$

$$\left[\begin{array}{cccc|c} 1 & 1 & 0 & 1 & 4 \\ 1 & 0 & 1 & 1 & 2 \\ 2 & 2 & 1 & 2 & 8 \\ 1 & -1 & 1 & -1 & -2 \end{array}\right] \Rightarrow \left[\begin{array}{cccc|c} 1 & 1 & 0 & 1 & 4 \\ 0 & -1 & 1 & 0 & -2 \\ 0 & 0 & 1 & 0 & 0 \\ 0 & -2 & 1 & -2 & -6 \end{array}\right]$$

$$R_1 + R_2 \Rightarrow R_1, \ -R_2 \Rightarrow R_2 \qquad\qquad -R_3 + R_1 \Rightarrow R_1, \ R_3 + R_2 \Rightarrow R_2$$
$$-2\,R_2 + R_4 \Rightarrow R_4 \qquad\qquad\qquad R_3 + R_4 \Rightarrow R_4$$

$$\left[\begin{array}{cccc|c} 1 & 0 & 1 & 1 & 2 \\ 0 & 1 & -1 & 0 & 2 \\ 0 & 0 & 1 & 0 & 0 \\ 0 & 0 & -1 & -2 & -2 \end{array}\right] \Rightarrow \left[\begin{array}{cccc|c} 1 & 0 & 0 & 1 & 2 \\ 0 & 1 & 0 & 0 & 2 \\ 0 & 0 & 1 & 0 & 0 \\ 0 & 0 & 0 & -2 & -2 \end{array}\right]$$

$$(1/2)\,R_4 + R_1 \Rightarrow R_1, \ -(1/2)\,R_4 \Rightarrow R_4$$

$$\left[\begin{array}{cccc|c} 1 & 0 & 0 & 0 & 1 \\ 0 & 1 & 0 & 0 & 2 \\ 0 & 0 & 1 & 0 & 0 \\ 0 & 0 & 0 & 1 & 1 \end{array}\right] \qquad \boxed{x = 1,\, y = 2,\, z = 0,\, t = 1}$$

45. Eliminate denominators first. $(12\,R_1,\ 6\,R_2,\ 24\,R_3)$

$(1/4)\,R_1 \Rightarrow R_1$ $\qquad\qquad\qquad -6\,R_1+R_2 \Rightarrow R_2,\ -4\,R_1+R_3 \Rightarrow R_3$

$$
\begin{bmatrix}
4 & 9 & -8 & -24 \\
6 & 3 & 2 & 6 \\
4 & -3 & -24 & 0
\end{bmatrix}
\Rightarrow
\begin{bmatrix}
1 & 9/4 & -2 & -6 \\
6 & 3 & 2 & 6 \\
4 & -3 & -24 & 0
\end{bmatrix}
\Rightarrow
\begin{bmatrix}
1 & 9/4 & -2 & -6 \\
0 & -21/2 & 14 & 42 \\
0 & -12 & -16 & 24
\end{bmatrix}
$$

$-(1/24)\,R_2 \Rightarrow R_2$ $\qquad\qquad -(9/4)\,R_2+R_1 \Rightarrow R_1$ $\qquad (1/8)\,R_3 \Rightarrow R_3$
$-(2/21)\,R_3 \Rightarrow R_3$ $\qquad\qquad -3\,R_2+R_3 \Rightarrow R_3$

$$
\begin{bmatrix}
1 & \frac{9}{4} & -2 & -6 \\
0 & 1 & -\frac{4}{3} & -4 \\
0 & 3 & 4 & -6
\end{bmatrix}
\Rightarrow
\begin{bmatrix}
1 & 0 & 1 & 3 \\
0 & 1 & -\frac{4}{3} & -4 \\
0 & 0 & 8 & 6
\end{bmatrix}
\Rightarrow
\begin{bmatrix}
1 & 0 & 1 & 3 \\
0 & 1 & -\frac{4}{3} & -4 \\
0 & 0 & 1 & \frac{3}{4}
\end{bmatrix}
$$

$-R_3+R_1 \Rightarrow R_1$
$(4/3)\,R_3+R_2 \Rightarrow R_2$

$$
\begin{bmatrix}
1 & 0 & 0 & \frac{9}{4} \\
0 & 0 & 0 & -3 \\
0 & 0 & 1 & \frac{3}{4}
\end{bmatrix}
\qquad
\boxed{x=\frac{9}{4},\ y=-3,\ z=\frac{3}{4}}
$$

47. Eliminate denominators first. $(4\,R_1,\ 12\,R_2,\ 3\,R_3)$

$\quad -4\,R_1+R_2 \Rightarrow R_2$ $\qquad\qquad\qquad R_1+R_2 \Rightarrow R_1,\ -R_2 \Rightarrow R_2$
$\quad -R_1+R_3 \Rightarrow R_3$ $\qquad\qquad\qquad -R_2+R_3 \Rightarrow R_3$

$$
\begin{bmatrix}
2 & 1 & -4 & 8 \\
8 & 3 & 6 & 18 \\
2 & 0 & 3 & -1
\end{bmatrix}
\Rightarrow
\begin{bmatrix}
2 & 1 & -4 & 8 \\
0 & -1 & 22 & -14 \\
0 & -1 & 7 & -9
\end{bmatrix}
\Rightarrow
\begin{bmatrix}
2 & 0 & 18 & -6 \\
0 & 1 & -22 & 14 \\
0 & 0 & -15 & 5
\end{bmatrix}
$$

$(1/2)\,R_1 \Rightarrow R_1$ $\qquad\qquad -9\,R_3+R_1 \Rightarrow R_1$
$-(1/15)\,R_3 \Rightarrow R_3$ $\qquad 22\,R_3+R_2 \Rightarrow R_2$

$$
\begin{bmatrix}
1 & 0 & 9 & -3 \\
0 & 1 & -22 & 14 \\
0 & 0 & 1 & -1/3
\end{bmatrix}
\Rightarrow
\begin{bmatrix}
1 & 0 & 0 & 0 \\
0 & 1 & 0 & 20/3 \\
0 & 0 & 1 & -1/3
\end{bmatrix}
\qquad
\boxed{x=0,\ y=\frac{20}{3},\ z=-\frac{1}{3}}
$$

224

49.

$$-3R_1 + R_2 \Rightarrow R_2 \qquad R_3 \Rightarrow R_4$$
$$-2R_1 + R_3 \Rightarrow R_3 \qquad -(1/4)R_2 \Rightarrow R_2$$
$$-R_1 + R_4 \Rightarrow R_4 \qquad -(1/2)R_4 \Rightarrow R_3$$

$$\begin{bmatrix} 1 & 1 & -2 \\ 3 & -1 & 6 \\ 2 & 2 & -4 \\ 1 & -1 & 4 \end{bmatrix} \Rightarrow \begin{bmatrix} 1 & 1 & -2 \\ 0 & -4 & 12 \\ 0 & 0 & 0 \\ 0 & -2 & 6 \end{bmatrix} \Rightarrow \begin{bmatrix} 1 & 1 & -2 \\ 0 & 1 & -3 \\ 0 & 1 & -3 \\ 0 & 0 & 0 \end{bmatrix}$$

$$-R_2 + R_1 \Rightarrow R_1$$
$$-R_2 + R_3 \Rightarrow R_3 \Rightarrow \begin{bmatrix} 1 & 0 & 1 \\ 0 & 1 & -3 \\ 0 & 0 & 0 \\ 0 & 0 & 0 \end{bmatrix} \qquad \boxed{x = 1,\ y = -3}$$

51.

$$-3R_1 + R_2 \Rightarrow R_2 \qquad\qquad\qquad -(1/7)R_2 \Rightarrow R_2$$

$$\begin{bmatrix} 1 & 2 & 1 & 4 \\ 3 & -1 & -1 & 2 \end{bmatrix} \Rightarrow \begin{bmatrix} 1 & 2 & 1 & 4 \\ 0 & -7 & -4 & -10 \end{bmatrix} \Rightarrow \begin{bmatrix} 1 & 2 & 1 & 4 \\ 0 & 1 & 4/7 & 10/7 \end{bmatrix}$$

$$-2R_2 + R_1 \Rightarrow R_1 \qquad \Rightarrow \begin{bmatrix} 1 & 0 & -\frac{1}{7} & \frac{8}{7} \\ 0 & 1 & \frac{4}{7} & \frac{10}{7} \end{bmatrix}$$

$$x - \frac{1}{7}z = \frac{8}{7} \Rightarrow \boxed{x = \frac{8}{7} + \frac{1}{7}z} \qquad y + \frac{4}{7}z = \frac{10}{7} \Rightarrow \boxed{y = \frac{10}{7} - \frac{4}{7}z}$$

53.

$$\begin{array}{cccc} w & x & y & z \end{array} \qquad R_3 \Leftrightarrow R_2$$

$$\begin{bmatrix} 1 & 1 & 0 & 0 & 1 \\ 1 & 0 & 1 & 0 & 0 \\ 0 & 1 & 0 & 1 & 0 \end{bmatrix} \Rightarrow \begin{bmatrix} 1 & 1 & 0 & 0 & 1 \\ 0 & 1 & 0 & 1 & 0 \\ 1 & 0 & 1 & 0 & 0 \end{bmatrix}$$

$$-R_1 + R_3 \Rightarrow R_3 \qquad\qquad -R_2 + R_1 \Rightarrow R_1,\ R_2 + R_3 \Rightarrow R_3$$

$$\begin{bmatrix} 1 & 1 & 0 & 0 & 1 \\ 0 & 1 & 0 & 1 & 0 \\ 0 & -1 & 1 & 0 & -1 \end{bmatrix} \Rightarrow \begin{bmatrix} 1 & 0 & 0 & -1 & 1 \\ 0 & 1 & 0 & 1 & 0 \\ 0 & 0 & 1 & 1 & -1 \end{bmatrix}$$

$$w - z = 1 \Rightarrow \boxed{w = 1 + z} \qquad x + z = 0 \Rightarrow \boxed{x = -z}$$

$$y + z = -1 \Rightarrow \boxed{y = -1 - z}$$

55.

$$-2\,R_1 + R_2 \Rightarrow R_2$$
$$-3\,R_1 + R_3 \Rightarrow R_3 \qquad\qquad -R_2 + R_3 \Rightarrow R_3$$

$$\begin{bmatrix} 1 & 1 & | & 3 \\ 2 & 1 & | & 1 \\ 3 & 2 & | & 2 \end{bmatrix} \Rightarrow \begin{bmatrix} 1 & 1 & | & 3 \\ 0 & -1 & | & -5 \\ 0 & -1 & | & -7 \end{bmatrix} \Rightarrow \begin{bmatrix} 1 & 1 & | & 3 \\ 0 & -1 & | & -5 \\ 0 & 0 & | & -2 \end{bmatrix}$$

The last row indicates $0x + 0y = -2$, which is impossible.
NO SOLUTION – THE SYSTEM IS INCONSISTENT.

57.

$$-2\,R_1 + R_2 \Rightarrow R_2$$
$$-R_1 + R_3 \Rightarrow R_3 \qquad\qquad R_3 \Leftrightarrow R_2$$

$$\begin{bmatrix} 1 & 1 & 1 & | & 14 \\ 2 & 3 & -2 & | & -7 \\ 1 & -5 & 1 & | & 8 \end{bmatrix} \Rightarrow \begin{bmatrix} 1 & 1 & 1 & | & 14 \\ 0 & 1 & -4 & | & -35 \\ 0 & -6 & 0 & | & -6 \end{bmatrix} \Rightarrow \begin{bmatrix} 1 & 1 & 1 & | & 14 \\ 0 & -6 & 0 & | & -6 \\ 0 & 1 & -4 & | & -35 \end{bmatrix}$$

$$6\,R_1 + R_2 \Rightarrow R_1$$
$$6\,R_3 + R_2 \Rightarrow R_3 \qquad\qquad R_3 + 4\,R_1 \Rightarrow R_1$$

$$\begin{bmatrix} 6 & 0 & 6 & | & 78 \\ 0 & -6 & 0 & | & -6 \\ 0 & 0 & -24 & | & -216 \end{bmatrix} \Rightarrow \begin{bmatrix} 24 & 0 & 0 & | & 96 \\ 0 & -6 & 0 & | & -6 \\ 0 & 0 & -24 & | & -216 \end{bmatrix}$$

$$(1/24)\,R_1 \Rightarrow R_1$$
$$-(1/6)\,R_2 \Rightarrow R_2 \qquad\qquad \Rightarrow \begin{bmatrix} 1 & 0 & 0 & | & 4 \\ 0 & 1 & 0 & | & 1 \\ 0 & 0 & 1 & | & 9 \end{bmatrix}$$
$$-(1/24)\,R_3 \Rightarrow R_3$$

$$x^2 = 4 \Rightarrow \boxed{x = \pm 2} \qquad y^2 = 1 \Rightarrow \boxed{y = \pm 1} \qquad z^2 = 9 \Rightarrow \boxed{z = \pm 3}$$

Exercise 7.3 (page 369)

1. $x = 2, \ y = 5$

3.
$$x + y = 3$$
$$3 + x = 4 \ \Rightarrow x = 1$$
$$5y = 10 \ \Rightarrow y = 2$$
$x = 1$ and $y = 2$ satisfy the first equation also.

5. $5A = 5 \begin{bmatrix} 3 & -3 \\ 0 & -2 \end{bmatrix} = \begin{bmatrix} 5 \cdot 3 & 5 \cdot (-3) \\ 5 \cdot 0 & 5 \cdot (-2) \end{bmatrix} = \begin{bmatrix} 15 & -15 \\ 0 & -10 \end{bmatrix}$

7. $5A = 5 \begin{bmatrix} 5 & 5 & -2 \\ -2 & -5 & 1 \end{bmatrix} = \begin{bmatrix} 5 \cdot 5 & 5 \cdot 5 & 5 \cdot (-2) \\ 5 \cdot (-2) & 5 \cdot (-5) & 5 \cdot 1 \end{bmatrix} = \begin{bmatrix} 25 & 25 & -10 \\ -10 & -25 & 5 \end{bmatrix}$

9. $A + B = \begin{bmatrix} 2 & 1 & -1 \\ -3 & 2 & 5 \end{bmatrix} + \begin{bmatrix} -3 & 1 & 2 \\ -3 & -2 & -5 \end{bmatrix}$

$= \begin{bmatrix} 2+(-3) & 1+1 & -1+(2) \\ -3+(-3) & 2+(-2) & 5+(-5) \end{bmatrix} = \begin{bmatrix} -1 & 2 & 1 \\ -6 & 0 & 0 \end{bmatrix}$

11. $A - B = \begin{bmatrix} -3 & 2 & -2 \\ -1 & 4 & -5 \end{bmatrix} - \begin{bmatrix} 3 & -3 & -2 \\ -2 & 5 & -5 \end{bmatrix}$

$= \begin{bmatrix} -3-3 & 2-(-3) & -2-(-2) \\ -1-(-2) & 4-5 & -5-(-5) \end{bmatrix} = \begin{bmatrix} -6 & 5 & 0 \\ 1 & -1 & 0 \end{bmatrix}$

13. $5A + 3B = 5\begin{bmatrix} 3 & 1 & -2 \\ -4 & 3 & -2 \end{bmatrix} + 3\begin{bmatrix} 1 & -2 & 2 \\ -5 & -5 & 3 \end{bmatrix}$

$= \begin{bmatrix} 15 & 5 & -10 \\ -20 & 15 & -10 \end{bmatrix} + \begin{bmatrix} 3 & -6 & 6 \\ -15 & -15 & 9 \end{bmatrix} = \begin{bmatrix} 18 & -1 & -4 \\ -35 & 0 & -1 \end{bmatrix}$

15. Additive inverse of $\begin{bmatrix} 5 & -2 & -7 \\ -5 & 0 & 3 \\ -2 & 3 & -5 \end{bmatrix} = \begin{bmatrix} -5 & 2 & 7 \\ 5 & 0 & -3 \\ 2 & -3 & 5 \end{bmatrix}$

17. $\begin{bmatrix} 2 & 3 \\ 3 & -2 \end{bmatrix}\begin{bmatrix} 1 & 2 \\ 0 & -2 \end{bmatrix} = \begin{bmatrix} 2(1)+3(0) & 2(2)+3(-2) \\ 3(1)+(-2)(0) & 3(2)+(-2)(-2) \end{bmatrix}$

$= \begin{bmatrix} 2 & -2 \\ 3 & 10 \end{bmatrix}$

19. $\begin{bmatrix} -4 & -2 \\ 21 & 0 \end{bmatrix}\begin{bmatrix} -5 & 6 \\ 21 & -1 \end{bmatrix} = \begin{bmatrix} -4(-5)+(-2)(21) & -4(6)+(-2)(-1) \\ 21(-5)+0(21) & 21(6)+0(-1) \end{bmatrix}$

$= \begin{bmatrix} -22 & -22 \\ -105 & 126 \end{bmatrix}$

21.
$$\begin{bmatrix} 2 & 1 & 3 \\ 1 & 2 & -1 \\ 0 & 1 & 0 \end{bmatrix}\begin{bmatrix} 1 & 2 & 3 \\ 2 & -2 & 1 \\ 0 & 0 & 1 \end{bmatrix}$$

$$= \begin{bmatrix} 2(1)+1(2)+3(0) & 2(2)+1(-2)+3(0) & 2(3)+1(1)+3(1) \\ 1(1)+2(2)+(-1)0 & 1(2)+2(-2)+(-1)0 & 1(3)+2(1)+(-1)1 \\ 0(1)+1(2)+0(0) & 0(2)+1(-2)+0(0) & 0(3)+1(1)+0(1) \end{bmatrix}$$

$$= \begin{bmatrix} 4 & 2 & 10 \\ 5 & -2 & 4 \\ 2 & -2 & 1 \end{bmatrix}$$

23.
$$\begin{bmatrix} 1 & -2 & -3 \\ 2 & 0 & 1 \end{bmatrix}\begin{bmatrix} 4 \\ -5 \\ -6 \end{bmatrix} = \begin{bmatrix} 1(4)+(-2)(-5)+(-3)(-6) \\ 2(4)+0(-5)+1(-6) \end{bmatrix} = \begin{bmatrix} 32 \\ 2 \end{bmatrix}$$

25. The multiplication cannot be performed.
{The number of columns of the first matrix $\neq$ the number of rows of the second.}

27.
$$\begin{bmatrix} 2 & 5 \\ -3 & 1 \\ 0 & -2 \\ 1 & -5 \end{bmatrix}\begin{bmatrix} 3 & -2 & 4 \\ -2 & -3 & 1 \end{bmatrix} = \begin{bmatrix} 6+(-10) & -4+(-15) & 8+5 \\ -9+(-2) & 6+(-3) & -12+1 \\ 0+4 & 0+6 & 0+(-2) \\ 3+10 & -2+15 & 4+(-5) \end{bmatrix}$$

$$= \begin{bmatrix} -4 & -19 & 13 \\ -11 & 3 & -11 \\ 4 & 6 & -2 \\ 13 & 13 & -1 \end{bmatrix}$$

29.
$$A - BC = \begin{bmatrix} 1 & 3 \\ 2 & 5 \end{bmatrix} - \begin{bmatrix} -1 \\ 3 \end{bmatrix}\begin{bmatrix} 3 & 2 \end{bmatrix} = \begin{bmatrix} 1 & 3 \\ 2 & 5 \end{bmatrix} - \begin{bmatrix} -3 & -2 \\ 9 & 6 \end{bmatrix}$$

$$= \begin{bmatrix} 4 & 5 \\ -7 & -1 \end{bmatrix}$$

31.
$$CB - AB = \begin{bmatrix} 3 & 2 \end{bmatrix}\begin{bmatrix} -1 \\ 3 \end{bmatrix} - \begin{bmatrix} 1 & 3 \\ 2 & 5 \end{bmatrix}\begin{bmatrix} -1 \\ 3 \end{bmatrix}$$

$$= [3] - \begin{bmatrix} 8 \\ 13 \end{bmatrix} \quad \text{cannot be performed -- dimensions do not match}$$

33. $ABC = \begin{bmatrix} 1 & 3 \\ 2 & 5 \end{bmatrix}\begin{bmatrix} -1 \\ 3 \end{bmatrix}\begin{bmatrix} 3 & 2 \end{bmatrix} = \begin{bmatrix} 8 \\ 13 \end{bmatrix}\begin{bmatrix} 3 & 2 \end{bmatrix} = \begin{bmatrix} 24 & 16 \\ 39 & 26 \end{bmatrix}$

35. $A^2B = \begin{bmatrix} 1 & 3 \\ 2 & 5 \end{bmatrix}\begin{bmatrix} 1 & 3 \\ 2 & 5 \end{bmatrix}\begin{bmatrix} -1 \\ 3 \end{bmatrix} = \begin{bmatrix} 7 & 18 \\ 12 & 31 \end{bmatrix}\begin{bmatrix} -1 \\ 3 \end{bmatrix} = \begin{bmatrix} 47 \\ 81 \end{bmatrix}$

37. $AB = \begin{bmatrix} 2.3 & -1.7 & 3.1 \\ -2 & 3.5 & 1 \\ -8 & 4.7 & 9.1 \end{bmatrix}\begin{bmatrix} -2.5 \\ 5.2 \\ -7 \end{bmatrix} = \begin{bmatrix} -36.29 \\ 16.2 \\ -19.26 \end{bmatrix}$

39. $A^2 = \begin{bmatrix} 2.3 & -1.7 & 3.1 \\ -2 & 3.5 & 1 \\ -8 & 4.7 & 9.1 \end{bmatrix}\begin{bmatrix} 2.3 & -1.7 & 3.1 \\ -2 & 3.5 & 1 \\ -8 & 4.7 & 9.1 \end{bmatrix} = \begin{bmatrix} -16.11 & 4.71 & 33.64 \\ -19.6 & 20.35 & 6.4 \\ -100.6 & 72.82 & 62.71 \end{bmatrix}$

41. $PQ = \begin{bmatrix} 217 & 23 & 319 \\ 347 & 24 & 340 \\ 3 & 97 & 750 \end{bmatrix}\begin{bmatrix} 0.75 \\ 1.00 \\ 1.25 \end{bmatrix} = \begin{bmatrix} 584.50 \\ 709.25 \\ 1036.75 \end{bmatrix}$

$584.50 was spent by adult males, $709.25 by adult females and $1036.75 by children on beverages at the game.

43. $A(B+C) = \begin{bmatrix} 1 & 2 \\ 1 & 3 \end{bmatrix} \cdot \left(\begin{bmatrix} 2 & 1 & -5 \\ 1 & 1 & 2 \end{bmatrix} + \begin{bmatrix} -2 & -1 & 6 \\ 0 & -1 & -1 \end{bmatrix} \right)$

$= \begin{bmatrix} 1 & 2 \\ 1 & 3 \end{bmatrix} \cdot \begin{bmatrix} 0 & 0 & 1 \\ 1 & 0 & 1 \end{bmatrix} = \boxed{\begin{bmatrix} 2 & 0 & 3 \\ 3 & 0 & 4 \end{bmatrix}}$

$AB + AC = \begin{bmatrix} 1 & 2 \\ 1 & 3 \end{bmatrix} \cdot \begin{bmatrix} 2 & 1 & -5 \\ 1 & 1 & 2 \end{bmatrix} + \begin{bmatrix} 1 & 2 \\ 1 & 3 \end{bmatrix} \cdot \begin{bmatrix} -2 & -1 & 6 \\ 0 & -1 & -1 \end{bmatrix}$

$= \begin{bmatrix} 4 & 3 & -1 \\ 5 & 4 & 1 \end{bmatrix} + \begin{bmatrix} -2 & -3 & 4 \\ -2 & -4 & 3 \end{bmatrix} = \boxed{\begin{bmatrix} 2 & 0 & 3 \\ 3 & 0 & 4 \end{bmatrix}}$

45. $3(AB) = 3\left(\begin{bmatrix} 1 & 2 \\ 1 & 3 \end{bmatrix} \cdot \begin{bmatrix} 2 & 1 & -5 \\ 1 & 1 & 2 \end{bmatrix}\right) = 3\begin{bmatrix} 4 & 3 & -1 \\ 5 & 4 & 1 \end{bmatrix}$

$$= \boxed{\begin{bmatrix} 12 & 9 & -3 \\ 15 & 12 & 3 \end{bmatrix}}$$

$(3A)B = \left(3\begin{bmatrix} 1 & 2 \\ 1 & 3 \end{bmatrix}\right) \cdot \begin{bmatrix} 2 & 1 & -5 \\ 1 & 1 & 2 \end{bmatrix}$

$$= \begin{bmatrix} 3 & 6 \\ 3 & 9 \end{bmatrix} \cdot \begin{bmatrix} 2 & 1 & -5 \\ 1 & 1 & 2 \end{bmatrix} = \boxed{\begin{bmatrix} 12 & 9 & -3 \\ 15 & 12 & 3 \end{bmatrix}}$$

47. $A^2 = \begin{bmatrix} 0 & 1 & 1 \\ 1 & 0 & 0 \\ 0 & 1 & 0 \end{bmatrix}\begin{bmatrix} 0 & 1 & 1 \\ 1 & 0 & 0 \\ 0 & 1 & 0 \end{bmatrix} = \begin{bmatrix} 1 & 1 & 0 \\ 0 & 1 & 1 \\ 1 & 0 & 0 \end{bmatrix}$

49. $A^2 = \begin{bmatrix} 0 & 2 & 1 & 0 \\ 2 & 0 & 1 & 0 \\ 1 & 1 & 0 & 2 \\ 0 & 0 & 2 & 0 \end{bmatrix}\begin{bmatrix} 0 & 2 & 1 & 0 \\ 2 & 0 & 1 & 0 \\ 1 & 1 & 0 & 2 \\ 0 & 0 & 2 & 0 \end{bmatrix} = \begin{bmatrix} 5 & 1 & 2 & 2 \\ 1 & 5 & 2 & 2 \\ 2 & 2 & 6 & 0 \\ 2 & 2 & 0 & 4 \end{bmatrix}$

$A + A^2 = \begin{bmatrix} 0 & 2 & 1 & 0 \\ 2 & 0 & 1 & 0 \\ 1 & 1 & 0 & 2 \\ 0 & 0 & 2 & 0 \end{bmatrix} + \begin{bmatrix} 5 & 1 & 2 & 2 \\ 1 & 5 & 2 & 2 \\ 2 & 2 & 6 & 0 \\ 2 & 2 & 0 & 4 \end{bmatrix} = \begin{bmatrix} 5 & 3 & 3 & 2 \\ 3 & 5 & 3 & 2 \\ 3 & 3 & 6 & 2 \\ 2 & 2 & 2 & 4 \end{bmatrix}$

A^2 represents the number of connections between cities which use another city in between. $A + A^2$ represents the total number of connections between two cities, either direct or through one other city.

51. $(AB)^2 \neq A^2B^2$. Let $A = \begin{bmatrix} 1 & 0 \\ 2 & 0 \end{bmatrix}$ and $B = \begin{bmatrix} 0 & 2 \\ 1 & 2 \end{bmatrix}$.

$(AB)^2 = \left(\begin{bmatrix} 1 & 0 \\ 2 & 0 \end{bmatrix}\begin{bmatrix} 0 & 2 \\ 1 & 2 \end{bmatrix}\right)^2 = \left(\begin{bmatrix} 0 & 2 \\ 0 & 4 \end{bmatrix}\right)^2 = \begin{bmatrix} 0 & 8 \\ 0 & 16 \end{bmatrix}$

$A^2B^2 = \left(\begin{bmatrix} 1 & 0 \\ 2 & 0 \end{bmatrix}\right)^2\left(\begin{bmatrix} 0 & 2 \\ 1 & 2 \end{bmatrix}\right)^2 = \begin{bmatrix} 1 & 0 \\ 2 & 0 \end{bmatrix}\begin{bmatrix} 2 & 4 \\ 2 & 6 \end{bmatrix}$

$$= \begin{bmatrix} 2 & 4 \\ 4 & 8 \end{bmatrix}$$

53. Let $A = \begin{bmatrix} 1 & 1 \\ 1 & 1 \end{bmatrix}$ and $B = \begin{bmatrix} 1 & 1 \\ -1 & -1 \end{bmatrix}$. Then $AB = \begin{bmatrix} 0 & 0 \\ 0 & 0 \end{bmatrix}$.

Exercise 7.4   (page 377)

1. $\begin{bmatrix} 3 & -4 & 1 & 0 \\ -2 & 3 & 0 & 1 \end{bmatrix} \Rightarrow \begin{bmatrix} 1 & -\frac{4}{3} & \frac{1}{3} & 0 \\ -2 & 3 & 0 & 1 \end{bmatrix} \Rightarrow \begin{bmatrix} 1 & -\frac{4}{3} & \frac{1}{3} & 0 \\ 0 & \frac{1}{3} & \frac{2}{3} & 1 \end{bmatrix}$

$\begin{bmatrix} 1 & 0 & 3 & 4 \\ 0 & \frac{1}{3} & \frac{2}{3} & 1 \end{bmatrix} \Rightarrow \begin{bmatrix} 1 & 0 & 3 & 4 \\ 0 & 3 & 2 & 3 \end{bmatrix}$ INVERSE: $\begin{bmatrix} 3 & 4 \\ 2 & 3 \end{bmatrix}$

3. $\begin{bmatrix} 3 & 7 & 1 & 0 \\ 2 & 5 & 0 & 1 \end{bmatrix} \Rightarrow \begin{bmatrix} 1 & \frac{7}{3} & \frac{1}{3} & 0 \\ 2 & 5 & 0 & 1 \end{bmatrix} \Rightarrow \begin{bmatrix} 1 & \frac{7}{3} & \frac{1}{3} & 0 \\ 0 & \frac{1}{3} & -\frac{2}{3} & 1 \end{bmatrix}$

$\begin{bmatrix} 1 & 0 & 5 & -7 \\ 0 & \frac{1}{3} & -\frac{2}{3} & 1 \end{bmatrix} \Rightarrow \begin{bmatrix} 1 & 0 & 5 & -7 \\ 0 & 3 & -2 & 3 \end{bmatrix}$ INVERSE: $\begin{bmatrix} 5 & -7 \\ -2 & 3 \end{bmatrix}$

5. $\begin{bmatrix} 1 & 0 & 3 & 1 & 0 & 0 \\ -1 & 1 & 3 & 0 & 1 & 0 \\ -2 & 1 & 1 & 0 & 0 & 1 \end{bmatrix} \Rightarrow \begin{bmatrix} 1 & 0 & 3 & 1 & 0 & 0 \\ 0 & 1 & 6 & 1 & 1 & 0 \\ 0 & 1 & 7 & 2 & 0 & 1 \end{bmatrix}$

$\begin{bmatrix} 1 & 0 & 3 & 1 & 0 & 0 \\ 0 & 1 & 6 & 1 & 1 & 0 \\ 0 & 0 & 1 & 1 & -1 & 1 \end{bmatrix} \Rightarrow \begin{bmatrix} 1 & 0 & 0 & -2 & 3 & -3 \\ 0 & 1 & 0 & -5 & 7 & -6 \\ 0 & 0 & 1 & 1 & -1 & 1 \end{bmatrix}$

INVERSE: $\begin{bmatrix} -2 & 3 & -3 \\ -5 & 7 & -6 \\ 1 & -1 & 1 \end{bmatrix}$

7. $\begin{bmatrix} 3 & 2 & 1 & | & 1 & 0 & 0 \\ 1 & 1 & -1 & | & 0 & 1 & 0 \\ 4 & 3 & 1 & | & 0 & 0 & 1 \end{bmatrix} \Rightarrow \begin{bmatrix} 1 & 1 & -1 & | & 0 & 1 & 0 \\ 3 & 2 & 1 & | & 1 & 0 & 0 \\ 4 & 3 & 1 & | & 0 & 0 & 1 \end{bmatrix}$

$\begin{bmatrix} 1 & 1 & -1 & | & 0 & 1 & 0 \\ 0 & -1 & 4 & | & 1 & -3 & 0 \\ 0 & -1 & 5 & | & 0 & -4 & 1 \end{bmatrix} \Rightarrow \begin{bmatrix} 1 & 1 & -1 & | & 0 & 1 & 0 \\ 0 & 1 & -4 & | & -1 & 3 & 0 \\ 0 & -1 & 5 & | & 0 & -4 & 1 \end{bmatrix}$

$\begin{bmatrix} 1 & 0 & 3 & | & 1 & -2 & 0 \\ 0 & 1 & -4 & | & -1 & 3 & 0 \\ 0 & 0 & 1 & | & -1 & -1 & 1 \end{bmatrix} \Rightarrow \begin{bmatrix} 1 & 0 & 0 & | & 4 & 1 & -3 \\ 0 & 1 & 0 & | & -5 & -1 & 4 \\ 0 & 0 & 1 & | & -1 & -1 & 1 \end{bmatrix}$

INVERSE: $\begin{bmatrix} 4 & 1 & -3 \\ -5 & -1 & 4 \\ -1 & -1 & 1 \end{bmatrix}$

9. $\begin{bmatrix} 1 & 3 & 5 & | & 1 & 0 & 0 \\ 0 & 1 & 6 & | & 0 & 1 & 0 \\ 1 & 4 & 11 & | & 0 & 0 & 1 \end{bmatrix} \Rightarrow \begin{bmatrix} 1 & 3 & 5 & | & 1 & 0 & 0 \\ 0 & 1 & 6 & | & 0 & 1 & 0 \\ 0 & 1 & 6 & | & -1 & 0 & 1 \end{bmatrix}$

$\begin{bmatrix} 1 & 0 & -13 & | & 1 & -3 & 0 \\ 0 & 1 & 6 & | & 0 & 1 & 0 \\ 0 & 0 & 0 & | & 1 & 1 & -1 \end{bmatrix}$ Since the original matrix cannot be changed into the identity matrix, there is no inverse matrix.

11. $\begin{bmatrix} 1 & 2 & 3 & | & 1 & 0 & 0 \\ 0 & 1 & 2 & | & 0 & 1 & 0 \\ 0 & 0 & 1 & | & 0 & 0 & 1 \end{bmatrix} \Rightarrow \begin{bmatrix} 1 & 0 & -1 & | & 1 & -2 & 0 \\ 0 & 1 & 2 & | & 0 & 1 & 0 \\ 0 & 0 & 1 & | & 0 & 0 & 1 \end{bmatrix}$

$\begin{bmatrix} 1 & 0 & 0 & | & 1 & -2 & 1 \\ 0 & 1 & 0 & | & 0 & 1 & -2 \\ 0 & 0 & 1 & | & 0 & 0 & 1 \end{bmatrix}$ INVERSE: $\begin{bmatrix} 1 & -2 & 1 \\ 0 & 1 & -2 \\ 0 & 0 & 1 \end{bmatrix}$

13. $\begin{bmatrix} 1 & 6 & 4 & | & 1 & 0 & 0 \\ 1 & -2 & -5 & | & 0 & 1 & 0 \\ 2 & 4 & -1 & | & 0 & 0 & 1 \end{bmatrix} \Rightarrow \begin{bmatrix} 1 & 6 & 4 & | & 1 & 0 & 0 \\ 0 & 8 & 9 & | & 1 & -1 & 0 \\ 0 & -8 & -9 & | & -2 & 0 & 1 \end{bmatrix}$

$\begin{bmatrix} 1 & 6 & 4 & | & 1 & 0 & 0 \\ 0 & 8 & 9 & | & 1 & -1 & 0 \\ 0 & 0 & 0 & | & -1 & -1 & 1 \end{bmatrix}$ Since the original matrix cannot be changed into the identity matrix, there is no inverse matrix.

15.
$$\begin{bmatrix} 1 & 2 & 3 & 4 & | & 1 & 0 & 0 & 0 \\ 0 & 1 & 2 & 3 & | & 0 & 1 & 0 & 0 \\ 0 & 0 & 1 & 2 & | & 0 & 0 & 1 & 0 \\ 0 & 0 & 0 & 1 & | & 0 & 0 & 0 & 1 \end{bmatrix} \Rightarrow \begin{bmatrix} 1 & 0 & -1 & -2 & | & 1 & -2 & 0 & 0 \\ 0 & 1 & 2 & 3 & | & 0 & 1 & 0 & 0 \\ 0 & 0 & 1 & 2 & | & 0 & 0 & 1 & 0 \\ 0 & 0 & 0 & 1 & | & 0 & 0 & 0 & 1 \end{bmatrix}$$

$$\begin{bmatrix} 1 & 0 & 0 & 0 & | & 1 & -2 & 1 & 0 \\ 0 & 1 & 0 & -1 & | & 0 & 1 & -2 & 0 \\ 0 & 0 & 1 & 2 & | & 0 & 0 & 1 & 0 \\ 0 & 0 & 0 & 1 & | & 0 & 0 & 0 & 1 \end{bmatrix} \Rightarrow \begin{bmatrix} 1 & 0 & 0 & 0 & | & 1 & -2 & 1 & 0 \\ 0 & 1 & 0 & 0 & | & 0 & 1 & -2 & 1 \\ 0 & 0 & 1 & 0 & | & 0 & 0 & 1 & -2 \\ 0 & 0 & 0 & 1 & | & 0 & 0 & 0 & 1 \end{bmatrix}$$

INVERSE:
$$\begin{bmatrix} 1 & -2 & 1 & 0 \\ 0 & 1 & -2 & 1 \\ 0 & 0 & 1 & -2 \\ 0 & 0 & 0 & 1 \end{bmatrix}$$

17. INVERSE:
$$\begin{bmatrix} 8 & -2 & -6 \\ -5 & 2 & 4 \\ 2 & 0 & -2 \end{bmatrix}$$

19. INVERSE:
$$\begin{bmatrix} -2.5 & 5 & 3 & 5.5 \\ 5.5 & -8 & -6 & -9.5 \\ -1 & 3 & 1 & 3 \\ -5.5 & 9 & 6 & 10.5 \end{bmatrix}$$

21.
$$\begin{bmatrix} 3 & -4 \\ -2 & 3 \end{bmatrix}\begin{bmatrix} x \\ y \end{bmatrix} = \begin{bmatrix} 1 \\ 5 \end{bmatrix}$$

$$\begin{bmatrix} 3 & 4 \\ 2 & 3 \end{bmatrix}\begin{bmatrix} 3 & -4 \\ -2 & 3 \end{bmatrix}\begin{bmatrix} x \\ y \end{bmatrix} = \begin{bmatrix} 3 & 4 \\ 2 & 3 \end{bmatrix}\begin{bmatrix} 1 \\ 5 \end{bmatrix}$$

$$\begin{bmatrix} 1 & 0 \\ 0 & 1 \end{bmatrix}\begin{bmatrix} x \\ y \end{bmatrix} = \begin{bmatrix} 23 \\ 17 \end{bmatrix} \quad x = 23, \; y = 17$$

23.
$$\begin{bmatrix} 3 & -4 \\ -2 & 3 \end{bmatrix}\begin{bmatrix} x \\ y \end{bmatrix} = \begin{bmatrix} 0 \\ 0 \end{bmatrix}$$

$$\begin{bmatrix} 3 & 4 \\ 2 & 3 \end{bmatrix}\begin{bmatrix} 3 & -4 \\ -2 & 3 \end{bmatrix}\begin{bmatrix} x \\ y \end{bmatrix} = \begin{bmatrix} 3 & 4 \\ 2 & 3 \end{bmatrix}\begin{bmatrix} 0 \\ 0 \end{bmatrix}$$

$$\begin{bmatrix} 1 & 0 \\ 0 & 1 \end{bmatrix}\begin{bmatrix} x \\ y \end{bmatrix} = \begin{bmatrix} 0 \\ 0 \end{bmatrix} \quad x = 0, \; y = 0$$

25.
$$\begin{bmatrix} 2 & 1 & -1 \\ 2 & 2 & -1 \\ -1 & -1 & 1 \end{bmatrix} \begin{bmatrix} x \\ y \\ z \end{bmatrix} = \begin{bmatrix} 2 \\ 4 \\ -1 \end{bmatrix}$$

$$\begin{bmatrix} 1 & 0 & 1 \\ -1 & 1 & 0 \\ 0 & 1 & 2 \end{bmatrix} \begin{bmatrix} 2 & 1 & -1 \\ 2 & 2 & -1 \\ -1 & -1 & 1 \end{bmatrix} \begin{bmatrix} x \\ y \\ z \end{bmatrix} = \begin{bmatrix} 1 & 0 & 1 \\ -1 & 1 & 0 \\ 0 & 1 & 2 \end{bmatrix} \begin{bmatrix} 2 \\ 4 \\ -1 \end{bmatrix}$$

$$\begin{bmatrix} 1 & 0 & 0 \\ 0 & 1 & 0 \\ 0 & 0 & 1 \end{bmatrix} \begin{bmatrix} x \\ y \\ z \end{bmatrix} = \begin{bmatrix} 1 \\ 2 \\ 2 \end{bmatrix} \qquad x = 1,\ y = 2,\ z = 2$$

27.
$$\begin{bmatrix} -2 & 1 & -3 \\ 2 & 3 & 0 \\ 1 & 0 & 1 \end{bmatrix} \begin{bmatrix} x \\ y \\ z \end{bmatrix} = \begin{bmatrix} 2 \\ -3 \\ 5 \end{bmatrix}$$

$$\begin{bmatrix} 3 & -1 & 9 \\ -2 & 1 & -6 \\ -3 & 1 & -8 \end{bmatrix} \begin{bmatrix} -2 & 1 & -3 \\ 2 & 3 & 0 \\ 1 & 0 & 1 \end{bmatrix} \begin{bmatrix} x \\ y \\ z \end{bmatrix} = \begin{bmatrix} 3 & -1 & 9 \\ -2 & 1 & -6 \\ -3 & 1 & -8 \end{bmatrix} \begin{bmatrix} 2 \\ -3 \\ 5 \end{bmatrix}$$

$$\begin{bmatrix} 1 & 0 & 0 \\ 0 & 1 & 0 \\ 0 & 0 & 1 \end{bmatrix} \begin{bmatrix} x \\ y \\ z \end{bmatrix} = \begin{bmatrix} 54 \\ -37 \\ -49 \end{bmatrix} \qquad \begin{array}{l} x = 54,\ y = -37 \\ z = -49 \end{array}$$

29.
$$\begin{bmatrix} 5 & 3 \\ -7 & 5 \end{bmatrix} \begin{bmatrix} x \\ y \end{bmatrix} = \begin{bmatrix} 13 \\ -9 \end{bmatrix}$$

$$\begin{bmatrix} 0.109 & -0.065 \\ 0.152 & 0.109 \end{bmatrix} \begin{bmatrix} 5 & 3 \\ -7 & 5 \end{bmatrix} \begin{bmatrix} x \\ y \end{bmatrix} = \begin{bmatrix} 0.109 & -0.065 \\ 0.152 & 0.109 \end{bmatrix} \begin{bmatrix} 13 \\ -9 \end{bmatrix}$$

$$\begin{bmatrix} 1 & 0 \\ 0 & 1 \end{bmatrix} \begin{bmatrix} x \\ y \end{bmatrix} = \begin{bmatrix} 2 \\ 1 \end{bmatrix} \qquad x = 2,\ y = 1$$

31.
$$\begin{bmatrix} 5 & 2 & 3 \\ 2 & 0 & 5 \\ 3 & 0 & 1 \end{bmatrix} \begin{bmatrix} x \\ y \\ z \end{bmatrix} = \begin{bmatrix} 12 \\ 7 \\ 4 \end{bmatrix}$$

$$\begin{bmatrix} 0 & -0.08 & 0.38 \\ 0.5 & -0.15 & -0.73 \\ 0 & 0.23 & -0.15 \end{bmatrix} \begin{bmatrix} 5 & 2 & 3 \\ 2 & 0 & 5 \\ 3 & 0 & 1 \end{bmatrix} \begin{bmatrix} x \\ y \\ z \end{bmatrix} = \begin{bmatrix} 0 & -0.08 & 0.38 \\ 0.5 & -0.15 & -0.73 \\ 0 & 0.23 & -0.15 \end{bmatrix} \begin{bmatrix} 12 \\ 7 \\ 4 \end{bmatrix}$$

$$\begin{bmatrix} 1 & 0 & 0 \\ 0 & 1 & 0 \\ 0 & 0 & 1 \end{bmatrix} \begin{bmatrix} x \\ y \\ z \end{bmatrix} = \begin{bmatrix} 1 \\ 2 \\ 1 \end{bmatrix} \qquad \begin{array}{l} x = 1,\ y = 2 \\ z = 1 \end{array}$$

33.

$$\begin{bmatrix} 23 & 27 \\ 21 & 22 \end{bmatrix} \begin{bmatrix} A \\ B \end{bmatrix} = \begin{bmatrix} 127 \\ 108 \end{bmatrix}$$

$$\begin{bmatrix} -0.361 & 0.443 \\ 0.344 & -0.377 \end{bmatrix} \begin{bmatrix} 23 & 27 \\ 21 & 22 \end{bmatrix} \begin{bmatrix} A \\ B \end{bmatrix} = \begin{bmatrix} -0.361 & 0.443 \\ 0.344 & -0.377 \end{bmatrix} \begin{bmatrix} 127 \\ 108 \end{bmatrix}$$

$$\begin{bmatrix} 1 & 0 \\ 0 & 1 \end{bmatrix} \begin{bmatrix} A \\ B \end{bmatrix} = \begin{bmatrix} 2 \\ 3 \end{bmatrix} \quad A = 2,\ B = 3$$

The available hours allow 2 Model A and 3 Model B to be manufactured.

35.
$$AX - 2IX = 0$$
$$(A - 2I)X = 0$$

$$\begin{bmatrix} 1 & 0 & 0 \\ -2 & -3 & -2 \\ 3 & 6 & 1 \end{bmatrix} \begin{bmatrix} x \\ y \\ z \end{bmatrix} = \begin{bmatrix} 0 \\ 0 \\ 0 \end{bmatrix}$$

$$\begin{bmatrix} x \\ y \\ z \end{bmatrix} = \begin{bmatrix} 0 \\ 0 \\ 0 \end{bmatrix}$$

37. Let $A = I$. Then $IB = BI = I$. But then $IB = I$ and $IB = B$. So $I = B$.

$-\frac{c}{a} R_1 + R_2 \Rightarrow R_2$

39.
$$\begin{bmatrix} a & b & 1 & 0 \\ c & d & 0 & 1 \end{bmatrix} \Rightarrow \begin{bmatrix} a & b & 1 & 0 \\ 0 & -\frac{cb}{a}+d & -\frac{c}{a} & 1 \end{bmatrix} \quad -\frac{c}{a} R_1 + R_2 \Rightarrow R_2$$

$$\Rightarrow \begin{bmatrix} a & b & 1 & 0 \\ 0 & ad-bc & -c & a \end{bmatrix} \quad a\, R_2 \Rightarrow R_2$$

$$\Rightarrow \begin{bmatrix} a & 0 & \dfrac{ad}{ad-bc} & -\dfrac{ab}{ad-bc} \\ 0 & ad-bc & -c & a \end{bmatrix} \quad \dfrac{-b}{ad-bc}R_2 + R_1 \Rightarrow R_1$$

$$\Rightarrow \begin{bmatrix} 1 & 0 & \dfrac{d}{ad-bc} & -\dfrac{b}{ad-bc} \\ 0 & 1 & \dfrac{-c}{ad-bc} & \dfrac{a}{ad-bc} \end{bmatrix} \quad \begin{array}{l} \frac{1}{a} R_1 \Rightarrow R_1 \\ \dfrac{1}{ad-bc} R_2 \Rightarrow R_2 \end{array}$$

The inverse exists if and only if $ad - bc = 0$.

41. No. Use $A = \begin{bmatrix} 1 & 2 \\ 0 & 1 \end{bmatrix}$ and $B = \begin{bmatrix} 1 & 0 \\ 3 & 1 \end{bmatrix}$.

43. Let $A = \begin{bmatrix} 1 & 2 \\ 0 & 1 \end{bmatrix}$, for example.

45. If they are inverse matrices, then their product equals the identity matrix.
$$(I - A)(I + A) = I^2 + IA - AI - A^2$$
$$= I + A - A - A^2$$
$$= I - A^2$$
$$= I \quad \{A^2 = 0 \text{ from } \#44\}$$

47. $(I - C)(I + C + C^2) = I^2 + IC + IC^2 - CI - C^2 - C^3$
$$= I + C + C^2 - C - C^2 - C^3$$
$$= I - C^3$$
$$= I \quad \{\text{since } C^3 = 0 \text{ in this problem}\}$$
Since their product is the identity matrix, they are inverse matrices, and $I - C$ is invertible.

Exercise 7.5 (page 389)

1. $\begin{vmatrix} 2 & 1 \\ -2 & 3 \end{vmatrix} = 2(3) - 1(-2)$
$= 6 + 2 = 8$

3. $\begin{vmatrix} 2 & -3 \\ -3 & 5 \end{vmatrix} = 2(5) - (-3)(-3)$
$= 10 - 9 = 1$

5. $M_{21} = \begin{vmatrix} -2 & 3 \\ 8 & 9 \end{vmatrix}$

7. $M_{33} = \begin{vmatrix} 1 & -2 \\ 4 & 5 \end{vmatrix}$

9. $C_{21} = -\begin{vmatrix} -2 & 3 \\ 8 & 9 \end{vmatrix}$ $(2 + 1$ is odd$)$

$= -[-2(9) - 3(8)]$
$= -(-42) = 42$

11. $C_{33} = +\begin{vmatrix} 1 & -2 \\ 4 & 5 \end{vmatrix}$ $(3 + 3$ is even$)$

$= 1(5) - (-2)(4)$
$= 5 + 8 = 13$

13. $\begin{vmatrix} 2 & -3 & 5 \\ -2 & 1 & 3 \\ 1 & 3 & -2 \end{vmatrix} = 2\begin{vmatrix} 1 & 3 \\ 3 & -2 \end{vmatrix} - (-3)\begin{vmatrix} -2 & 3 \\ 1 & -2 \end{vmatrix} + 5\begin{vmatrix} -2 & 1 \\ 1 & 3 \end{vmatrix}$

$= 2(-11) + 3(1) + 5(-7) = -54$

15. $\begin{vmatrix} 1 & -1 & 2 \\ 2 & 1 & 3 \\ 1 & 1 & -1 \end{vmatrix} = 1\begin{vmatrix} 1 & 3 \\ 1 & -1 \end{vmatrix} - (-1)\begin{vmatrix} 2 & 3 \\ 1 & -1 \end{vmatrix} + 2\begin{vmatrix} 2 & 1 \\ 1 & 1 \end{vmatrix}$

$= 1(-4) + 1(-5) + 2(1) = -7$

17. $\begin{vmatrix} 2 & 1 & -1 \\ 1 & 3 & 5 \\ 2 & -5 & 3 \end{vmatrix} = 2\begin{vmatrix} 3 & 5 \\ -5 & 3 \end{vmatrix} - 1\begin{vmatrix} 1 & 5 \\ 2 & 3 \end{vmatrix} + (-1)\begin{vmatrix} 1 & 3 \\ 2 & -5 \end{vmatrix}$

$$= 2(34) - 1(-7) - 1(-11) = 86$$

19. $\begin{vmatrix} 0 & 1 & -3 \\ -3 & 5 & 2 \\ 2 & -5 & 3 \end{vmatrix} = 0\begin{vmatrix} 5 & 2 \\ -5 & 3 \end{vmatrix} - 1\begin{vmatrix} -3 & 2 \\ 2 & 3 \end{vmatrix} + (-3)\begin{vmatrix} -3 & 5 \\ 2 & -5 \end{vmatrix}$

$$= 0(25) - 1(-13) - 3(5) = -2$$

21. $\begin{vmatrix} 0 & 0 & 1 & 0 \\ -2 & 1 & 0 & 1 \\ 1 & 0 & 1 & 2 \\ 2 & 0 & 1 & 2 \end{vmatrix} = 0(***) - 0(***) + 1\begin{vmatrix} -2 & 1 & 1 \\ 1 & 0 & 2 \\ 2 & 0 & 2 \end{vmatrix} - 0(***)$

$$= -2\begin{vmatrix} 0 & 2 \\ 0 & 2 \end{vmatrix} - 1\begin{vmatrix} 1 & 2 \\ 2 & 2 \end{vmatrix} + 1\begin{vmatrix} 1 & 0 \\ 2 & 0 \end{vmatrix}$$

$$= -2(0) - 1(-2) + 1(0) = 2$$

23. $\begin{vmatrix} 1 & 2 & 1 & 3 \\ -2 & 1 & -3 & 1 \\ -1 & 0 & 1 & -2 \\ 2 & -1 & -1 & 3 \end{vmatrix} = \begin{vmatrix} 1 & 2 & 1 & 3 \\ 0 & 0 & -4 & 4 \\ 0 & 2 & 2 & 1 \\ 0 & -5 & -3 & -3 \end{vmatrix}$
$\begin{array}{l} R_2 + R_4 \\ \\ R_1 + R_3 \\ -2\,R_1 + R_4 \end{array}$

$$= 1\begin{vmatrix} 0 & -4 & 4 \\ 2 & 2 & 1 \\ -5 & -3 & -3 \end{vmatrix} \quad \text{(expand along first column)}$$

$$= 0\begin{vmatrix} 2 & 1 \\ -3 & -3 \end{vmatrix} - (-4)\begin{vmatrix} 2 & 1 \\ -5 & -3 \end{vmatrix} + 4\begin{vmatrix} 2 & 2 \\ -5 & -3 \end{vmatrix}$$

$$= 0 + 4(-1) + 4(4) = 12$$

25. True. Switching R_1 with R_2 multiplies the determinant by (-1).

27. True. Two rows have been multiplied by (-1). Each such multiplication multiplies the determinant by (-1), but $(-1)(-1)\det = \det$.

29. Switching R_1 and R_2 multiplies the determinant by (-1), and multiplying R_3 by (-1) multiplies the determinant by (-1) again. The value of the resulting determinant is then $(-1)(-1)(3) = 3$.

31. Adding $R_1 + R_3$ leaves the value of the determinant alone. It is still 3.

33. $x = \dfrac{\begin{vmatrix} 7 & 2 \\ -4 & -3 \end{vmatrix}}{\begin{vmatrix} 3 & 2 \\ 2 & -3 \end{vmatrix}} = \dfrac{-13}{-13} = 1$
  $\qquad y = \dfrac{\begin{vmatrix} 3 & 7 \\ 2 & -4 \end{vmatrix}}{\begin{vmatrix} 3 & 2 \\ 2 & -3 \end{vmatrix}} = \dfrac{-26}{-13} = 2$

35. $x = \dfrac{\begin{vmatrix} 3 & -1 \\ 9 & -7 \end{vmatrix}}{\begin{vmatrix} 1 & -1 \\ 3 & -7 \end{vmatrix}} = \dfrac{-12}{-4} = 3$
  $\qquad y = \dfrac{\begin{vmatrix} 1 & 3 \\ 3 & 9 \end{vmatrix}}{\begin{vmatrix} 1 & -1 \\ 3 & -7 \end{vmatrix}} = \dfrac{0}{-4} = 0$

37. $x = \dfrac{\begin{vmatrix} 2 & 2 & 1 \\ 2 & -1 & 1 \\ 4 & 1 & 3 \end{vmatrix}}{\begin{vmatrix} 1 & 2 & 1 \\ 1 & -1 & 1 \\ 1 & 1 & 3 \end{vmatrix}} = \dfrac{-6}{-6} = 1$
  $\qquad y = \dfrac{\begin{vmatrix} 1 & 2 & 1 \\ 1 & 2 & 1 \\ 1 & 4 & 3 \end{vmatrix}}{\begin{vmatrix} 1 & 2 & 1 \\ 1 & -1 & 1 \\ 1 & 1 & 3 \end{vmatrix}} = \dfrac{0}{-6} = 0$

$z = \dfrac{\begin{vmatrix} 1 & 2 & 2 \\ 1 & -1 & 2 \\ 1 & 1 & 4 \end{vmatrix}}{\begin{vmatrix} 1 & 2 & 1 \\ 1 & -1 & 1 \\ 1 & 1 & 3 \end{vmatrix}} = \dfrac{-6}{-6} = 1$

39. $x = \dfrac{\begin{vmatrix} 5 & -1 & 1 \\ 10 & -3 & 2 \\ 0 & 3 & 1 \end{vmatrix}}{\begin{vmatrix} 2 & -1 & 1 \\ 3 & -3 & 2 \\ 1 & 3 & 1 \end{vmatrix}} = \dfrac{-5}{-5} = 1$
  $\qquad y = \dfrac{\begin{vmatrix} 2 & 5 & 1 \\ 3 & 10 & 2 \\ 1 & 0 & 1 \end{vmatrix}}{\begin{vmatrix} 2 & -1 & 1 \\ 3 & -3 & 2 \\ 1 & 3 & 1 \end{vmatrix}} = \dfrac{5}{-5} = -1$

$z = \dfrac{\begin{vmatrix} 2 & -1 & 5 \\ 3 & -3 & 10 \\ 1 & 3 & 0 \end{vmatrix}}{\begin{vmatrix} 2 & -1 & 1 \\ 3 & -3 & 2 \\ 1 & 3 & 1 \end{vmatrix}} = \dfrac{-10}{-5} = 2$

41. Rewrite system:
$$3x + 2y + 3z = 66$$
$$2x + 6y - z = 36$$
$$3x + y + 6z = 96$$

$$x = \frac{\begin{vmatrix} 66 & 2 & 3 \\ 36 & 6 & -1 \\ 96 & 1 & 6 \end{vmatrix}}{\begin{vmatrix} 3 & 2 & 3 \\ 2 & 6 & -1 \\ 3 & 1 & 6 \end{vmatrix}} = \frac{198}{33} = 6$$

$$y = \frac{\begin{vmatrix} 3 & 66 & 3 \\ 2 & 36 & -1 \\ 3 & 96 & 6 \end{vmatrix}}{\begin{vmatrix} 3 & 2 & 3 \\ 2 & 6 & -1 \\ 3 & 1 & 6 \end{vmatrix}} = \frac{198}{33} = 6$$

$$z = \frac{\begin{vmatrix} 3 & 2 & 66 \\ 2 & 6 & 36 \\ 3 & 1 & 96 \end{vmatrix}}{\begin{vmatrix} 3 & 2 & 3 \\ 2 & 6 & -1 \\ 3 & 1 & 6 \end{vmatrix}} = \frac{396}{33} = 12$$

43.
$$p = \frac{\begin{vmatrix} 0 & -1 & 3 & -1 \\ -1 & 1 & 0 & -1 \\ 2 & 0 & -1 & 0 \\ -8 & -2 & 0 & -3 \end{vmatrix}}{\begin{vmatrix} 2 & -1 & 3 & -1 \\ 1 & 1 & 0 & -1 \\ 3 & 0 & -1 & 0 \\ 1 & -2 & 0 & -3 \end{vmatrix}} = \frac{5}{6}$$

$$q = \frac{\begin{vmatrix} 2 & 0 & 3 & -1 \\ 1 & -1 & 0 & -1 \\ 3 & 2 & -1 & 0 \\ 1 & -8 & 0 & -3 \end{vmatrix}}{\begin{vmatrix} 2 & -1 & 3 & -1 \\ 1 & 1 & 0 & -1 \\ 3 & 0 & -1 & 0 \\ 1 & -2 & 0 & -3 \end{vmatrix}} = \frac{2}{3}$$

$$r = \frac{\begin{vmatrix} 2 & -1 & 0 & -1 \\ 1 & 1 & -1 & -1 \\ 3 & 0 & 2 & 0 \\ 1 & -2 & -8 & -3 \end{vmatrix}}{\begin{vmatrix} 2 & -1 & 3 & -1 \\ 1 & 1 & 0 & -1 \\ 3 & 0 & -1 & 0 \\ 1 & -2 & 0 & -3 \end{vmatrix}} = \frac{1}{2}$$

$$s = \frac{\begin{vmatrix} 2 & -1 & 3 & 0 \\ 1 & 1 & 0 & -1 \\ 3 & 0 & -1 & 2 \\ 1 & -2 & 0 & -8 \end{vmatrix}}{\begin{vmatrix} 2 & -1 & 3 & -1 \\ 1 & 1 & 0 & -1 \\ 3 & 0 & -1 & 0 \\ 1 & -2 & 0 & -3 \end{vmatrix}} = \frac{5}{2}$$

45.
$$\begin{vmatrix} x & y & 1 \\ 0 & 0 & 1 \\ 4 & 6 & 1 \end{vmatrix} = 0$$

$$x\begin{vmatrix} 0 & 1 \\ 6 & 1 \end{vmatrix} - y\begin{vmatrix} 0 & 1 \\ 4 & 1 \end{vmatrix} + 1\begin{vmatrix} 0 & 0 \\ 4 & 6 \end{vmatrix} = 0$$

$$-6x + 4y + 0 = 0$$
$$6x + 4y = 0 \implies 3x + 2y = 0$$

47.
$$\begin{vmatrix} x & y & 1 \\ -2 & 3 & 1 \\ 5 & -3 & 1 \end{vmatrix} = 0$$

$$x\begin{vmatrix} 3 & 1 \\ -3 & 1 \end{vmatrix} - y\begin{vmatrix} -2 & 1 \\ 5 & 1 \end{vmatrix} + 1\begin{vmatrix} -2 & 3 \\ 5 & -3 \end{vmatrix} = 0$$

$$6x + 7y - 9 = 0$$
$$6x + 7y = 9$$

49. $\pm\dfrac{1}{2}\begin{vmatrix} 0 & 0 & 1 \\ 12 & 0 & 1 \\ 12 & 5 & 1 \end{vmatrix} = \pm\dfrac{1}{2}(60) = 30$ square units

51. $\pm\dfrac{1}{2}\begin{vmatrix} 2 & 3 & 1 \\ 10 & 8 & 1 \\ 0 & 20 & 1 \end{vmatrix} = \pm\dfrac{1}{2}(146) = 73$ square units

FOR 53 – 55, use $\begin{vmatrix} a & b \\ c & d \end{vmatrix} = ad - bc.$

53. $\begin{vmatrix} b & a \\ d & c \end{vmatrix} = bc - ad = -(ad - bc) = -\begin{vmatrix} a & b \\ c & d \end{vmatrix}$

55. $\begin{vmatrix} a & ak+b \\ c & ck+d \end{vmatrix} = a(ck+d) - (ak+b)c = ack + ad - akc - bc$
$$= ad - bc$$
$$= \begin{vmatrix} a & b \\ c & d \end{vmatrix}$$

57. $\begin{vmatrix} 1 & 3 & 4 \\ 0 & 5 & 2 \\ 0 & 0 & 2 \end{vmatrix} = 10$

59. $\begin{vmatrix} 1 & 2 & 2 & 3 \\ 0 & 2 & 2 & 1 \\ 0 & 0 & 3 & 2 \\ 0 & 0 & 0 & 4 \end{vmatrix} = 24$

THE DETERMINANT = THE PRODUCT OF THE DIAGONAL ENTRIES.

61. $\begin{vmatrix} 3 & x \\ 1 & 2 \end{vmatrix} = \begin{vmatrix} 2 & -1 \\ x & -5 \end{vmatrix}$
$$6 - x = -10 + x$$
$$-2x = -16$$
$$x = 8$$

63. $\begin{vmatrix} 3 & x & 1 \\ x & 0 & -2 \\ 4 & 0 & 1 \end{vmatrix} = \begin{vmatrix} 2 & x \\ x & 4 \end{vmatrix}$
$$-x(x+8) = 8 - x^2$$
$$-x^2 - 8x = 8 - x^2$$
$$-8x = 8$$
$$x = -1$$

65. Examples will vary.

67. The domain is the set of square $n \times n$ matrices, and the range is the set of real numbers.

69. Yes. By #65, $|AB| = |A| \cdot |B|$. Then, $|AB| = 0 \Rightarrow |A| \cdot |B| = 0$. Then, either $|A| = 0$ or $|B| = 0$.

71.
$$\begin{vmatrix} x-2 & -3 \\ -2 & x-3 \end{vmatrix} = 0$$
$$(x-2)(x-3) - 6 = 0$$
$$x^2 - 5x + 6 - 6 = 0$$
$$x^2 - 5x = 0$$
$$x(x-5) = 0 \Rightarrow x = 0 \text{ or } x = 5$$

73.
$$\begin{vmatrix} 2.3 & 5.7 & 6.1 \\ 3.4 & 6.2 & 8.3 \\ 5.8 & 8.2 & 9.2 \end{vmatrix} = 21.468$$

Exercise 7.6 (page 397)

In the text, the problems are worked by converting to a common denominator and then equating the numerators. In this manual, the same work is achieved by multiplying the equation through by the common denominator.

1. Find A and B such that:
$$\frac{3x-1}{x(x-1)} = \frac{A}{x} + \frac{B}{x-1}$$
$$3x - 1 = A(x-1) + Bx$$
$$3x - 1 = Ax - A + Bx$$
$$3x - 1 = (A+B)x - A$$

$$\begin{array}{rcl} A + B & = & 3 \\ -A & = & -1 \end{array}$$

$$A = 1, B = 2$$

$$\frac{3x-1}{x(x-1)} = \frac{1}{x} + \frac{2}{x-1}$$

3. Find A and B such that:
$$\frac{2x-15}{x(x-3)} = \frac{A}{x} + \frac{B}{x-3}$$
$$2x - 15 = A(x-3) + Bx$$
$$2x - 15 = Ax - 3A + Bx$$
$$2x - 15 = (A+B)x - 3A$$

$$\begin{array}{rcl} A + B & = & 2 \\ -3A & = & -15 \end{array}$$

$$A = 5, B = -3$$

$$\frac{2x-15}{x(x-3)} = \frac{5}{x} + \frac{-3}{x-3} = \frac{5}{x} - \frac{3}{x-3}$$

5. Find A and B such that:

$$\frac{3x+1}{(x+1)(x-1)} = \frac{A}{x+1} + \frac{B}{x-1}$$

$$3x+1 = A(x-1) + B(x+1)$$
$$3x+1 = Ax - A + Bx + B$$
$$3x+1 = (A+B)x + (-A+B)$$

$$A + B = 3$$
$$-A + B = 1$$

$$A = 1, B = 2 \quad \Rightarrow \quad \frac{3x+1}{(x+1)(x-1)} = \frac{1}{x+1} + \frac{2}{x-1}$$

7. Factor, and then find A and B such that:

$$\frac{-2x+11}{(x+2)(x-3)} = \frac{A}{x+2} + \frac{B}{x-3}$$

$$-2x+11 = A(x-3) + B(x+2)$$
$$-2x+11 = Ax - 3A + Bx + 2B$$
$$-2x+11 = (A+B)x + (-3A+2B)$$

$$A + B = -2$$
$$-3A + 2B = 11$$

$$A = -3, B = 1 \quad \Rightarrow \quad \frac{-2x+11}{x^2-x-6} = \frac{-3}{x+2} + \frac{1}{x-3} = \frac{1}{x-3} - \frac{3}{x+2}$$

9. Factor, and then find A and B such that:

$$\frac{3x-23}{(x+3)(x-1)} = \frac{A}{x+3} + \frac{B}{x-1}$$

$$3x-23 = A(x-1) + B(x+3)$$
$$3x-23 = Ax - A + Bx + 3B$$
$$3x-23 = (A+B)x + (-A+3B)$$

$$A + B = 3$$
$$-A + 3B = -23$$

$$A = 8, B = -5 \quad \Rightarrow \quad \frac{3x-23}{x^2+2x-3} = \frac{8}{x+3} + \frac{-5}{x-1} = \frac{8}{x+3} - \frac{5}{x-1}$$

11. Factor, and then find A and B such that:

$$\frac{9x - 31}{(2x - 3)(x - 5)} = \frac{A}{2x - 3} + \frac{B}{x - 5}$$

$$9x - 31 = A(x - 5) + B(2x - 3)$$
$$9x - 31 = Ax - 5A + 2Bx - 3B$$
$$9x - 31 = (A + 2B)x + (-5A - 3B)$$

$$A + 2B = 9$$
$$-5A - 3B = -31$$

$$A = 5, B = 2 \Rightarrow \frac{9x - 31}{2x^2 - 13x + 15} = \frac{5}{2x - 3} + \frac{2}{x - 5}$$

13. Find A, B and C such that:

$$\frac{5x^2 + 9x + 3}{x(x + 1)^2} = \frac{A}{x} + \frac{B}{x + 1} + \frac{C}{(x + 1)^2}$$

$$5x^2 + 9x + 3 = A(x + 1)^2 + Bx(x + 1) + Cx$$
$$5x^2 + 9x + 3 = A(x^2 + 2x + 1) + Bx^2 + Bx + Cx$$
$$5x^2 + 9x + 3 = Ax^2 + 2Ax + A + Bx^2 + Bx + Cx$$
$$5x^2 + 9x + 3 = (A + B)x^2 + (2A + B + C)x + A$$

$$\left.\begin{array}{lllll} A & + & B & & = 5 \\ 2A & + & B & + C & = 9 \\ A & & & & = 3 \end{array}\right\} \quad A = 3, B = 2, C = 1$$

$$\frac{5x^2 + 9x + 3}{x(x + 1)^2} = \frac{3}{x} + \frac{2}{x + 1} + \frac{1}{(x + 1)^2}$$

15. Find A, B and C such that:

$$\frac{-2x^2 + x - 2}{x^2(x - 1)} = \frac{A}{x} + \frac{B}{x^2} + \frac{C}{(x - 1)}$$

$$-2x^2 + x - 2 = Ax(x - 1) + B(x - 1) + Cx^2$$
$$-2x^2 + x - 2 = Ax^2 - Ax + Bx - B + Cx^2$$
$$-2x^2 + x - 2 = (A + C)x^2 + (-A + B)x - B$$

$$\left.\begin{array}{lllll} A & & & + C & = -2 \\ -A & + & B & & = 1 \\ & - & B & & = -2 \end{array}\right\} \quad A = 1, B = 2, C = -3$$

$$\frac{-2x^2 + x - 2}{x^2(x - 1)} = \frac{1}{x} + \frac{2}{x^2} + \frac{-3}{x - 1} = \frac{1}{x} + \frac{2}{x^2} - \frac{3}{x - 1}$$

17. Factor, and then find A, B and C such that:

$$\frac{4x^2 + 4x - 2}{x(x+1)(x-1)} = \frac{A}{x} + \frac{B}{x+1} + \frac{C}{x-1}$$

$$4x^2 + 4x - 2 = A(x+1)(x-1) + Bx(x-1) + Cx(x+1)$$
$$4x^2 + 4x - 2 = A(x^2 - 1) + Bx^2 - Bx + Cx^2 + Cx$$
$$4x^2 + 4x - 2 = Ax^2 - A + Bx^2 - Bx + Cx^2 + Cx$$
$$4x^2 + 4x - 2 = (A + B + C)x^2 + (-B + C)x - A$$

$$\left.\begin{array}{rcl} A + B + C &=& 4 \\ - B + C &=& 4 \\ -A &=& -2 \end{array}\right\} \quad A = 2, B = -1, C = 3$$

$$\frac{4x^2 + 4x - 2}{x(x^2 - 1)} = \frac{2}{x} + \frac{-1}{x+1} + \frac{3}{x-1} = \frac{2}{x} - \frac{1}{x+1} + \frac{3}{x-1}$$

19. Factor, and then find A, B and C such that:

$$\frac{3x^2 - 13x + 18}{x(x-3)^2} = \frac{A}{x} + \frac{B}{x-3} + \frac{C}{(x-3)^2}$$

$$3x^2 - 13x + 18 = A(x-3)^2 + Bx(x-3) + Cx$$
$$3x^2 - 13x + 18 = A(x^2 - 6x + 9) + Bx^2 - 3Bx + Cx$$
$$3x^2 - 13x + 18 = Ax^2 - 6Ax + 9A + Bx^2 - 3Bx + Cx$$
$$3x^2 - 13x + 18 = (A + B)x^2 + (-6A - 3B + C)x + 9A$$

$$\left.\begin{array}{rcl} A + B &=& 3 \\ -6A - 3B + C &=& -13 \\ 9A &=& 18 \end{array}\right\} \quad A = 2, B = 1, C = 2$$

$$\frac{3x^2 - 13x + 18}{x^3 - 6x^2 + 9x} = \frac{2}{x} + \frac{1}{x-3} + \frac{2}{(x-3)^2}$$

21. Find A, B and C such that:

$$\frac{x^2 - 2x - 3}{(x-1)^3} = \frac{A}{x-1} + \frac{B}{(x-1)^2} + \frac{C}{(x-1)^3}$$

$$x^2 - 2x - 3 = A(x-1)^2 + B(x-1) + C$$
$$x^2 - 2x - 3 = A(x^2 - 2x + 1) + Bx - B + C$$
$$x^2 - 2x - 3 = Ax^2 - 2Ax + A + Bx - B + C$$
$$x^2 - 2x - 3 = Ax^2 + (-2A + B)x + (A - B + C)$$

$$\left.\begin{array}{rcl} A &=& 1 \\ -2A + B &=& -2 \\ A - B + C &=& -3 \end{array}\right\} \quad A = 1, B = 0, C = -4$$

$$\frac{x^2 - 2x - 3}{(x-1)^3} = \frac{1}{x-1} + \frac{0}{(x-1)^2} + \frac{-4}{(x-1)^3} = \frac{1}{x-1} - \frac{4}{(x-1)^3}$$

23. Find A, B and C such that:

$$\frac{x^2+x+3}{x(x^2+3)} = \frac{A}{x} + \frac{Bx+C}{x^2+3}$$

$$x^2+x+3 = A(x^2+3)+(Bx+C)x$$
$$x^2+x+3 = Ax^2+3A+Bx^2+Cx$$
$$x^2+x+3 = (A+B)x^2+Cx+3A$$

$$\left.\begin{array}{rcl} A+B & = & 1 \\ C & = & 1 \\ 3A & = & 3 \end{array}\right\} \quad A=1, B=0, C=1$$

$$\frac{x^2+x+3}{x(x^2+3)} = \frac{1}{x} + \frac{0x+1}{x^2+3} = \frac{1}{x} + \frac{1}{x^2+3}$$

25. Factor, and then find A, B and C such that:

$$\frac{5x^2+2x+2}{x(x^2+1)} = \frac{A}{x} + \frac{Bx+C}{x^2+1}$$

$$5x^2+2x+2 = A(x^2+1)+(Bx+C)x$$
$$5x^2+2x+2 = Ax^2+A+Bx^2+Cx$$
$$5x^2+2x+2 = (A+B)x^2+Cx+A$$

$$\left.\begin{array}{rcl} A+B & = & 5 \\ C & = & 2 \\ A & = & 2 \end{array}\right\} \quad A=2, B=3, C=2$$

$$\frac{5x^2+2x+2}{x^3+x} = \frac{2}{x} + \frac{3x+2}{x^2+1}$$

27. Find A, B, C and D such that:

$$\frac{3x^3+5x^2+3x+1}{x^2(x^2+x+1)} = \frac{A}{x} + \frac{B}{x^2} + \frac{Cx+D}{x^2+x+1}$$

$$3x^3+5x^2+3x+1 = Ax(x^2+x+1)+B(x^2+x+1)+(Cx+D)x^2$$
$$3x^3+5x^2+3x+1 = Ax^3+Ax^2+Ax+Bx^2+Bx+B+Cx^3+Dx^2$$
$$3x^3+5x^2+3x+1 = (A+C)x^3+(A+B+D)x^2+(A+B)x+B$$

$$\left.\begin{array}{rcl} A+C & = & 3 \\ A+B+D & = & 5 \\ A+B & = & 3 \\ B & = & 1 \end{array}\right\} \quad A=2, B=1, C=1, D=2$$

$$\frac{3x^3+5x^2+3x+1}{x^2(x^2+x+1)} = \frac{2}{x} + \frac{1}{x^2} + \frac{x+2}{x^2+x+1}$$

29. Factor, and then find A, B, C and D such that:

$$\frac{2x^2 + 1}{x^2(x^2 + 1)} = \frac{A}{x} + \frac{B}{x^2} + \frac{Cx + D}{x^2 + 1}$$

$$2x^2 + 1 = Ax(x^2 + 1) + B(x^2 + 1) + (Cx + D)x^2$$
$$2x^2 + 1 = Ax^3 + Ax + Bx^2 + B + Cx^3 + Dx^2$$
$$2x^2 + 1 = (A + C)x^3 + (B + D)x^2 + Ax + B$$

$$\left.\begin{array}{rcl}
A \qquad\quad + C & = & 0 \\
\quad B \qquad\quad + D & = & 2 \\
A & = & 0 \\
\quad B & = & 1
\end{array}\right\} \quad A = 0, B = 1, C = 0, \ D = 1$$

$$\frac{2x^2 + 1}{x^4 + x^2} = \frac{0}{x} + \frac{1}{x^2} + \frac{0x + 1}{x^2 + 1} = \frac{1}{x^2} + \frac{1}{x^2 + 1}$$

31. Factor, and then find A, B and C such that:

$$\frac{-x^2 - 3x - 5}{(x + 1)(x^2 + 2)} = \frac{A}{x + 1} + \frac{Bx + C}{x^2 + 2}$$

$$-x^2 - 3x - 5 = A(x^2 + 2) + (Bx + C)(x + 1)$$
$$-x^2 - 3x - 5 = Ax^2 + 2A + Bx^2 + Bx + Cx + C$$
$$-x^2 - 3x - 5 = (A + B)x^2 + (B + C)x + (2A + C)$$

$$\left.\begin{array}{rcl}
A \ + B & = & -1 \\
\quad B \ + C & = & -3 \\
2A \qquad + C & = & -5
\end{array}\right\} \quad A = -1, B = 0, C = -3$$

$$\frac{-2x^2 - 3x - 5}{x^3 + x^2 + 2x + 2} = \frac{-1}{x + 1} + \frac{0x - 3}{x^2 + 2} = -\frac{1}{x + 1} - \frac{3}{x^2 + 2}$$

33. Find A, B, C and D such that:

$$\frac{x^3 + 4x^2 + 3x + 6}{(x^2 + 2)(x^2 + x + 2)} = \frac{Ax + B}{x^2 + 2} + \frac{Cx + D}{x^2 + x + 2}$$

$$x^3 + 4x^2 + 3x + 6 = (Ax + B)(x^2 + x + 2) + (Cx + D)(x^2 + 2)$$
$$x^3 + 4x^2 + 3x + 6 = Ax^3 + Ax^2 + 2Ax + Bx^2 + Bx + 2B + Cx^3 + 2Cx + Dx^2 + 2D$$
$$x^3 + 4x^2 + 3x + 6 = (A + C)x^3 + (A + B + D)x^2 + (2A + B + 2C)x + (2B + 2D)$$

$$\left.\begin{array}{rcl}
A \qquad\quad + C & = & 1 \\
A \ + B \qquad\quad + D & = & 4 \\
2A \ + B \ + 2C & = & 3 \\
\quad 2B \qquad\quad + 2D & = & 6
\end{array}\right\} \quad A = 1, B = 1, C = 0, \ D = 2$$

$$\frac{x^3 + 4x^2 + 3x + 6}{(x^2 + 2)(x^2 + x + 2)} = \frac{1x + 1}{x^2 + 2} + \frac{0x + 2}{x^2 + x + 2} = \frac{x + 1}{x^2 + 2} + \frac{2}{x^2 + x + 2}$$

35. Find A, B, C, D and E such that:

$$\frac{2x^4 + 6x^3 + 20x^2 + 22x + 25}{x(x^2 + 2x + 5)^2} = \frac{A}{x} + \frac{Bx + C}{x^2 + 2x + 5} + \frac{Dx + E}{(x^2 + 2x + 5)^2}$$

$$
\begin{aligned}
2x^4 + 6x^3 + 20x^2 + 22x + 25 \\
&= A(x^2 + 2x + 5)^2 + (Bx + C)(x)(x^2 + 2x + 5) + (Dx + E)x \\
&= Ax^4 + 4Ax^3 + 14Ax^2 + 20Ax + 25A + Bx^4 + 2Bx^3 + 5Bx^2 + Cx^3 \\
&\quad + 2Cx^2 + 5Cx + Dx^2 + Ex \\
&= (A + B)x^4 + (4A + 2B + C)x^3 + (14A + 5B + 2C + D)x^2 \\
&\quad + (20A + 5C + E)x + 25A
\end{aligned}
$$

$$
\begin{array}{llllll}
A & + B & & & & = 2 \\
4A & + 2B & + C & & & = 6 \\
14A & + 5B & + 2C & + D & & = 20 \\
20A & & + 5C & & + E & = 22 \\
25A & & & & & = 25
\end{array}
$$

$A = 1$, $B = 1$, $C = 0$, $D = 1$, $E = 2$

$$\frac{2x^4 + 6x^3 + 20x^2 + 22x + 25}{x(x^2 + 2x + 5)^2} = \frac{1}{x} + \frac{1x + 0}{x^2 + 2x + 5} + \frac{1x + 2}{(x^2 + 2x + 5)^2}$$

37. First, simplify using long division: $\quad \dfrac{x^3}{x^2 + 3x + 2} = x - 3 + \dfrac{7x + 6}{x^2 + 3x + 2}$

Factor, and find A and B so that:

$$\frac{7x + 6}{(x + 2)(x + 1)} = \frac{A}{x + 2} + \frac{B}{x + 1}$$

$$
\begin{aligned}
7x + 6 &= A(x + 1) + B(x + 2) \\
7x + 6 &= Ax + A + Bx + 2B \\
7x + 6 &= (A + B)x + (A + 2B)
\end{aligned}
$$

$$A + B = 7$$
$$A + 2B = 6$$

$A = 8, B = -1 \implies \dfrac{x^3}{x^2 + 3x + 2} = x - 3 + \dfrac{8}{x + 2} - \dfrac{1}{x + 1}$

39. First, simplify using long division: $\dfrac{3x^3 + 3x^2 + 6x + 4}{3x^3 + x^2 + 3x + 1} = 1 + \dfrac{2x^2 + 3x + 3}{3x^3 + x^2 + 3x + 1}$

Factor, and find A, B and C so that:

$$\frac{2x^2 + 3x + 3}{(3x+1)(x^2+1)} = \frac{A}{3x+1} + \frac{Bx+C}{x^2+1}$$

$$2x^2 + 3x + 3 = A(x^2+1) + (Bx+C)(3x+1)$$
$$2x^2 + 3x + 3 = Ax^2 + A + 3Bx^2 + Bx + 3Cx + C$$
$$2x^2 + 3x + 3 = (A+3B)x^2 + (B+3C)x + (A+C)$$

$$\left.\begin{array}{lllll} A & + & 3B & & = 2 \\ & & B & + 3C & = 3 \\ A & & & + C & = 3 \end{array}\right\} \quad A = 2, B = 0, C = 1$$

$$\frac{3x^3 + 3x^2 + 6x + 4}{3x^3 + x^2 + 3x + 1} = 1 + \frac{2}{3x+1} + \frac{1}{x^2+1}$$

41. First, simplify using long division: $\dfrac{x^3 + 3x^2 + 2x + 1}{x^3 + x^2 + x} = 1 + \dfrac{2x^2 + x + 1}{x^3 + x^2 + x}$

Factor, and find

$$\frac{2x^2 + x + 1}{x(x^2 + x + 1)} = \frac{A}{x} + \frac{Bx + C}{x^2 + x + 1}$$

$$2x^2 + x + 1 = A(x^2 + x + 1) + (Bx + C)x$$
$$2x^2 + x + 1 = Ax^2 + Ax + A + Bx^2 + Cx$$
$$2x^2 + x + 1 = (A+B)x^2 + (A+C)x + A$$

$$\left.\begin{array}{lllll} A & + & B & & = 2 \\ A & & & + C & = 1 \\ A & & & & = 1 \end{array}\right\} \quad A = 1, B = 1, C = 0$$

$$\frac{x^3 + 3x^2 + 2x + 1}{x^3 + x^2 + x} = 1 + \frac{1}{x} + \frac{x}{x^2 + x + 1}$$

43. Multiply the denominator and then perform long division to simplify:

$$\frac{2x^4 + 2x^3 + 3x^2 - 1}{(x^2 - x)(x^2 + 1)} = \frac{2x^4 + 2x^3 + 3x^2 - 1}{x^4 - x^3 + x^2 - x} = 2 + \frac{4x^3 + x^2 + 2x - 1}{(x^2 - x)(x^2 + 1)}$$

Then factor, and find A, B, C and D so that:

$$\frac{4x^3 + x^2 + 2x - 1}{x(x - 1)(x^2 + 1)} = \frac{A}{x} + \frac{B}{x - 1} + \frac{Cx + D}{x^2 + 1}$$

$$4x^3 + x^2 + 2x - 1 = A(x - 1)(x^2 + 1) + Bx(x^2 + 1) + (Cx + D)(x)(x - 1)$$
$$= Ax^3 - Ax^2 + Ax - A + Bx^3 + Bx + Cx^3 - Cx^2 + Dx^2 - Dx$$
$$= (A + B + C)x^3 + (-A - C + D)x^2 + (A + B - D)x + (-A)$$

$$\left.\begin{array}{rcl}
A \quad + B \quad + C \qquad\qquad &=& 4 \\
-A \qquad\qquad - C \quad + D &=& 1 \\
A \quad + B \qquad\quad - D &=& 2 \\
-A \qquad\qquad\qquad\qquad &=& -1
\end{array}\right\} \quad A = 1, B = 3, C = 0, D = 2$$

$$\frac{2x^4 + 2x^3 + 3x^2 - 1}{(x^2 - x)(x^2 + 1)} = 2 + \frac{1}{x} + \frac{3}{x - 1} + \frac{2}{x^2 + 1}$$

Exercise 7.7   (page 402)

1. $2x + 3y < 12$

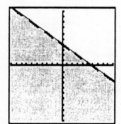

3. $x < 3$

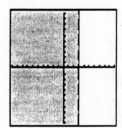

5. $4x - y > 4$

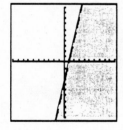

7. $y > 2x$

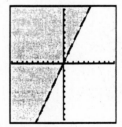

9. $y \leq \frac{1}{2}x + 1$

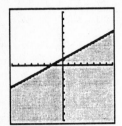

11. $2y \geq 3x - 2$

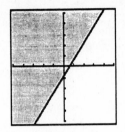

13. $y < 3,\ x \geq 2$

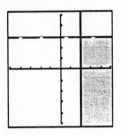

15. $y \geq 1,\ x < 2$

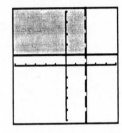

17. $y \leq x - 2,\ y \geq 2x + 1$

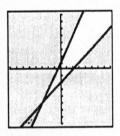

19. $x + y < 2,\ x + y \leq 1$

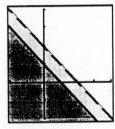

21. $x + 2y < 3,\ 2x + 4y < 8$

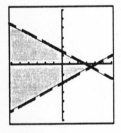

23. $3x - 3y \geq 6,\ 3x + 2y < 6$

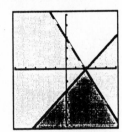

25. $2x - y \leq 0$
$x + 2y \leq 10,\ y \geq 0$

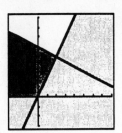

27. $x - 2y \geq 0$
$x - y \leq 2,\ x \geq 0$

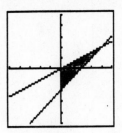

29. $x + y \leq 4$
$x - y \leq 4$
$x \geq 0,\ y \geq 0$

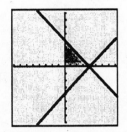

31. $3x - 2y \leq 6$
$x + 2y \leq 10$
$x \geq 0,\ y \geq 0$

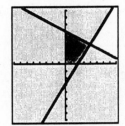

Exercise 7.8   (page 410)

1.

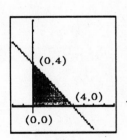

point	$P = 2x + 3y$
$(0, 0)$	$= 0$
$(0, 4)$	$= 12$ ⟸ MAXIMUM
$(4, 0)$	$= 8$

3.

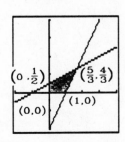

point	$P = y + \frac{1}{2}x$
$(0, 0)$	$= 0$
$\left(0, \frac{1}{2}\right)$	$= \frac{1}{2}$
$\left(\frac{5}{3}, \frac{4}{3}\right)$	$= \frac{13}{6}$ ⟸ MAXIMUM
$(1, 0)$	$= \frac{1}{2}$

5.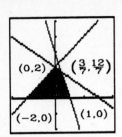

point	$P = 2x + y$
$(-2, 0)$	$= -4$
$(1, 0)$	$= 2$
$(0, 2)$	$= 2$
$\left(\dfrac{3}{7}, \dfrac{12}{7}\right)$	$= \dfrac{18}{7}$ ⟵ MAXIMUM

7.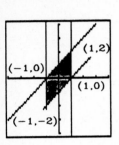

point	$P = 3x - 2y$
$(-1, 0)$	$= -3$
$(1, 2)$	$= -1$
$(1, 0)$	$= 3$ ⟵ MAXIMUM
$(-1, -2)$	$= 1$

9.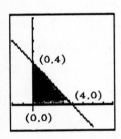

point	$P = 5x + 12y$
$(0, 0)$	$= 0$ ⟵ MINIMUM
$(0, 4)$	$= 12$
$(4, 0)$	$= 8$

11.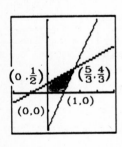

point	$P = y + \dfrac{1}{2}x$
$(0, 0)$	$= 0$ ⟵ MINIMUM
$\left(0, \dfrac{1}{2}\right)$	$= \dfrac{1}{2}$
$\left(\dfrac{5}{3}, \dfrac{4}{3}\right)$	$= \dfrac{13}{6}$
$(1, 0)$	$= \dfrac{1}{2}$

13.

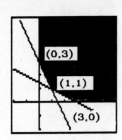

point	$P = 6x + 2y$
(3, 0)	= 18
(1, 1)	= 8
(0, 3)	= 6 ⇐ MINIMUM

15.

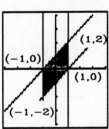

point	$P = 2x - 2y$
(−1, 0)	= −2 ⇐ MINIMUM
(1, 2)	= −2 ⇐ MINIMUM
(1, 0)	= 2
(−1, −2)	= 2

17.

	chair	table	available
Tom's time	3	2	42
Carlos' time	2	6	42
profit ($)	80	100	
# made	x	y	

Maximize $P = 80x + 100y$ with:
$$3x + 2y \le 42$$
$$2x + 6y \le 42$$
$$x \ge 0, \, y \ge 0$$

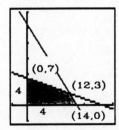

point	$P = 80x + 100y$
(0, 0)	= 0
(0, 7)	= 700
(12, 3)	= 1260 ⇐ max
(14, 0)	= 1120

They should make 12 chairs and 3 tables, with a profit of $1260.

19. Let $x =$ the number of IBM compatible computers and $y =$ the number of Macintosh computers.

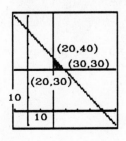

Maximize $C = 50x + 40y$ with:
$$x + y \le 60$$
$$x \ge 20, \, x \le 30$$
$$y \ge 30, \, y \le 50$$

point	$C = 50x + 40y$
(20, 30)	= 2200
(20, 40)	= 2600
(30, 30)	= 2700 ⇐ max

She should stock 30 of each type, with her commission equaling $2700.

21.

	A	B	need
protein	2	8	360
carb.	20	40	2000
% fat	10	30	
amt. used	x	y	

Minimize $F = 0.10x + 0.30y$ with
$$2x + 8y \geq 360$$
$$20x + 40y \geq 2000$$
$$x \geq 0, y \geq 0$$

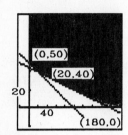

point	$F = 0.10x + 0.30y$	
(0, 50)	= 15	
(20, 40)	= 14	$\Leftarrow$ min
(180, 0)	= 18	

Batches should be made with 20 oz of A's hamburger and 40 oz of B's hamburger. Each batch will contain 14 oz of fat.

23.

	chew	bubble	available
A time	2	4	168
B time	8	2	168
profit	200	160	
cases	x	y	

Maximize $P = 200x + 160y$ with
$$2x + 4y \leq 168$$
$$8x + 2y \leq 168$$
$$x \geq 0, y \geq 0$$

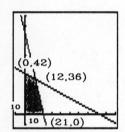

point	$P = 200x + 160y$	
(0, 42)	= 6720	
(12, 36)	= 8160	$\Leftarrow$ max
(21, 0)	= 4200	

Make 12 cases of chewing gum and 36 cases of bubble gum, for a profit of $8160.

25. Let x = amount in stocks and y = amount in bonds.

Maximize $I = 0.09x + 0.07y$ with
$$x + y \leq 200{,}000$$
$$x \geq 100{,}000, \ y \geq 50{,}000$$

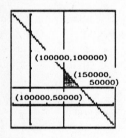

point	$I = 0.09x + 0.07y$	
(100,000, 50,000)	= 12,500	
(100,000, 100,000)	= 16,000	
(150,000, 50,000)	= 17,000	$\Leftarrow$ max

She should invest $150,000 in stocks and $50,000 in bonds, for interest of $17,000.

27.

	bus	truck	need
students	40	10	100
instruments	3	6	18
cost	350	200	
# used	x	y	

Minimize $C = 350x + 200y$ with
$$40x + 10y \geq 100$$
$$3x + 6y \geq 18$$
$$x \geq 0, y \geq 0$$

point	$C = 350x + 200y$	
(0, 10)	= 2000	
(2, 2)	= 1100	$\Leftarrow$ min
(0, 6)	= 1200	

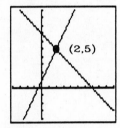

The band should use 2 buses and 2 trucks, at a cost of $1100.

Chapter 7 Review Exercises (page 413)

1. $2x - y = -1$
$x + y = 7$

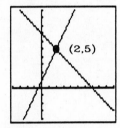

solution: $(2, 5)$

3. $y = 5x + 7$
$x = y - 7$

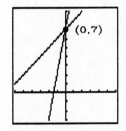

solution: $(0, 7)$

5. $y = 3x + 2$
$y = 5x$

$$3x + 2 = 5x$$
$$2 = 2x$$
$$x = 1$$

$y = 5x = 5(1) = 5$
solution: $(1, 5)$

7. $2x + y = -3 \Rightarrow y = -2x - 3$
$x - y = 3$

$$x - (-2x - 3) = 3$$
$$3x = 0$$
$$x = 0$$

$y = -2x - 3 = -2(0) - 3 = -3$
solution: $(0, -3)$

9.

$$x + 5y = 7$$
$$3x + y = -7$$

$$-3x - 15y = -21$$
$$\underline{3x + y = -7}$$
$$-14y = -28$$
$$y = 2 \Rightarrow x = -3, \quad (-3, 2)$$

11. $2(x + y) - x = 0$
$$3(x + y) + 2y = 1$$

$$x + 2y = 0$$
$$3x + 5y = 1$$

$$-3x - 6y = 0$$
$$\underline{3x + 5y = 1}$$
$$-y = 1$$
$$y = -1 \Rightarrow x = 2, \quad (2, -1)$$

13.

$$3x + 2y - z = 2 \quad \textbf{(1)}$$
$$x + y - z = 0 \quad \textbf{(2)}$$
$$2x + 3y - z = 1 \quad \textbf{(3)}$$

Add $-$ **(1)** and **(2)**:
$$-3x - 2y + z = -2$$
$$\underline{x + y - z = 0}$$
$$-2x - y = -2 \quad \textbf{(4)}$$

Add $-$ **(1)** and **(3)**:
$$-3x - 2y + z = -2$$
$$\underline{2x + 3y - z = 1}$$
$$-x + y = -1 \quad \textbf{(5)}$$

Add **(4)** and **(5)**:

$$-2x - y = -2$$
$$\underline{-x + y = -1}$$
$$-3x = -3 \Rightarrow x = 1$$

Substitute $x = 1$ into
(4) and solve for y:
$$-2x - y = -2$$
$$-2(1) - y = -2$$
$$y = 0$$

Substitute $x = 1$ and $y = 0$
into **(2)** and solve for z:
$$x + y - z = 0$$
$$1 + 0 - z = 0$$
$$z = 1$$

The solution is $x = 1$, $y = 0$ and $z = 1$.

15.

$$2x - y + z = 1 \quad (1)$$
$$x - y + 2z = 3 \quad (2)$$
$$x - y + z = 1 \quad (3)$$

Add $-(1)$ and (2):

$$-2x + y - z = -1$$
$$\underline{x - y + 2z = 3}$$

$$-x \quad\quad + z = 2 \quad (4)$$

Add $-(1)$ and (3):

$$-2x + y - z = -1$$
$$\underline{x - y + z = 1}$$

$$-x \quad\quad\quad = 0 \quad (5)$$

Substitute $x = 0$ into
(4) and solve for z:

$$-x + z = 2$$
$$-(0) + z = 2$$
$$z = 2$$

Substitute $x = 0$ and $z = 2$
into (2) and solve for y:

$$x - y + 2z = 3$$
$$0 - y + 2(2) = 3$$
$$y = 1$$

The solution is $x = 0$, $y = 1$ and $z = 2$.

17. Let $x =$ the cost of a fur coat and $y =$ the cost of a leather coat.

$$25x + 15y = 9300 \quad \Rightarrow \quad -50x - 30y = -18{,}600$$
$$10x + 30y = 12{,}600 \quad \Rightarrow \quad \underline{10x + 30y = 12{,}600}$$

$$-40x = -6000 \Rightarrow x = 150, \ y = 370$$

He pays $20 \cdot 150 + 20 \cdot 370 = \$10{,}400$ for 20 of each.

19.

$$-3/2\ R_2 + R_1 \Rightarrow R_2 \qquad (1/2)\ R_1 \Rightarrow R_1, \ -(2/17)\ R_2 \Rightarrow R_2$$

$$\begin{bmatrix} 2 & 5 & | & 7 \\ 3 & -1 & | & 2 \end{bmatrix} \Rightarrow \begin{bmatrix} 2 & 5 & | & 7 \\ 0 & -17/2 & | & -17/2 \end{bmatrix} \Rightarrow \begin{bmatrix} 1 & 5/2 & | & 7/2 \\ 0 & 1 & | & 1 \end{bmatrix}$$

From R_2 of the last matrix, $y = 1$. Substitute into R_1:

$$x + (5/2)y = 7/2$$
$$x + (5/2) = 7/2$$
$$x = 1$$

$$\boxed{x = 1, \ y = 1}$$

21.

$$-2\ R_1 + R_2 \Rightarrow R_2 \qquad\qquad -(1/7)\ R_3 \Rightarrow R_2$$
$$-3\ R_1 + R_3 \Rightarrow R_3 \qquad\qquad R_2 \Rightarrow R_3$$

$$\begin{bmatrix} 1 & 3 & 1 & | & 3 \\ 2 & -1 & 1 & | & -11 \\ 3 & 2 & 3 & | & 2 \end{bmatrix} \Rightarrow \begin{bmatrix} 1 & 3 & 1 & | & 3 \\ 0 & -7 & -1 & | & -17 \\ 0 & -7 & 0 & | & -7 \end{bmatrix} \Rightarrow \begin{bmatrix} 1 & 3 & 1 & | & 3 \\ 0 & 1 & 0 & | & 1 \\ 0 & -7 & -1 & | & -17 \end{bmatrix}$$

$$-3\ R_2 + R_1 \Rightarrow R_1 \qquad\qquad R_3 + R_1 \Rightarrow R_1$$
$$7\ R_2 + R_3 \Rightarrow R_3 \qquad\qquad -R_3 \Rightarrow R_3$$

$$\begin{bmatrix} 1 & 0 & 1 & | & 0 \\ 0 & 1 & 0 & | & 1 \\ 0 & 0 & -1 & | & -10 \end{bmatrix} \Rightarrow \begin{bmatrix} 1 & 0 & 0 & | & -10 \\ 0 & 1 & 0 & | & 1 \\ 0 & 0 & 1 & | & 10 \end{bmatrix} \qquad \boxed{x = -10, \ y = 1, \ z = 10}$$

23.
$$-R_1 + R_3 \Rightarrow R_3, \quad -R_1 + R_4 \Rightarrow R_4$$

$$\begin{bmatrix} 1 & 1 & -3 & 1 & | & 2 \\ 0 & 1 & -1 & 2 & | & 0 \\ 1 & 0 & 1 & -1 & | & 2 \\ 1 & -1 & 0 & 2 & | & -3 \end{bmatrix} \Rightarrow \begin{bmatrix} 1 & 1 & -3 & 1 & | & 2 \\ 0 & 1 & -1 & 2 & | & 0 \\ 0 & -1 & 4 & -2 & | & 0 \\ 0 & -2 & 3 & 1 & | & -5 \end{bmatrix}$$

$$-R_2 + R_1 \Rightarrow R_1, \quad R_2 + R_3 \Rightarrow R_3 \qquad R_3 \Leftrightarrow R_4$$
$$2R_2 + R_4 \Rightarrow R_4$$

$$\begin{bmatrix} 1 & 0 & -2 & -1 & | & 2 \\ 0 & 1 & -1 & 2 & | & 0 \\ 0 & 0 & 3 & 0 & | & 0 \\ 0 & 0 & 1 & 5 & | & -5 \end{bmatrix} \Rightarrow \begin{bmatrix} 1 & 0 & -2 & -1 & | & 2 \\ 0 & 1 & -1 & 2 & | & 0 \\ 0 & 0 & 1 & 5 & | & -5 \\ 0 & 0 & 3 & 0 & | & 0 \end{bmatrix}$$

$$2R_3 + R_1 \Rightarrow R_1 \qquad\qquad (9/15)R_4 + R_1 \Rightarrow R_1$$
$$R_3 + R_2 \Rightarrow R_2 \qquad\qquad (7/15)R_4 + R_2 \Rightarrow R_2$$
$$-3R_3 + R_4 \Rightarrow R_4 \qquad\quad (5/15)R_4 + R_3 \Rightarrow R_3, \quad -(1/15)R_4 \Rightarrow R_4$$

$$\begin{bmatrix} 1 & 0 & 0 & 9 & | & -8 \\ 0 & 1 & 0 & 7 & | & -5 \\ 0 & 0 & 1 & 5 & | & -5 \\ 0 & 0 & 0 & -15 & | & 15 \end{bmatrix} \Rightarrow \begin{bmatrix} 1 & 0 & 0 & 0 & | & 1 \\ 0 & 1 & 0 & 0 & | & 2 \\ 0 & 0 & 1 & 0 & | & 0 \\ 0 & 0 & 0 & 1 & | & -1 \end{bmatrix}$$

$$\boxed{w = 1, \; x = 2, \; y = 0, \; z = -1}$$

25. $\begin{bmatrix} 3 & 2 & 1 \\ 3 & 2 & 1 \end{bmatrix} + \begin{bmatrix} -2 & 1 & 3 \\ 1 & -2 & 1 \end{bmatrix} = \begin{bmatrix} 1 & 3 & 4 \\ 4 & 0 & 2 \end{bmatrix}$

27. $\begin{bmatrix} 2 & 3 \\ -1 & 2 \end{bmatrix}\begin{bmatrix} 1 & -2 \\ -3 & 1 \end{bmatrix} = \begin{bmatrix} -7 & -1 \\ -7 & 4 \end{bmatrix}$

29. $\begin{bmatrix} 1 & -3 & 2 \end{bmatrix} \begin{bmatrix} 2 \\ 1 \\ 3 \end{bmatrix} = [5]$

31. $\begin{bmatrix} 1 \\ 2 \\ 1 \\ 5 \end{bmatrix} \begin{bmatrix} 2 & -1 & 1 & 3 \end{bmatrix} = \begin{bmatrix} 2 & -1 & 1 & 3 \\ 4 & -2 & 2 & 6 \\ 2 & -1 & 1 & 3 \\ 10 & -5 & 5 & 15 \end{bmatrix}$

33. $\begin{bmatrix} 1 & -3 & 2 \end{bmatrix} \begin{bmatrix} 2 \\ 1 \\ -5 \end{bmatrix} + \begin{bmatrix} 1 & -3 \end{bmatrix} \begin{bmatrix} 2 \\ 5 \end{bmatrix} = [-11] + [-13] = [-24]$

35. $\begin{bmatrix} 1 & 3 & | & 1 & 0 \\ -3 & 5 & | & 0 & 1 \end{bmatrix} \Rightarrow \begin{bmatrix} 1 & 3 & | & 1 & 0 \\ 0 & 14 & | & 3 & 1 \end{bmatrix} \Rightarrow \begin{bmatrix} 1 & 3 & | & 1 & 0 \\ 0 & 1 & | & \frac{3}{14} & \frac{1}{14} \end{bmatrix}$

$\begin{bmatrix} 1 & 0 & | & \frac{5}{14} & -\frac{3}{14} \\ 0 & 1 & | & \frac{3}{14} & \frac{1}{14} \end{bmatrix}$ INVERSE: $\begin{bmatrix} \frac{5}{14} & -\frac{3}{14} \\ \frac{3}{14} & \frac{1}{14} \end{bmatrix}$

37. $\begin{bmatrix} 1 & 3 & -5 & | & 1 & 0 & 0 \\ 1 & 4 & 4 & | & 0 & 1 & 0 \\ 0 & 0 & 1 & | & 0 & 0 & 1 \end{bmatrix} \Rightarrow \begin{bmatrix} 1 & 3 & -5 & | & 1 & 0 & 0 \\ 0 & 1 & 9 & | & -1 & 1 & 0 \\ 0 & 0 & 1 & | & 0 & 0 & 1 \end{bmatrix}$

$\begin{bmatrix} 1 & 0 & -32 & | & 4 & -3 & 0 \\ 0 & 1 & 9 & | & -1 & 1 & 0 \\ 0 & 0 & 1 & | & 0 & 0 & 1 \end{bmatrix} \Rightarrow \begin{bmatrix} 1 & 0 & 0 & | & 4 & -3 & 32 \\ 0 & 1 & 0 & | & -1 & 1 & -9 \\ 0 & 0 & 1 & | & 0 & 0 & 1 \end{bmatrix}$

INVERSE: $\begin{bmatrix} 4 & -3 & 32 \\ -1 & 1 & -9 \\ 0 & 0 & 1 \end{bmatrix}$

$$(-3 R_3 + R_2 \Rightarrow R_2)$$

39. $\begin{bmatrix} 1 & 0 & 8 & | & 1 & 0 & 0 \\ 3 & 7 & 6 & | & 0 & 1 & 0 \\ 1 & 2 & 3 & | & 0 & 0 & 1 \end{bmatrix} \Rightarrow \begin{bmatrix} 1 & 0 & 8 & | & 1 & 0 & 0 \\ 0 & 1 & -3 & | & 0 & 1 & -3 \\ 0 & 2 & -5 & | & -1 & 0 & 1 \end{bmatrix}$

$\begin{bmatrix} 1 & 0 & 8 & | & 1 & 0 & 0 \\ 0 & 1 & -3 & | & 0 & 1 & -3 \\ 0 & 0 & 1 & | & -1 & -2 & 7 \end{bmatrix} \Rightarrow \begin{bmatrix} 1 & 0 & 0 & | & 9 & 16 & -56 \\ 0 & 1 & 0 & | & -3 & -5 & 18 \\ 0 & 0 & 1 & | & -1 & -2 & 7 \end{bmatrix}$

INVERSE: $\begin{bmatrix} 9 & 16 & -56 \\ -3 & -5 & 18 \\ -1 & -2 & 7 \end{bmatrix}$

41.
$$\left[\begin{array}{ccc|ccc} -1 & 1 & 0 & 1 & 0 & 0 \\ -2 & 1 & 0 & 0 & 1 & 0 \\ 3 & -1 & -1 & 0 & 0 & 1 \end{array}\right] \Rightarrow \left[\begin{array}{ccc|ccc} 1 & -1 & 0 & -1 & 0 & 0 \\ 0 & -1 & 0 & -2 & 1 & 0 \\ 0 & 2 & -1 & 3 & 0 & 1 \end{array}\right]$$

$$\left[\begin{array}{ccc|ccc} 1 & 0 & 0 & 1 & -1 & 0 \\ 0 & 1 & 0 & 2 & -1 & 0 \\ 0 & 0 & -1 & -1 & 2 & 1 \end{array}\right] \Rightarrow \left[\begin{array}{ccc|ccc} 1 & 0 & 0 & 1 & -1 & 0 \\ 0 & 1 & 0 & 2 & -1 & 0 \\ 0 & 0 & 1 & 1 & -2 & -1 \end{array}\right]$$

INVERSE: $\left[\begin{array}{ccc} 1 & -1 & 0 \\ 2 & -1 & 0 \\ 1 & -2 & -1 \end{array}\right]$

43.
$$\left[\begin{array}{ccc} 4 & -1 & 2 \\ 1 & 1 & 2 \\ 1 & 0 & 1 \end{array}\right]\left[\begin{array}{c} x \\ y \\ z \end{array}\right] = \left[\begin{array}{c} 0 \\ 1 \\ 0 \end{array}\right]$$

$$\left[\begin{array}{ccc} 1 & 1 & -4 \\ 1 & 2 & -6 \\ -1 & -1 & 5 \end{array}\right]\left[\begin{array}{ccc} 4 & -1 & 2 \\ 1 & 1 & 2 \\ 1 & 0 & 1 \end{array}\right]\left[\begin{array}{c} x \\ y \\ z \end{array}\right] = \left[\begin{array}{ccc} 1 & 1 & -4 \\ 1 & 2 & -6 \\ -1 & -1 & 5 \end{array}\right]\left[\begin{array}{c} 0 \\ 1 \\ 0 \end{array}\right]$$

$$\left[\begin{array}{ccc} 1 & 0 & 0 \\ 0 & 1 & 0 \\ 0 & 0 & 1 \end{array}\right]\left[\begin{array}{c} x \\ y \\ z \end{array}\right] = \left[\begin{array}{c} 1 \\ 2 \\ -1 \end{array}\right] \quad x = 1, \, y = 2, \, z = -1$$

45. $\begin{vmatrix} 3 & -2 \\ 1 & -3 \end{vmatrix} = 3(-3) - (-2)(1) = -9 + 2 = -7$

47. $\begin{vmatrix} 1 & 3 & -1 \\ 1 & 2 & 1 \\ 1 & 0 & 2 \end{vmatrix} = 1\begin{vmatrix} 2 & 1 \\ 0 & 2 \end{vmatrix} - 3\begin{vmatrix} 1 & 1 \\ 1 & 2 \end{vmatrix} + (-1)\begin{vmatrix} 1 & 2 \\ 1 & 0 \end{vmatrix}$

$$= 1(4) - 3(1) - 1(-2) = 3$$

49. $x = \dfrac{\begin{vmatrix} -5 & 3 \\ -4 & 1 \end{vmatrix}}{\begin{vmatrix} 1 & 3 \\ -2 & 1 \end{vmatrix}} = \dfrac{7}{7} = 1 \qquad\qquad y = \dfrac{\begin{vmatrix} 1 & -5 \\ -2 & -4 \end{vmatrix}}{\begin{vmatrix} 1 & 3 \\ -2 & 1 \end{vmatrix}} = \dfrac{-14}{7} = -2$

51.

$$x = \frac{\begin{vmatrix} 7 & -3 & 1 \\ -9 & 1 & -3 \\ 3 & 1 & 1 \end{vmatrix}}{\begin{vmatrix} 1 & -3 & 1 \\ 1 & 1 & -3 \\ 1 & 1 & 1 \end{vmatrix}} = \frac{16}{16} = 1 \qquad y = \frac{\begin{vmatrix} 1 & 7 & 1 \\ 1 & -9 & -3 \\ 1 & 3 & 1 \end{vmatrix}}{\begin{vmatrix} 1 & -3 & 1 \\ 1 & 1 & -3 \\ 1 & 1 & 1 \end{vmatrix}} = \frac{-16}{16} = -1$$

$$z = \frac{\begin{vmatrix} 1 & -3 & 7 \\ 1 & 1 & -9 \\ 1 & 1 & 3 \end{vmatrix}}{\begin{vmatrix} 1 & -3 & 1 \\ 1 & 1 & -3 \\ 1 & 1 & 1 \end{vmatrix}} = \frac{48}{16} = 3$$

53. Factor, and then find A and B such that:

$$\frac{7x+3}{x(x+1)} = \frac{A}{x} + \frac{B}{x+1}$$

$$7x + 3 = A(x+1) + Bx$$
$$7x + 3 = Ax + A + Bx$$
$$7x + 3 = (A+B)x + A$$

$$A + B = 7$$
$$A \quad\;\; = 3$$

$$A = 3, B = 4 \;\Rightarrow\; \frac{7x+3}{x^2+x} = \frac{3}{x} + \frac{4}{x+1}$$

55. Factor, and then find A, B and C such that:

$$\frac{x^2+5}{x(x^2+x+5)} = \frac{A}{x} + \frac{Bx+C}{x^2+x+5}$$

$$x^2 + 5 = A(x^2+x+5) + (Bx+C)x$$
$$x^2 + 5 = Ax^2 + Ax + 5A + Bx^2 + Cx$$
$$x^2 + 5 = (A+B)x^2 + (A+C)x + 5A$$

$$\left.\begin{aligned} A \;+\; B \qquad\qquad &= 1 \\ A \qquad\quad +\; C &= 0 \\ 5A \qquad\qquad\quad &= 5 \end{aligned}\right\} \quad A = 1, B = 0, C = -1$$

$$\frac{x^2+5}{x^3+x^2+5x} = \frac{1}{x} + \frac{0x-1}{x^2+x+5} = \frac{1}{x} - \frac{1}{x^2+x+5}$$

57.

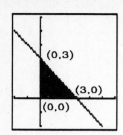

point	$P = 2x + y$	
(0, 0)	= 0	
(0, 3)	= 3	
(3, 0)	= 6	⟸ MAX

59.

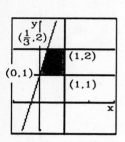

point	$P = 3x - y$	
(0, 1)	= -1	
(1, 1)	= 2	⟸ MAX
(1, 2)	= 1	
(1/3, 2)	= -1	

61. Maximize $P = 6x + 5y$ with
$$6x + 10y \leq 20000$$
$$8x + 6y \leq 16400$$
$$6x + 4y \leq 12000$$
$$x \geq 0, \ y \geq 0$$

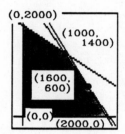

point	$P = 6x + 5y$
(0, 2000)	= 10,000
(1000, 1400)	= 13,000
(1600, 600)	= 12,600
(2000, 0)	= 12,000

The company should use 1000 bags of x and 1400 bags of y, for a profit of $13,000.

Chapter 7 Test (page 416)

1. $x - 3y = -5$
$2x - y = 0$
solution: (1, 2)

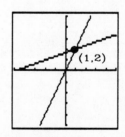

3.
$$3x + y = 0$$
$$3x - 5y = 18$$

$$3x + y = 0$$
$$-3x + 5y = -18$$

$$6y = -18$$

$$y = -3, \quad x = 1$$

5. Let $x =$ liters of 20% solution and $y =$ liters of 45% solution.

$$x + y = 10$$
$$0.20x + 0.45y = 0.30(10)$$

$$-2x - 2y = -20$$
$$2x + 4.5y = 30$$

$$2.5y = 10$$
$$y = 4, \quad x = 6$$

Mix 6 liters of the 20% solution with 4 liters of the 45% solution.

7. $\begin{bmatrix} 3 & -2 & | & 4 \\ 2 & 3 & | & 7 \end{bmatrix} \Rightarrow \begin{bmatrix} 3 & -2 & | & 4 \\ 0 & 13/3 & | & 13/3 \end{bmatrix} \Rightarrow \begin{bmatrix} 1 & -2/3 & | & 4/3 \\ 0 & 1 & | & 1 \end{bmatrix}$

$y = 1 \Rightarrow x = 2.$

9. $\begin{bmatrix} 1 & 2 & 1 & | & 0 \\ 3 & -2 & -2 & | & 7 \\ 4 & 0 & -1 & | & 7 \end{bmatrix} \Rightarrow \begin{bmatrix} 1 & 2 & 1 & | & 0 \\ 0 & -8 & -5 & | & 7 \\ 0 & -8 & -5 & | & 7 \end{bmatrix} \Rightarrow \begin{bmatrix} 1 & 2 & 1 & | & 0 \\ 0 & 1 & 5/8 & | & -7/8 \\ 0 & 0 & 0 & | & 0 \end{bmatrix}$

$x = \frac{1}{4}z + \frac{7}{4}, \quad y = -\frac{5}{8}z - \frac{7}{8}$

$\begin{bmatrix} 1 & 0 & -1/4 & | & 7/4 \\ 0 & 1 & 5/8 & | & -7/8 \\ 0 & 0 & 0 & | & 0 \end{bmatrix}$

11. $3\begin{bmatrix} 2 & -3 & 5 \\ 0 & 3 & -1 \end{bmatrix} - 5\begin{bmatrix} -2 & 1 & -1 \\ 0 & 3 & 2 \end{bmatrix}$

$= \begin{bmatrix} 6 & -9 & 15 \\ 0 & 9 & -3 \end{bmatrix} - \begin{bmatrix} -10 & 5 & -5 \\ 0 & 15 & 10 \end{bmatrix}$

$= \begin{bmatrix} 16 & -14 & 20 \\ 0 & -6 & -13 \end{bmatrix}$

13. $\begin{bmatrix} 5 & 19 & | & 1 & 0 \\ 2 & 7 & | & 0 & 1 \end{bmatrix} \Rightarrow \begin{bmatrix} 1 & 5 & | & 1 & -2 \\ 0 & -3 & | & -2 & 5 \end{bmatrix} \Rightarrow \begin{bmatrix} 1 & 5 & | & 1 & -2 \\ 0 & 1 & | & 2/3 & -5/3 \end{bmatrix}$

$\begin{bmatrix} 1 & 0 & | & -7/3 & 19/3 \\ 0 & 1 & | & 2/3 & -5/3 \end{bmatrix}$ INVERSE: $\begin{bmatrix} -\frac{7}{3} & \frac{19}{3} \\ \frac{2}{3} & -\frac{5}{3} \end{bmatrix}$

15.

$$\begin{bmatrix} 5 & 19 \\ 2 & 7 \end{bmatrix}\begin{bmatrix} x \\ y \end{bmatrix}=\begin{bmatrix} 3 \\ 2 \end{bmatrix}$$

$$\begin{bmatrix} -\frac{7}{3} & \frac{19}{3} \\ \frac{2}{3} & -\frac{5}{3} \end{bmatrix}\begin{bmatrix} 5 & 19 \\ 2 & 7 \end{bmatrix}\begin{bmatrix} x \\ y \end{bmatrix}=\begin{bmatrix} -\frac{7}{3} & \frac{19}{3} \\ \frac{2}{3} & -\frac{5}{3} \end{bmatrix}\begin{bmatrix} 3 \\ 2 \end{bmatrix}$$

$$\begin{bmatrix} 1 & 0 \\ 0 & 1 \end{bmatrix}\begin{bmatrix} x \\ y \end{bmatrix}=\begin{bmatrix} 17/3 \\ -4/3 \end{bmatrix} \quad x=\frac{17}{3}, y=-\frac{4}{3}$$

17. $\begin{vmatrix} 3 & -5 \\ -3 & 1 \end{vmatrix}=(3)(1)-(-5)(-3)=-12$

19. $x=\dfrac{\begin{vmatrix} 3 & -5 \\ 2 & 1 \end{vmatrix}}{\begin{vmatrix} 3 & -5 \\ -3 & 1 \end{vmatrix}}=\dfrac{13}{-12}=-\dfrac{13}{12} \qquad y=\dfrac{\begin{vmatrix} 3 & 3 \\ -3 & 2 \end{vmatrix}}{\begin{vmatrix} 3 & -5 \\ -3 & 1 \end{vmatrix}}=\dfrac{15}{-12}=-\dfrac{5}{4}$

21. Find A and B such that:

$$\frac{5x}{(2x-3)(x+1)}=\frac{A}{2x-3}+\frac{B}{x+1}$$

$$5x=A(x+1)+B(2x-3)$$
$$5x=Ax+A+2Bx-3B$$
$$5x=(A+2B)x+(A-3B)$$

$$A+2B=5$$
$$A-3B=0$$

$$A=3, B=1 \implies \frac{5x}{(2x-3)(x+1)}=\frac{3}{2x-3}+\frac{1}{x+1}$$

23. $x-3y\geq 3$
$x+3y\leq 3$

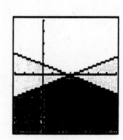

264

25.

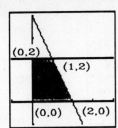

point	$P = 3x + 2y$	
(0, 0)	$= 0$	
(0, 2)	$= 4$	
(1, 2)	$= 7$	⟵ MAX
(2, 0)	$= 6$	

Exercise 8.1 (page 426)

1.
$$(x - h)^2 + (y - k)^2 = r^2$$
$$(x - 0)^2 + (y - 0)^2 = 7^2$$
$$x^2 + y^2 = 49$$

3. $r = \sqrt{(3 - 2)^2 + (2 - - 2)^2} = \sqrt{17}$
$$(x - h)^2 + (y - k)^2 = r^2$$
$$(x - 2)^2 + (y - - 2)^2 = (\sqrt{17})^2$$
$$(x - 2)^2 + (y + 2)^2 = 17$$

5. Find center:
$$3x + y = 1 \Rightarrow 9x + 3y = 3$$
$$-2x - 3y = 4 \Rightarrow -2x - 3y = 4$$
$$7x = 7$$

$$x = 1, y = -2$$

$$(x - 1)^2 + (y + 2)^2 = 6^2$$
$$(x - 1)^2 + (y + 2)^2 = 36$$

7. $x^2 + y^2 = 4$
center $(0, 0)$, $r = 2$

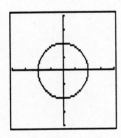

9.
$$3x^2 + 3y^2 - 12x - 6y = 12$$
$$x^2 - 4x + y^2 - 2y = 4$$
$$x^2 - 4x + 4 + y^2 - 2y + 1 = 4 + 4 + 1$$
$$(x - 2)^2 + (y - 1)^2 = 9$$
center $(2, 1)$, $r = 3$

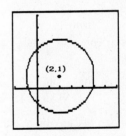

11. Check the coordinate $(50, 70)$: $50^2 + 70^2 = 2500 + 4900 = 7400$.
The city can receive the signal.

13. original center: $(0, 0)$
original range $= \sqrt{1600} = 40$
translator:
$$x^2 + y^2 - 70y + 600 = 0$$
$$x^2 + y^2 - 70y + 1225 = -600 + 1225$$
$$(x - 0)^2 + (y - 35)^2 = 625$$
center: $(0, 35)$; range $= \sqrt{625} = 25$
From the figure, it can be seen that
the signal can be received 60 miles
from the main transmitter.

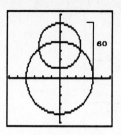

15. The center is $(4, 0)$ and the radius is 4: $(x - 4)^2 + y^2 = 16$.

17. If a point is on the circle, then it must satisfy the equation of the circle.
Substitute each point into the equation of a circle:

$$(x - h)^2 + (y - k)^2 = r^2 \qquad\qquad (x - h)^2 + (y - k)^2 = r^2$$
$$(0 - h)^2 + (8 - k)^2 = r^2 \qquad\qquad (5 - h)^2 + (3 - k)^2 = r^2$$
$$h^2 + 64 - 16k + k^2 = r^2 \qquad 25 - 10h + h^2 + 9 - 6k + k^2 = r^2$$
$$h^2 + k^2 - 16k + 64 = r^2 \qquad\qquad h^2 - 10h + k^2 - 6k + 34 = r^2$$

$$(x - h)^2 + (y - k)^2 = r^2$$
$$(4 - h)^2 + (6 - k)^2 = r^2$$
$$16 - 8h + h^2 + 36 - 12k + k^2 = r^2$$
$$h^2 - 8h + k^2 - 12k + 52 = r^2$$

Write each equation in the form $h^2 + k^2 - r^2 = \,??????$

$$h^2 + k^2 - r^2 = 16k - 64$$
$$h^2 + k^2 - r^2 = 10h + 6k - 34$$
$$h^2 + k^2 - r^2 = 8h + 12k - 52$$

$$16k - 64 = 10h + 6k - 34 \qquad\qquad 16k - 64 = 8h + 12k - 52$$
$$10k - 10h = 30 \qquad\qquad\qquad\qquad 4k - 8h = 12$$

$$10k - 10h = 30 \;\Rightarrow\quad k - h = 3$$
$$4k - 8h = 12 \;\Rightarrow\quad k - 2h = 3 \quad \Rightarrow \; k = 3, h = 0$$

Substitute into one of the above equations to get $r = 5$.

EQUATION: $(x - 0)^2 + (y - 3)^2 = 5^2 \;\Rightarrow\; x^2 + (y - 3)^2 = 25$

19. vertical parabola (up), $p = 3$
$$(x - h)^2 = 4p(y - k)$$
$$(x - 0)^2 = 4(3)(y - 0)$$
$$x^2 = 12y$$

21. horizontal parabola (left), $p = 3$
$$(y - k)^2 = -4p(x - h)$$
$$(y - 0)^2 = -4(3)(x - 0)$$
$$y^2 = -12x$$

23. vertical parabola (down), $p = 3$

$$(x - h)^2 = -4p(y - k)$$
$$(x - 3)^2 = -4(3)(y - 5)$$
$$(x - 3)^2 = -12(y - 5)$$

25. vertical parabola (down), $p = 7$

$$(x - h)^2 = -4p(y - k)$$
$$(x - 3)^2 = -4(7)(y - 5)$$
$$(x - 3)^2 = -28(y - 5)$$

27.
$$(x - 2)^2 = 4p(y - 2)$$
$$(0 - 2)^2 = 4p(0 - 2)$$
$$4 = -8p$$
$$-2 = 4p$$
$$(x - 2)^2 = -2(y - 2)$$

or

$$(y - 2)^2 = 4p(x - 2)$$
$$(0 - 2)^2 = 4p(0 - 2)$$
$$4 = -8p$$
$$-2 = 4p$$
$$(y - 2)^2 = -2(x - 2)$$

or

29.
$$(x + 4)^2 = 4p(y - 6)$$
$$(0 + 4)^2 = 4p(3 - 6)$$
$$16 = -12p$$
$$-\frac{16}{3} = 4p$$
$$(x + 4)^2 = -\frac{16}{3}(y - 6)$$

or

$$(y - 6)^2 = 4p(x + 4)$$
$$(3 - 6)^2 = 4p(0 + 4)$$
$$9 = 16p$$
$$\frac{9}{4} = 4p$$
$$(y - 6)^2 = \frac{9}{4}(x + 4)$$

or

31.
$$(x - 6)^2 = 4p(y - 8)$$
$$(5 - 6)^2 = 4p(10 - 8)$$
$$1 = 8p$$
$$\frac{1}{2} = 4p$$
$$(x - 6)^2 = \frac{1}{2}(y - 8)$$

or

$$(y - 8)^2 = 4p(x - 6)$$
$$(10 - 8)^2 = 4p(5 - 6)$$
$$4 = -4p$$
$$-4 = 4p$$
$$(y - 8)^2 = -4(x - 6)$$

or

Check to see which equation is satisfied by (5, 6) as well. The second equation is the solution: $(y - 8)^2 = -4(x - 6)$

33.
$$(x - 3)^2 = 4p(y - 1)$$
$$(4 - 3)^2 = 4p(3 - 1)$$
$$1 = 8p$$
$$\frac{1}{2} = 4p$$
$$(x - 3)^2 = \frac{1}{2}(y - 1)$$

or

$$(y - 1)^2 = 4p(x - 3)$$
$$(3 - 1)^2 = 4p(4 - 3)$$
$$4 = 4p$$
$$(y - 1)^2 = 4(x - 3)$$

or

Check to see which equation is satisfied by (2, 3) as well. The first equation is the solution:
$$(x - 3)^2 = \frac{1}{2}(y - 1)$$

35.
$$y = x^2 + 4x + 5$$
$$y - 5 = x^2 + 4x$$
$$y - 5 + 4 = x^2 + 4x + 4$$
$$y - 1 = (x + 2)^2$$

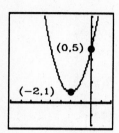

37.
$$y^2 + 4x - 6y = -1$$
$$y^2 - 6y = -4x - 1$$
$$y^2 - 6y + 9 = -4x - 1 + 9$$
$$(y - 3)^2 = -4(x - 2)$$

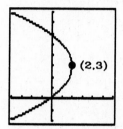

39.
$$y^2 + 2x - 2y = 5$$
$$y^2 - 2y = -2x + 5$$
$$y^2 - 2y + 1 = -2x + 5 + 1$$
$$(y - 1)^2 = -2(x - 3)$$

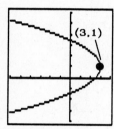

41.
$$x^2 - 6y + 22 = -4x$$
$$x^2 + 4x = 6y - 22$$
$$x^2 + 4x + 4 = 6y - 22 + 4$$
$$(x + 2)^2 = 6(y - 3)$$

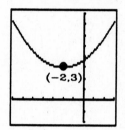

43.
$$4x^2 - 4x + 32y = 47$$
$$4x^2 - 4x = -32y + 47$$
$$4\left(x^2 - x + \frac{1}{4}\right) = -32y + 47 + 1$$
$$4\left(x - \frac{1}{2}\right)^2 = -32y + 48$$
$$\left(x - \frac{1}{2}\right)^2 = -8y + 12$$
$$\left(x - \frac{1}{2}\right)^2 = -8\left(y - \frac{3}{2}\right)$$

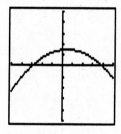

45. Find the distance to the focus:
$$4p = 8 \Rightarrow y = 2$$
It will be hottest 2 feet from the vertex of the parabola.

47. The point $(15, -10)$ is on the curve.
$$(x - h)^2 = -4p(y - k)$$
$$(15 - 0)^2 = -4p(-10 - 0)$$
$$225 = -4p(-10)$$
$$\frac{225}{10} = -4p$$
$$x^2 = -\frac{45}{2}y$$

49. Income = price · # rented

$$= -45\left(\frac{n}{32} - \frac{1}{2}\right) \cdot n$$

$$I = -\frac{45}{32} n^2 + \frac{45}{2} n$$

$$-\frac{32}{45} I = -\frac{32}{45}\left(-\frac{45}{32} n^2 + \frac{45}{2} n\right)$$

$$-\frac{32}{45} I = n^2 - 16n$$

$$-\frac{32}{45} I + 64 = n^2 - 16n + 64$$ If I is graphed along the y-axis, this is a vertical parabola opening up. The

$$-\frac{32}{45} I + 64 = (n - 8)^2$$ maximum income is then at the vertex.

She should build 8 cabins.

51. Consider the vertex to be at $(0, 0)$. Assume the parabola is horizontal and opening along the positive x-axis. Find its equation, and substitute $x = 10$.

$$(y - k)^2 = 4p(x - h)$$
$$(y - 0)^2 = 4(1)(x - 0)$$
$$y^2 = 4x$$
$$y^2 = 4(10)$$
$$y^2 = 40 \Rightarrow y = \pm\sqrt{40} = \pm 2\sqrt{10}$$

The diameter is $2(2\sqrt{10})$, or $4\sqrt{10}$, or about 12.6 cm.

53. Find the equation of the parabola: NOTE: $(315, -630)$ is on the curve.

$$(x - h)^2 = 4p(y - k)$$
$$x^2 = 4py$$
$$315^2 = 4p(-630)$$

$$x^2 = -\frac{315^2}{630} y$$

$$\frac{315^2}{-630} = 4p$$

$$= -\frac{315^2}{630} (-430)$$

$$\approx 67725$$
$$x \approx 260$$

The width is about $2(260)$, or 520 feet.

55. Its maximum height occurs at the vertex, while it hits the ground when $s = 0$.

HEIGHT: GROUND:

$$s = 16t^2 - 128t \qquad\qquad s = 16t^2 - 128t$$
$$s = 16(t^2 - 8t) \qquad\qquad 0 = 16t(t - 8)$$
$$s + 256 = 16(t^2 - 8t + 16) \qquad t = 0 \text{ (start)}, \quad \text{or } t = 8$$
$$s + 256 = 16(t - 4)^2$$

It reaches the vertex after 4 seconds, and hits the ground after a total of 8 seconds, or 4 seconds after reaching its maximum height.

57.
$$y = ax^2 + bx + c \qquad\qquad y = ax^2 + bx + c \qquad\qquad y = ax^2 + bx + c$$
$$8 = a(1)^2 + b(1) + c \qquad -1 = a(-2)^2 + b(-2) + c \qquad 15 = a(2)^2 + b(2) + c$$
$$8 = a + b + c \qquad\qquad -1 = 4a - 2b + c \qquad\qquad 15 = 4a + 2b + c$$

$$\left.\begin{array}{l} a + b + c = 8 \\ 4a - 2b + c = -1 \\ 4a + 2b + c = 15 \end{array}\right\} \quad a = 1, \, b = 4, \, c = 3$$

$$y = x^2 + 4x + 3$$

Exercise 8.2   (page 436)

1. $c = 3, a = 5$, horizontal
$$c^2 = a^2 - b^2$$
$$9 = 25 - b^2 \;\Rightarrow\; b^2 = 16$$
$$\frac{(x-0)^2}{25} + \frac{(y-0)^2}{16} = 1$$
$$\frac{x^2}{25} + \frac{y^2}{16} = 1$$

3. $c = 1, b = \frac{4}{3}$, vertical
$$c^2 = a^2 - b^2$$
$$1^2 = a^2 - \left(\frac{4}{3}\right)^2 \;\Rightarrow\; a^2 = \frac{25}{9}$$
$$\frac{(x-0)^2}{16/9} + \frac{(y-0)^2}{25/9} = 1$$
$$\frac{9x^2}{16} + \frac{9y^2}{25} = 1$$

5. $c = 3, a = 4$, vertical
$$c^2 = a^2 - b^2$$
$$3^2 = 4^2 - b^2 \;\Rightarrow\; b^2 = 7$$
$$\frac{x^2}{7} + \frac{y^2}{16} = 1$$

7. vertical ellipse
$$\frac{(x-3)^2}{4} + \frac{(y-4)^2}{9} = 1$$

9. horizontal ellipse
$$\frac{(x-3)^2}{9} + \frac{(y-4)^2}{4} = 1$$

11. $C: \left(\dfrac{-2+8}{2}, \dfrac{4+4}{2}\right) = (3, 4)$

horizontal ellipse, $c = 5$
$$c^2 = a^2 - b^2 \Rightarrow 5^2 = a^2 - 4^2 \Rightarrow a^2 = 41$$
$$\frac{(x-3)^2}{41} + \frac{(y-4)^2}{16} = 1$$

13. $C: \left(\dfrac{-4+4}{2}, \dfrac{4+4}{2}\right) = (0, 4)$

horizontal ellipse, $c = 4, \, a = 6$
$$c^2 = a^2 - b^2 \Rightarrow 4^2 = 6^2 - b^2$$
$$\Rightarrow b^2 = 20$$
$$\frac{x^2}{36} + \frac{(y-4)^2}{20} = 1$$

15. $C: \left(\dfrac{6 + -6}{2}, \dfrac{0+0}{2}\right) = (0, 0)$

horizontal, $c = \frac{3}{5}a, \, c = 6 \Rightarrow a = 10$
$$c^2 = a^2 - b^2 \Rightarrow b^2 = 10^2 - 6^2 = 64$$
$$\frac{x^2}{100} + \frac{y^2}{64} = 1$$

17. $\dfrac{x^2}{25} + \dfrac{y^2}{49} = 1$

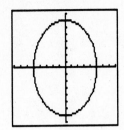

19. $\dfrac{x^2}{16} + \dfrac{(y+2)^2}{36} = 1$

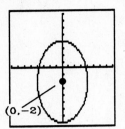

(0,−2)

21.

$$x^2 + 4y^2 - 4x + 8y + 4 = 0$$
$$x^2 - 4x + 4(y^2 + 2y) = -4$$
$$x^2 - 4x + 4 + 4(y^2 + 2y + 1) = -4 + 4 + 4$$
$$(x-2)^2 + 4(y+1)^2 = 4$$

$$\dfrac{(x-2)^2}{4} + \dfrac{(y+1)^2}{1} = 1$$

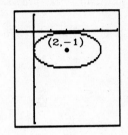

(2,−1)

23.

$$16x^2 + 25y^2 - 160x - 200y + 400 = 0$$
$$16(x^2 - 10x) + 25(y^2 - 8y) = -400$$
$$16(x^2 - 10x + 25) + 25(y^2 - 8y + 16) = -400 + 400 + 400$$
$$16(x-5)^2 + 25(y-4)^2 = 400$$

$$\dfrac{(x-5)^2}{25} + \dfrac{(y-4)^2}{16} = 1$$

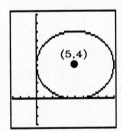

(5,4)

25. $a = 378{,}000 \div 2 = 189{,}000.$ $\dfrac{c}{a} = \dfrac{11}{200} \Rightarrow c = \dfrac{11a}{200} = 10{,}395.$

The farthest distance is $a + c$, or 199,395 miles.

27. Let the center $= (0, 0)$. $a = 50$, $b = 30$. Then $c^2 = 50^2 - 30^2 = 40^2$ and $c = 40$.

$$\frac{x^2}{50^2} + \frac{y^2}{30^2} = 1 \qquad \text{To find half the focal width, let } x = 40 \text{ and solve for } y.$$

$$\frac{40^2}{50^2} + \frac{y^2}{30^2} = 1$$

$$\frac{y^2}{30^2} = 1 - \frac{40^2}{50^2}$$

$$y^2 = 30^2\left(1 - \frac{40^2}{50^2}\right) = 324 \quad \Rightarrow \quad y = \pm 18$$

The focal width is $2(18) = 36$ meters.

29. $a = 24, b = 12 \Rightarrow a^2 = 576,\ b^2 = 144,\ c^2 = 432.$

$$\frac{x^2}{144} + \frac{y^2}{576} = 1$$

$$\frac{x^2}{144} = 1 - \frac{y^2}{576}$$

$$x^2 = 144\left(1 - \frac{y^2}{576}\right)$$

$$x = \sqrt{144\left(1 - \frac{y^2}{576}\right)} \qquad \text{Let } y = 12.$$

$$x \approx 10.4 \qquad \text{The whole width is then about 20.8 inches.}$$

31. $FB = \sqrt{(0 - c)^2 + (b - 0)^2} = \sqrt{c^2 + b^2} = \sqrt{a^2} = a$

33.
$$\frac{x^2}{a^2} + \frac{y^2}{b^2} = 1 \qquad \text{Note: } c^2 = a^2 - b^2, \text{ so } b^2 = a^2 - c^2.$$

$$\frac{x^2}{a^2} + \frac{y^2}{a^2 - c^2} = 1 \qquad\qquad\qquad \text{THEN: } \quad y = \sqrt{\frac{a^4 - 2a^2c^2 + c^4}{a^2}}$$

$$\frac{c^2}{a^2} + \frac{y^2}{a^2 - c^2} = 1$$

$$y^2 = (a^2 - c^2)\left(1 - \frac{c^2}{a^2}\right) \qquad\qquad\qquad = \frac{\sqrt{(a^2 - c^2)^2}}{a}$$

$$= a^2 - c^2 - c^2 + \frac{c^4}{a^2} \qquad\qquad\qquad = \frac{a^2 - c^2}{a} = \frac{b^2}{a}$$

$$= a^2 - 2c^2 + \frac{c^4}{a^2} \qquad\qquad \text{So the length } = \frac{2b^2}{a}$$

35. Let $(0, 0)$ be the center between the thumbtacks. Then $c = 1$. Consider when the pencil is in line with the thumbtacks. Then the string reaches from one thumbtack, to a vertex of the ellipse, and back to the other thumbtack. If you examine this, you will see that the length of the string $= 2a$, the length of the major axis. Thus $2a = 6$, or $a = 3$. Then, $b^2 = a^2 - c^2 = 9 - 1 = 8$.

$$\frac{x^2}{9} + \frac{y^2}{8} = 1.$$

37.

$$\frac{(x - h)^2}{a^2} + \frac{(y - k)^2}{b^2} = 1$$

$$b^2(x - h)^2 + a^2(y - k)^2 = a^2 b^2$$
$$b^2(x^2 - 2hx + h^2) + a^2(y^2 - 2ky + k^2) = a^2 b^2$$
$$b^2 x^2 - 2b^2 hx + b^2 h^2 + a^2 y^2 - 2a^2 ky + a^2 k^2 = a^2 b^2$$

$$(b^2)x^2 + (a^2)y^2 + (-2b^2 h)x + (-2a^2 k)y = \boxed{a^2 b^2 - a^2 y^2 - a^2 k^2 - b^2 h^2}$$

The boxed expression is simply a constant, and this is a general second degree equation.

Exercise 8.3   (page 443)

1. C: $(0, 0)$, horizontal, $a = 5$, $c = 7$
$c^2 = a^2 + b^2 \Rightarrow b^2 = 24$

$$\frac{x^2}{25} - \frac{y^2}{24} = 1$$

3. $\dfrac{(x - 2)^2}{4} - \dfrac{(y - 4)^2}{9} = 1$

5. vertical, $a = 3$

$$\frac{(y - 3)^2}{9} - \frac{(x - 5)^2}{b^2} = 1$$

$$\frac{(8 - 3)^2}{9} - \frac{(1 - 5)^2}{b^2} = 1$$

$$\frac{25}{9} - \frac{16}{b^2} = 1$$

$$\frac{16}{b^2} = \frac{16}{9}$$

$$b^2 = 9$$

$$\frac{(y - 3)^2}{9} - \frac{(x - 5)^2}{9} = 1$$

7. C: $(0, 0)$, vertical, $a = 3$

$$c = \frac{5}{3}a \Rightarrow c = 5, \quad b^2 = c^2 - a^2 = 16$$

$$\frac{y^2}{9} - \frac{x^2}{16} = 1$$

9. $\dfrac{(x-1)^2}{4} - \dfrac{(y+3)^2}{16} = 1$, or $\dfrac{(y+3)^2}{4} - \dfrac{(x-1)^2}{16} = 1$

11.

$$\dfrac{x^2}{a^2} - \dfrac{y^2}{b^2} = 1$$

$$\dfrac{4^2}{a^2} - \dfrac{2^2}{b^2} = 1$$

$$\dfrac{16}{a^2} - \dfrac{4}{b^2} = 1$$

$$\dfrac{16}{a^2} = 1 + \dfrac{4}{b^2}$$

$$\dfrac{64}{a^2} = 4 + \dfrac{16}{b^2} \quad \Rightarrow$$

$$\boxed{\dfrac{x^2}{10} - \dfrac{3y^2}{20} = 1}$$

$$\dfrac{x^2}{a^2} - \dfrac{y^2}{b^2} = 1$$

$$\dfrac{8^2}{a^2} - \dfrac{(-6)^2}{b^2} = 1$$

$$\dfrac{64}{a^2} - \dfrac{36}{b^2} = 1$$

$$4 + \dfrac{16}{b^2} - \dfrac{36}{b^2} = 1$$

$$\dfrac{20}{b^2} = 3$$

$$b^2 = \dfrac{20}{3} \quad \Rightarrow \quad a^2 = 10$$

13. $4(x-1)^2 - 9(y+2)^2 = 36 \Rightarrow \dfrac{(x-1)^2}{9} - \dfrac{(y+2)^2}{4} = 1$; $a = 3$, $b = 2$; $(2a)(2b) = 24$

15.

$$x^2 + 6x - y^2 + 2y = -11$$
$$x^2 + 6x - (y^2 - 2y) = -11$$
$$x^2 + 6x + 9 - (y^2 - 2y + 1) = -11 + 9 - 1$$
$$(x+3)^2 - (y-1)^2 = -3$$
$$(y-1)^2 - (x+3)^2 = 3$$

$$\dfrac{(y-1)^2}{3} - \dfrac{(x+3)^2}{3} = 1 \Rightarrow a = \sqrt{3}, \ b = \sqrt{3} \ ; \ (2a)(2b) = 12$$

17. $(2a)(2b) = 36$

$4(2b) = 36$

$b = 9/2$

$b^2 = 81/4$

$\dfrac{(x+2)^2}{4} - \dfrac{4(y+4)^2}{81} = 1$, or

$\dfrac{(y+4)^2}{4} - \dfrac{4(x+2)^2}{81} = 1$

19. C: $(0, 0)$, $a = 6$, $b = 5/4$, horizontal

$$\dfrac{x^2}{36} - \dfrac{16y^2}{25} = 1$$

274

21. $\dfrac{x^2}{9} - \dfrac{y^2}{4} = 1$; horizontal

$a = 3,\ b = 2,\ c = \sqrt{13}$

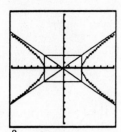

23. $4x^2 - 3y^2 = 36 \Rightarrow \dfrac{x^2}{9} - \dfrac{y^2}{12} = 1$

horizontal; $a = 3,\ b = \sqrt{12},\ c = \sqrt{21}$

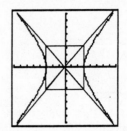

25. $\dfrac{y^2}{1} - \dfrac{x^2}{1} = 1$; vertical

$a = 1,\ b = 1,\ c = \sqrt{2}$

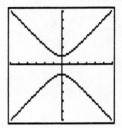

27.

$$4x^2 - 2y^2 + 8x - 8y = 8$$
$$4(x^2 + 2x) - 2(y^2 + 4y) = 8$$
$$4(x^2 + 2x + 1) - 2(y^2 + 4y + 4) = 8 + 4 - 8$$
$$4(x + 1)^2 - 2(y + 2)^2 = 4$$

$$\dfrac{(x + 1)^2}{1} - \dfrac{(y + 2)^2}{2} = 1$$

$a = 1,\ b = \sqrt{2},\ c = \sqrt{3}$; horizontal

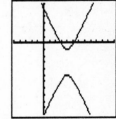

29.

$$y^2 - 4x^2 + 6y + 32x = 59$$
$$y^2 + 6y - 4(x^2 - 8x) = 59$$
$$y^2 + 6y + 9 - 4(x^2 - 8x + 16) = 59 + 9 - 64$$
$$(y + 3)^2 - 4(x - 4)^2 = 4$$

$$\dfrac{(y + 3)^2}{4} - \dfrac{(x - 4)^2}{1} = 1$$

$a = 2,\ b = 1,\ c = \sqrt{5}$; vertical

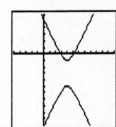

31. $-xy = 6$

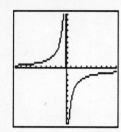

33. $a = 100,000,000$; horizontal; C: $(0, 0)$

$$\frac{b}{a} = 2 \Rightarrow b = 200,000,000$$

$$\frac{x^2}{100,000,000^2} - \frac{y^2}{200,000,000} = 1$$

35. Treat the transmitters as foci of a hyperbola. Then 24 is the constant difference between the distances from the ship to the transmitters. The ship's position (x, y) then satisfies the following equation:

$$\sqrt{(x - 13)^2 + (y - 0)^2} - \sqrt{(x + 13)^2 + (y - 0)^2} = 24$$

From the textbook's development of the standard form of the equation of a hyperbola, $2a$ must equal 24. From the figure, $c = 13$. $a = 12$ and $c = 13$ indicate that $b^2 = 13^2 - 12^2 = 25$. The hyperbola is:

$$\frac{x^2}{144} - \frac{y^2}{25} = 1$$

37. $(-2, 1)$ and $(8, 1)$ are the foci of a hyperbola. As in #35, $2a$ = the common difference of 6. Then $a = 3$, $c = 5$ (from the coordinates of the foci) and $b^2 = 5^2 - 3^2 = 4^2$. The center is $(3, 1)$. The equation is:

$$\frac{(x - 3)^2}{9} - \frac{(y - 1)^2}{16} = 1$$

39. The distance between P and the line $y = -2$ is the difference between the y-coordinates, or $y + 2$.

$$\sqrt{(x - 0)^2 + (y - 3)^2} = \frac{3}{2}(y + 2)$$

$$x^2 + (y - 3)^2 = \frac{9}{4}(y + 2)^2$$

$$4x^2 + 4(y - 3)^2 = 9(y + 2)^2$$
$$4x^2 + 4(y^2 - 6y + 9) = 9(y^2 + 4y + 4)$$
$$4x^2 + 4y^2 - 24y + 36 = 9y^2 + 36y + 36$$
$$4x^2 - 5y^2 - 60y = 0$$

41-43. Answers will vary.

Exercise 8.4 (page 448)

1. $8x^2 + 32y^2 = 256$
$x = 2y$

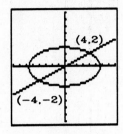

$(-4, -2), (4, 2)$

3. $x^2 + y^2 = 90$
$y = x^2$

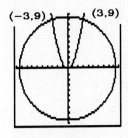

$(-3, 9), (3, 9)$

5. $x^2 + y^2 = 25$
$12x^2 + 64y^2 = 768$

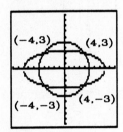

$(4, 3), (4, -3),$
$(-4, 3), (-4, -3)$

7. $x^2 - 13 = -y^2$
$y = 2x - 4$

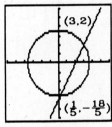

$(3, 2), \left(\frac{1}{5}, -\frac{18}{5}\right)$

9. $x^2 - 6x - y = -5$
$x^2 - 6x + y = -5$

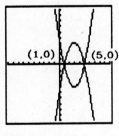

$(1, 0), (5, 0)$

11. $y = x + 1$
$x = x^2 + x$

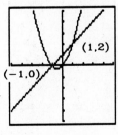

$(1, 2), (-1, 0)$

13. $6x^2 + 9y^2 = 10 \Rightarrow y = \pm \sqrt{(10 - 6x^2)/9}$
$3y - 2x = 0 \Rightarrow y = 2x/3$

$(1, 0.67), (-1, -0.67)$

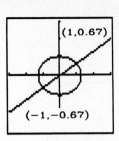

15. $5x + 3y = 15 \Rightarrow y = \dfrac{15 - 5x}{3}$

$$25x^2 + 9y^2 = 225$$
$$25x^2 + 9\left(\dfrac{15 - 5x}{3}\right)^2 = 225$$
$$25x^2 + (15 - 5x)^2 = 225$$
$$25x^2 + 225 - 150x + 25x^2 = 225$$
$$50x^2 - 150x = 0$$
$$50x(x - 3) = 0$$

$x = 0$ or $x = 3$
Substitute into $y = \dfrac{15 - 5x}{3}$.
$(0, 5)$ or $(3, 0)$

17. $x + y = 2 \Rightarrow y = 2 - x$

$$x^2 + y^2 = 2$$
$$x^2 + (2 - x)^2 = 2$$
$$x^2 + 4 - 4x + x^2 = 2$$
$$2x^2 - 4x + 2 = 0$$
$$2(x^2 - 2x + 1) = 0$$
$$2(x - 1)(x - 1) = 0$$

$x = 1$
Substitute into $y = 2 - x$.
$(1, 1)$

19. $x + y = 3 \Rightarrow y = 3 - x$

$$x^2 + y^2 = 5$$
$$x^2 + (3 - x)^2 = 5$$
$$x^2 + 9 - 6x + x^2 = 5$$
$$2x^2 - 6x + 4 = 0$$
$$2(x - 2)(x - 1) = 0$$

$x = 2$ or $x = 1$
Substitute to get $y = 1$ or $y = 2$.
$(2, 1), (1, 2)$

21. $y = x^2 - 1 \Rightarrow x^2 = y + 1$

$$x^2 + y^2 = 13$$
$$y + 1 + y^2 = 13$$
$$y^2 + y - 12 = 0$$
$$(y + 4)(y - 3) = 0$$

$y = -4$ or $y = 3$
Substitute to get $x^2 = -3$ or $x^2 = 4$.
The only real solutions are $x = \pm 2$.
$(2, 3), (-2, 3)$

23. $y = x^2$

$$x^2 + y^2 = 30$$
$$y + y^2 = 30$$
$$y^2 + y - 30 = 0$$
$$(y + 6)(y - 5) = 0$$

$y = -6$ or $y = 5$
Substitute and get the only real
solutions for x from $y = 5$.
$(\sqrt{5}, 5), (-\sqrt{5}, 5)$

25.
$$x^2 + y^2 = 13$$
$$x^2 - y^2 = 5$$

$$2x^2 = 18$$
$$x^2 = 9 \Rightarrow x = \pm 3$$

Substitute and get $y^2 = 4$, or $y = \pm 2$.
$(3, 2), (-3, 2), (3, -2), (-3, -2)$

27. $x^2 + y^2 = 20$
$x^2 - y^2 = -12$

$2x^2 = 8$
$x^2 = 4 \Rightarrow x = \pm 2$
Substitute and get $y^2 = 16$.
$(2, 4), (2, -4), (-2, 4), (-2, -4)$

29. $y^2 = 40 - x^2 \Rightarrow x^2 = 40 - y^2$
$y = x^2 - 10$
$y = 40 - y^2 - 10$
$y^2 + y - 30 = 0$
$(y + 6)(y - 5) = 0 \Rightarrow y = -6, 5$
$y = -6 \Rightarrow x^2 = 4$
$y = 5 \Rightarrow x^2 = 15$
$(2, -6), (-2, -6)$
$(\sqrt{15}, 5), (-\sqrt{15}, 5)$

31. $y = x^2 - 4 \Rightarrow x^2 = y + 4$
$x^2 - y^2 = -16$
$y + 4 - y^2 = -16$
$-y^2 + y + 20 = 0$
$-(y^2 - y - 20) = 0$
$-(y - 5)(y + 4) = 0$
$y = 5 \Rightarrow x^2 = 9 \Rightarrow x = \pm 3$
$y = -4 \Rightarrow x^2 = 0 \Rightarrow x = 0$
$(3, 5), (-3, 5), (0, 4)$

33. $x^2 - y^2 = -5$
$3x^2 + 2y^2 = 30$

$2x^2 - 2y^2 = -10$
$3x^2 + 2y^2 = 30$

$5x^2 = 20 \Rightarrow x^2 = 4$
$x^2 = 4 \Rightarrow y^2 = 9$
$(2, 3), (2, -3), (-2, 3), (-2, -3)$

35. $\frac{1}{x} + \frac{2}{y} = 1$

$\frac{2}{x} - \frac{1}{y} = \frac{1}{3}$

$\frac{1}{x} + \frac{2}{y} = 1$

$\frac{4}{x} - \frac{2}{y} = \frac{2}{3}$

$\frac{5}{x} = \frac{5}{3}$

$x = 3, y = 3 \Rightarrow (3, 3)$

37. $3y^2 = xy \Rightarrow x = 3y$ or $y = 0$
$2x^2 + xy - 84 = 0$
$2(3y)^2 + (3y)y - 84 = 0$
$18y^2 + 3y^2 = 84$
$21y^2 = 84$
$y^2 = 4 \Rightarrow y = \pm 2$
$y = 2, x = 6; \ y = -2, x = -6$
$y = 0 \Rightarrow 2x^2 = 84 \Rightarrow x = \pm \sqrt{42}$
$(2, 6), (-2, -6), (\sqrt{42}, 0), (-\sqrt{42}, 0)$

39. $xy = \frac{1}{6} \Rightarrow y = \frac{1}{6x}$

$y + x = 5xy$

$\frac{1}{6x} + x = 5x\left(\frac{1}{6x}\right)$

$\frac{1}{6x} + x = \frac{5}{6}$

$1 + 6x^2 = 5x$
$6x^2 - 5x + 1 = 0$
$(2x - 1)(3x - 1) = 0$

$x = \frac{1}{2}$ or $x = \frac{1}{3} \Rightarrow \left(\frac{1}{2}, \frac{1}{3}\right), \left(\frac{1}{3}, \frac{1}{2}\right)$

41. Let $l =$ the length and $w =$ the width.
$lw = 63$
$2l + 2w = 32 \Rightarrow w = 16 - l$

$lw = 63$
$l(16 - l) = 63$
$-l^2 + 16l - 63 = 0$
$l^2 - 16l + 63 = 0$
$(l - 9)(l - 7) = 0$
The dimensions are 9 cm by 7 cm.

43. Let c = her investment and r = her rate. Then $c + 150$ represent's John's investment, at a rate of $r + 0.015$.

$$cr = 67.5 \quad \Rightarrow \quad r = \frac{67.5}{c}$$
$$(c + 150)(r + 0.015) = 94.5$$
$$cr + 0.015c + 150r + 2.25 = 94.5$$
$$67.5 + 0.015c + 150 \cdot \frac{67.5}{c} + 2.25 = 94.5$$
$$0.015c - 24.75 + \frac{10125}{c} = 0$$
$$0.015c^2 - 24.75c + 10125 = 0$$
$$15c^2 - 24750c + 10125000 = 0$$
$$c^2 - 1650c + 675000 = 0$$
$$(c - 750)(c - 900) = 0 \quad c = 750 \text{ or } c = 900$$

$c = 750 \Rightarrow r = 0.09, \quad c = 900 \Rightarrow r = 0.075$

She invests either $750 at 9% or $900 at 7.5%.

45. Road: $y = \frac{100}{200}x = \frac{1}{2}x;$ Station limit: $(x - 120)^2 + y^2 = 100^2$

$$(x - 120)^2 + y^2 = 100^2$$
$$x^2 - 240x + 120^2 + \left(\frac{x}{2}\right)^2 = 100^2$$
$$x^2 + \frac{x^2}{4} - 240x + 4400 = 0$$
$$4x^2 + x^2 - 960x + 17600 = 0$$
$$5x^2 - 960x + 17600 = 0$$
$$5(x^2 - 192x + 3520) = 0 \quad \Rightarrow \quad x \approx 20.5 \text{ or } x \approx 171.5$$

The first time would be when $x \approx 20.5$. Find the y-coordinate. $y \approx 10.25$.
Find the distance from $(0, 0)$ to $(20.5, 10.25) \approx 23$.
The driver will pick up the station starting about 23 miles from Collinsville.

Chapter 8 Review Exercises (page 450)

1. $r = \sqrt{(5 - 0)^2 + (5 - 0)^2} = \sqrt{50}$

$$x^2 + y^2 = 50$$

3. $C:\left(\dfrac{-2 + 12}{2}, \dfrac{4 + 16}{2}\right) = (5, 10)$

$$r = \sqrt{(12 - 5)^2 + (16 - 10)^2} = \sqrt{85}$$
$$(x - 5)^2 + (y - 10)^2 = 85$$

5.
$$x^2 + y^2 - 6x + 4y = 3$$
$$x^2 - 6x + 9 + y^2 + 4y + 4 = 3 + 9 + 4$$
$$(x - 3)^2 + (y + 2)^2 = 16$$

$C: (3, -2); \quad r = 4$

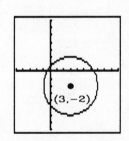

(3, −2)

7. horizontal

$$(y-k)^2 = 4p(x-h)$$

$$(4-0)^2 = 4p(-8-0)$$
$$16 = 4p(-8)$$
$$-2 = 4p$$
$$y^2 = -2x$$

9.

$$y = ax^2 + bx + c$$
$$ax^2 + bx + c = y$$
$$a\left(x^2 + \tfrac{b}{a}x\right) = y - c$$
$$a\left[x^2 + \tfrac{b}{a}x + \left(\tfrac{b}{2a}\right)^2\right] = y - c + \tfrac{b^2}{4a}$$
$$a\left(x + \tfrac{b}{2a}\right)^2 = y + \tfrac{b^2 - 4ac}{4a}$$
$$\left(x + \tfrac{b}{2a}\right)^2 = \tfrac{1}{a}\left(y + \tfrac{b^2 - 4ac}{4a}\right)$$

vertex: $\left(-\tfrac{b}{2a},\ \tfrac{4ac - b^2}{4a}\right)$

11.

$$x^2 - 4y - 2x = -9$$
$$x^2 - 2x = 4y - 9$$
$$x^2 - 2x + 1 = 4y - 9 + 1$$
$$(x-1)^2 = 4(y-2)$$

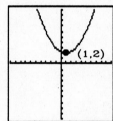

13.

$$y^2 - 4x - 2y = -13$$
$$y^2 - 2y = 4x - 13$$
$$y^2 - 2y + 1 = 4x - 13 + 1$$
$$(y-1)^2 = 4(x-3)$$

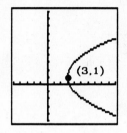

15. $a = 6, b = 4;\ \dfrac{x^2}{36} + \dfrac{y^2}{16} = 1$

17.

$$4x^2 + y^2 - 16x + 2y = -13$$
$$4(x^2 - 4x) + y^2 + 2y = -13$$
$$4(x^2 - 4x + 4) + y^2 + 2y + 1 = -13 + 16 + 1$$
$$4(x-2)^2 + (y+1)^2 = 4$$
$$\dfrac{(y+1)^2}{4} + \dfrac{(x-2)^2}{1} = 1$$

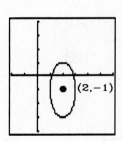

19. $C:\ (0,3),\ a = 3,\ c = 5 \Rightarrow b = 4;\ \dfrac{x^2}{9} - \dfrac{(y-3)^2}{16} = 1$

21.
$$9x^2 - 4y^2 - 16y - 18x = 43$$
$$9(x^2 - 2x) - 4(y^2 + 4y) = 43$$
$$9(x^2 - 2x + 1) - 4(y^2 + 4y + 4) = 43 + 9 - 16$$
$$9(x - 1)^2 - 4(y + 2)^2 = 36$$
$$\frac{(x - 1)^2}{4} - \frac{(y + 2)^2}{9} = 1$$

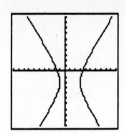

23. $3x^2 + y^2 = 52$
$x^2 - y^2 = 12$

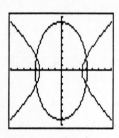

$(4, 2), (4, -2), (-4, 2), (-4, -2)$

25.
$$-\sqrt{3}\, y + 4\sqrt{3} = 3x$$
$$\frac{-\sqrt{3}\, y + 4\sqrt{3}}{3} = x$$
$$x^2 + y^2 = 16$$
$$\left(\frac{-\sqrt{3}\, y + 4\sqrt{3}}{3}\right)^2 + y^2 = 16$$
$$\frac{3y^2 - 24y + 48}{9} + y^2 = 16$$
$$3y^2 - 24y + 48 + 9y^2 = 144$$
$$12y^2 - 24y - 964 = 0$$
$$12(y^2 - 2y - 8) = 0$$
$$12(y - 4)(y + 2) = 0$$
$$y = 4 \Rightarrow x = 0$$
$$y = -2 \Rightarrow x = 2\sqrt{3}$$
$$(0, 4) \text{ or } (2\sqrt{3}, -2)$$

27. $\dfrac{x^2}{16} + \dfrac{y^2}{12} = 1$

$x^2 - \dfrac{y^2}{3} = 1$

$(2, 3), (2, -3), (-2, 3), (-2, -3)$

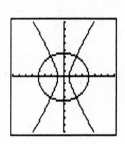

Chapter 8 Test (page 451)

1. $(x - 2)^2 + (y - 3)^2 = 9$

3. $r = \sqrt{(2 - 7)^2 + (-5 - 7)^2} = \sqrt{169}$
$(x - 2)^2 + (y + 5)^2 = 169$

5. vertical, $p = 4$; $(x - 3)^2 = 16(y - 2)$

7.

$$(x-h)^2 = 4p(y-k)$$
$$(0-2)^2 = 4p(0+3)$$
$$4 = 4p(3)$$
$$\frac{4}{3} = 4p$$
$$(x-2)^2 = \frac{4}{3}(y+3)$$

or

$$(y-k)^2 = 4p(x-h)$$
$$(0+3)^2 = 4p(0-2)$$
$$9 = 4p(-2)$$
$$-\frac{9}{2} = 4p$$
$$(y+3)^2 = -\frac{9}{2}(x-2)$$

9. $a = 10$, $c = 6$, horizontal, $b^2 = 10^2 - 6^2 = 64$; $\quad \frac{x^2}{100} + \frac{y^2}{64} = 1$

11.

$$\frac{(x-2)^2}{a^2} + \frac{(y-3)^2}{b^2} = 1$$
$$\frac{(2-2)^2}{a^2} + \frac{(9-3)^2}{b^2} = 1$$
$$\frac{36}{b^2} = 1$$
$$b^2 = 36$$

$$\frac{(x-2)^2}{a^2} + \frac{(y-3)^2}{b^2} = 1$$
$$\frac{(0-2)^2}{a^2} + \frac{(3-3)^2}{b^2} = 1$$
$$\frac{4}{a^2} = 1$$
$$a^2 = 4$$

$$\frac{(y-3)^2}{36} + \frac{(x-2)^2}{4} = 1$$

13. horizontal, $c = 13$, $a = 5$
$b^2 = c^2 - a^2 = 144$

$$\frac{x^2}{25} - \frac{y^2}{144} = 1$$

15. $c = 10$, $a = 8$
$b^2 = c^2 - a^2 = 36$

$$\frac{(x-2)^2}{64} - \frac{(y+1)^2}{36} = 1$$

17. $y = x^2 - 3 \Rightarrow x^2 = y + 3$
$$x^2 + y^2 = 23$$
$$y + 3 + y^2 = 23$$
$$y^2 + y - 20 = 0$$
$$(y+5)(y-4) = 0$$
$$y = -5 \Rightarrow x^2 = -2$$
$$y = 4 \Rightarrow x^2 = 7$$
$$(\sqrt{7}, 4), (-\sqrt{7}, 4)$$

19.
$$y^2 - 4y - 6x - 14 = 0$$
$$y^2 - 4y = 6x + 14$$
$$y^2 - 4y + 4 = 6x + 14 + 4$$
$$(y-2)^2 = 6(x+3)$$

parabola

Exercise 9.1 (page 458)

1. $4! = 4 \cdot 3 \cdot 2 \cdot 1 = 24$

3. $3! \cdot 6! = 6 \cdot 720 = 4320$

5. $6! + 6! = 720 \cdot 720 = 1440$

7. $\dfrac{9!}{12!} = \dfrac{9!}{12 \cdot 11 \cdot 10 \cdot 9!} = \dfrac{1}{12 \cdot 11 \cdot 10}$

$$= \frac{1}{1320}$$

9. $\dfrac{5!\,7!}{9!} = \dfrac{5\cdot\cancel{4}\cdot\cancel{3}\cdot\cancel{2}\cdot1\cdot7!}{\underset{3}{\cancel{9}}\cdot\cancel{8}\cdot7!} = \dfrac{5}{3}$

11. $\dfrac{18!}{6!\,(18-6)!} = \dfrac{18\cdot17\cdot16\cdot15\cdot14\cdot13\cdot12!}{6\cdot5\cdot4\cdot3\cdot2\cdot1\cdot12!} = 18{,}564$

13. $(a+b)^3 = a^3 + \dfrac{3!}{1!\,2!}\,a^2b + \dfrac{3!}{2!\,1!}\,ab^2 + b^3 = a^3 + 3a^2b + 3ab^2 + b^3$

15. $(a-b)^5 = a^5 + \dfrac{5!}{1!4!}a^4(-b) + \dfrac{5!}{2!3!}a^3(-b)^2 + \dfrac{5!}{3!2!}a^2(-b)^3 + \dfrac{5!}{4!1!}a(-b)^4 + (-b)^5$

$\qquad = a^5 - 5a^4b + 10a^3b^2 - 10a^2b^3 + 5ab^4 - b^5$

17. $(2x+y)^3 = (2x)^3 + \dfrac{3!}{1!2!}(2x)^2y + \dfrac{3!}{2!1!}(2x)y^2 + y^3 = 8x^3 + 12x^2y + 6xy^2 + y^3$

19. $(x-2y)^3 = x^3 + \dfrac{3!}{1!2!}x^2(-2y) + \dfrac{3!}{2!1!}x(-2y)^2 + (-2y)^3 = x^3 - 6x^2y + 12xy^2 - 8y^3$

21. $(2x+3y)^4 = (2x)^4 + \dfrac{4!}{1!3!}(2x)^3(3y) + \dfrac{4!}{2!2!}(2x)^2(3y)^2 + \dfrac{4!}{3!1!}(2x)(3y)^3 + (3y)^4$

$\qquad = 16x^4 + 96x^3y + 216x^2y^2 + 216xy^3 + 81y^4$

23. $(x-2y)^4 = x^4 + \dfrac{4!}{1!3!}x^3(-2y) + \dfrac{4!}{2!2!}x^2(-2y)^2 + \dfrac{4!}{3!1!}x(-2y)^3 + (-2y)^4$

$\qquad = x^4 - 8x^3y^2 + 24x^2y^2 - 32xy^3 + 16y^4$

25. $(x-3y)^5$

$\qquad = x^5 + \dfrac{5!}{1!4!}x^4(-3y) + \dfrac{5!}{2!3!}x^3(-3y)^2 + \dfrac{5!}{3!2!}x^2(-3y)^3 + \dfrac{5!}{4!1!}x(-3y)^4 + (-3y)^5$

$\qquad = x^5 - 15x^4y + 90x^3y^2 - 270x^2y^3 + 405xy^4 - 243y^5$

27. $\left(\dfrac{x}{2}+y\right)^4 = \left(\dfrac{x}{2}\right)^4 + \dfrac{4!}{1!3!}\left(\dfrac{x}{2}\right)^3y + \dfrac{4!}{2!2!}\left(\dfrac{x}{2}\right)^2y^2 + \dfrac{4!}{3!1!}\left(\dfrac{x}{2}\right)y^3 + y^4$

$\qquad = \dfrac{x^4}{16} + \dfrac{x^3y}{2} + \dfrac{3x^2y^2}{2} + 2xy^3 + y^4$

29. The third term will have a $b^2 \Rightarrow \dfrac{4!}{2!2!}a^2b^2 = 6a^2b^2$

31. The fifth term will have a $b^4 \Rightarrow \dfrac{7!}{4!3!}a^3b^4 = 35a^3b^4$

33. The sixth term will have a $(-b)^5 \Rightarrow \dfrac{5!}{5!0!}a^0(-b)^5 = -b^5$

35. The fifth term will have a $b^4 \Rightarrow \dfrac{17!}{4!13!}a^{13}b^4 = 2380a^{13}b^4$

37. The second term will have a $(-\sqrt{2})^1 \Rightarrow \dfrac{4!}{1!3!}a^3(-\sqrt{2})^1 = -(4\sqrt{2})a^3$

39. The fifth term will have a $(\sqrt{3}\, b)^4 \Rightarrow \dfrac{9!}{4!5!}a^5(\sqrt{3}\, b)^4 = 1134a^5b^4$

41. The third term will have a $y^2 \Rightarrow \dfrac{4!}{2!2!}\left(\dfrac{x}{2}\right)^2 y^2 = \dfrac{3x^2y^2}{2}$

43. The tenth term will have a $\left(-\dfrac{s}{2}\right)^9 \Rightarrow \dfrac{11!}{9!2!}\left(\dfrac{r}{2}\right)^2\left(-\dfrac{s}{2}\right)^9 = 55 \cdot \dfrac{r^2}{4} \cdot \left(-\dfrac{s^9}{512}\right)$

$$= -\dfrac{55r^2s^9}{2048}$$

45. The fourth term will have a $b^3 \Rightarrow \dfrac{n!}{3!(n-3)!}\, a^{n-3}b^3$

47. The rth term will have a $b^{r-1} \Rightarrow \dfrac{n!}{(r-1)!(n-r+1)!}\, a^{n-r+1}b^{r-1}$

49. The sum of the nth row is 2^{n-1}.

51. Each term will contain $a^n\left(-\dfrac{1}{a}\right)^{10-n}$, or $a^n(-a)^{n-10}$. The constant term will occur when this product is 1, or when the sum of the exponents is 0.
$n + n - 10 = 0 \Rightarrow n = 5 \Rightarrow$ Find the term when $n = 5$.

$$\dfrac{10!}{5!5!}a^5\left(-\dfrac{1}{a}\right)^5 = -252$$

53. $\dfrac{n!}{0!(n-0)!} = \dfrac{n!}{1 \cdot n!} = 1$

Exercise 9.2 (page 464)

1. $f(1) = 5(1)(1-1) = 0$, $f(2) = 5(2)(2-1) = 10$, $f(3) = 5(3)(3-1) = 30$
$f(4) = 5(4)(4-1) = 60$, $f(5) = 5(5)(5-1) = 100$, $f(6) = 5(6)(6-1) = 150$
0, 10, 30, 60, 100, 150

3. 1, 6, 11, 16,... Add 5 to a term to get the next term. The next term is 21.

5. $a, a+d, a+2d, a+3d, \dots$ Add d to get the next term. The next term is $a+4d$.

7. 1, 3, 6, 10, ... The difference between terms is increasing. Add 5 to get the next term, which is then 15.

9. $1 + 2 + 3 + 4 + 5 = 15$ **11.** $3 + 3 + 3 + 3 + 3 = 15$

13. $2\left(\frac{1}{3}\right)^1 + 2\left(\frac{1}{3}\right)^2 + 2\left(\frac{1}{3}\right)^3 + 2\left(\frac{1}{3}\right)^4 + 2\left(\frac{1}{3}\right)^5 = \frac{2}{3} + \frac{2}{9} + \frac{2}{27} + \frac{2}{81} + \frac{2}{243} = \frac{242}{243}$

15. $3(1) - 2 + 3(2) - 2 + 3(3) - 2 + 3(4) - 2 + 3(5) - 2 = 35$

17. $a_2 = 2a_1 + 1 = 2(3) + 1 = 7$
$a_3 = 2a_2 + 1 = 2(7) + 1 = 15$
$a_4 = 2a_3 + 1 = 2(15) + 1 = 31$
3, 7, 15, 31

19. $a_2 = \frac{a_1}{2} = \frac{-4}{2} = -2$

$a_3 = \frac{a_2}{2} = \frac{-2}{2} = -1$

$a_4 = \frac{a_3}{2} = \frac{-1}{2} = -\frac{1}{2}$

$-4, \ -2, \ -1, \ -\frac{1}{2}$

21. $a_2 = a_1{}^2 = k^2$

$a_3 = a_2{}^2 = (k^2)^2 = k^4$

$a_4 = a_3{}^2 = (k^4)^2 = k^8$

k, k^2, k^4, k^8

23. $a_2 = \frac{2a_1}{k} = \frac{2(8)}{k} = \frac{16}{k}$

$a_3 = \frac{2a_2}{k} = \frac{2 \cdot \frac{16}{k}}{k} = \frac{32}{k^2}$

$a_4 = \frac{2a_3}{k} = \frac{2 \cdot \frac{32}{k^2}}{k} = \frac{64}{k^3}$

$8, \ \frac{16}{k}, \ \frac{32}{k^2}, \ \frac{64}{k^3}$

25. yes   **27.** no

29. $\displaystyle\sum_{k=1}^{5} 2k = 2 \sum_{k=1}^{5} k = 2(1 + 2 + 3 + 4 + 5) = 2(15) = 30$

31. $\displaystyle\sum_{k=1}^{4} (-2k^2) = -2 \sum_{k=1}^{4} k^2 = -2(1^2 + 2^2 + 3^2 + 4^2) = -2(30) = -60$

33. $\displaystyle\sum_{k=1}^{5}(3k-1) = \sum_{k=1}^{5}3k + \sum_{k=1}^{5}(-1) = 3\left(\sum_{k=1}^{5}k\right) - 5 = 3(15) - 5 = 40$

35. $\displaystyle\sum_{k=1}^{1000}\frac{1}{2} = 1000\cdot\frac{1}{2} = 500$ **37.** $\displaystyle\sum_{x=3}^{4}\frac{1}{x} = \frac{1}{3} + \frac{1}{4} = \frac{7}{12}$

39. $\displaystyle\sum_{x=1}^{4}(4x+1)^2 - \sum_{x=1}^{4}(4x-1)^2 = \sum_{x=1}^{4}(16x^2 + 8x + 1) - \sum_{x=1}^{4}(16x^2 - 8x + 1)$

$$= \sum_{x=1}^{4}(16x^2 + 8x + 1 - 16x^2 + 8x - 1)$$

$$= \sum_{x=1}^{4}16x = 16\sum_{x=1}^{4}x = 16(1+2+3+4) = 160$$

41. $\displaystyle\sum_{x=6}^{8}(5x-1)^2 + \sum_{x=6}^{8}(10x-1) = \sum_{x=6}^{8}(25x^2 - 10x + 1) + \sum_{x=6}^{8}(10x - 1)$

$$= \sum_{x=6}^{8}(25x^2 - 10x + 1 + 10x - 1)$$

$$= \sum_{x=6}^{8}25x^2 = 25\sum_{x=6}^{8}x^2 = 25(6^2 + 7^2 + 8^2) = 3725$$

43. Let $f(k) = k$ and $g(k) = k$. If the statement were true, then:
$1^2 + 2^2 + 3^2 + 4^2 + ... = (1 + 2 + 3 + 4 + ...)(1 + 2 + 3 + 4 + ...)$
This is obviously false (check with $n = 2$).

Exercise 9.3   (page 472)

1. 1, 3, 5, 7, 9, 11   **3.** $-8, -3, 2, 7, 12$

5. $9, \frac{23}{2}, 14, \frac{33}{2}, 19, \frac{43}{2}$ **7.** $a = 5, d = 2$
15th term $= 5 + 14\cdot 2 = 33$

$$\text{sum} = \frac{15(5+33)}{2} = 15\cdot 19 = 285$$

9. $\frac{27}{2}, \frac{30}{2}, \frac{33}{2}, \ldots$ $a = \frac{27}{2}, d = \frac{3}{2}$

20th term $= \frac{27}{2} + 19 \cdot \frac{3}{2} = 42$

$$\text{sum} = \frac{20 \cdot \left(\frac{27}{2} + 42\right)}{2}$$

$$= \frac{20 \cdot \frac{111}{2}}{2}$$

$$= \frac{1110}{2} = 555$$

11. $a + 24\left(\frac{1}{2}\right) = 10 \Rightarrow a = -2$

30th term $= -2 + 29\left(\frac{1}{2}\right) = \frac{25}{2}$

$$\text{sum} = \frac{30 \cdot \left(-2 + \frac{25}{2}\right)}{2}$$

$$= \frac{30 \cdot \frac{21}{2}}{2}$$

$$= \frac{315}{2} = 157\frac{1}{2}$$

13. $a + 4d = 14$
$a + d = 5$
Solve the system to get:
$d = 3, a = 2$
15th term $= 2 + 14 \cdot 3 = 44$

15. $20 = 10 + 4d \Rightarrow d = \frac{5}{2}$

$10, \frac{25}{2}, 15, \frac{35}{2}, 20$

17. $\frac{2}{3} = -7 + 5d \Rightarrow d = \frac{23}{15}$

$-7, -\frac{82}{15}, -\frac{59}{15}, -\frac{36}{15}, -\frac{13}{15}, \frac{2}{3}$

19. $10, 20, 40, 80$

21. $-2, -6, -18, -54$

23. $3, 3\sqrt{2}, 6, 6\sqrt{2}$

25. $2r^3 = 54 \Rightarrow r = 3$
$2, 6, 18, 54$

27. 5th term $= 4 \cdot 2^4 = 64$

$$\text{sum} = \frac{4 - 4 \cdot 2^5}{1 - 2} = \frac{4 - 128}{-1} = 124$$

29. 10th term: $2 \cdot (-3)^9 = -39{,}366$

$$\text{sum} = \frac{2 - 2 \cdot (-3)^{10}}{1 - (-3)}$$

$$= \frac{2 - 118{,}098}{4} = -29{,}524$$

31. $3, \frac{9}{2}, \frac{27}{4}, \frac{81}{8}, \ldots$ $a = 3, r = \frac{3}{2}$

6th term: $3 \cdot \left(\frac{3}{2}\right)^5 = \frac{729}{32}$

$$\text{sum} = \frac{3 - 3 \cdot \left(\frac{3}{2}\right)^6}{1 - \frac{3}{2}}$$

$$= \frac{3 - \frac{2187}{64}}{-\frac{1}{2}} = \frac{1995}{32}$$

33. $6 + 4 + \frac{8}{3} + \dots \quad a = 6, \; r = \frac{2}{3}$

$$S = \frac{6}{1 - \frac{2}{3}} = \frac{6}{\frac{1}{3}} = 18$$

35. $12, \; -6, \; 3, \; -\frac{3}{2}, \; \dots \; a = 12, \; r = -\frac{1}{2}$

$$S = \frac{12}{1 - -\frac{1}{2}} = \frac{12}{\frac{3}{2}} = 8$$

37. $20 = 10 \cdot r^4 \Rightarrow r = \sqrt[4]{2}$
$10, \; 10 \cdot \sqrt[4]{2}, \; 10 \cdot \sqrt[4]{4},$
$10 \cdot \sqrt[4]{8}, \; 20$

39. $2048 = 2 \cdot r^5 \Rightarrow r = \sqrt[5]{1024} = 4$
$2, \; 8, \; 32, \; 128, \; 512, \; 2048$

41. $\frac{5}{10} + \frac{5}{100} + \frac{5}{1000} + \dots$

$a = \frac{1}{2}, \; r = \frac{1}{10}$

$$S = \frac{\frac{1}{2}}{1 - \frac{1}{10}} = \frac{\frac{1}{2}}{\frac{9}{10}} = \frac{5}{9}$$

43. $\frac{25}{100} + \frac{25}{10,000} + \frac{25}{1,000,000} + \dots$

$a = \frac{1}{4}, \; r = \frac{1}{100}$

$$S = \frac{\frac{1}{4}}{1 - \frac{1}{100}} = \frac{\frac{1}{4}}{\frac{99}{100}} = \frac{25}{99}$$

Exercise 9.4 (page 475)

1. Let a = the original enrollment. Next year's enrollment is then $a + 0.10a$, or $1.1a$. The enrollments will be a geometric sequence with $r = 1.1$.
   in 8 years: $ar^8 = 623 \cdot (1.1)^7 \approx 1335.5$
   number of professors $\approx 1335.5/60 \approx 22.3$ 23 professors will be needed.

3. After 1st: $0.95(10)$; after 2nd: $0.95(0.95(10)) = (0.95)^2(10)$;
   height after 13th $= (0.95)^{13} \cdot 10 \approx 5.13$ meters

5. after 30 minutes: 2; after 60 minutes: 4; after 90 minutes: 8; ...

   $30 \Rightarrow 2^{30/30} = 2, \; 60 \Rightarrow 2^{60/30} = 4, \; 90 \Rightarrow 2^{90/30} = 8, \; \dots$

   after 10 hours is after 600 minutes $\Rightarrow 2^{600/30} = 2^{20} = 1,048,576$

7. compounded annually: interest $= 1000 \cdot 0.075 = 75$, amount $= \$1075$
   compounded daily: rate $= 0.0725/365$
   $\qquad\qquad$ after nth day, amount $= 1000 \cdot (1 + 0.0725/365)^n$
   $\qquad\qquad$ after 365 days, amount $= 1000 \cdot (1 + 0.0725/365)^{365} = \$1075.19$
   The 7.25% investment will yield an additional 19 cents.

9. $1990 + n \cdot 30 = 3010 \Rightarrow n = 34$ Thus there will be 34 doubling periods.
   population in 3010 $= 5,000,000,000 \cdot 2^{34} \approx 8.6 \times 10^{19}$

11. amount $= 50,000 \cdot 1.06^{n - 1988} = 50,000 \cdot 1.06^{22} \approx \$180,176.87$

13. $1000 \cdot (1 + 0.07/4)^{40} = \$2001.60$ **15.** $1000 \cdot (1 + 0.07/365)^{3650} = \$2013.62$

17. $2000 \cdot (1 + 0.11/4)^{180} = \$264,094.58$

19. $1 + 2 + 4 + \ldots + 2^{63} = \dfrac{1 - 1 \cdot 2^{64}}{1 - 2} \approx 1.84 \times 10^{19}$ grains

21. No. $1 - 0.999999 = 0.000001$. If they were equal, the difference would be 0.

Exercise 9.5   (page 481)

1. $n = 1$:
$$5(1) = \frac{5(1)(2)}{2}$$
$$5 = 5$$

$n = 2$:
$$5 + 5(2) = \frac{5(2)(2+1)}{2}$$
$$15 = 15$$

$n = 3$:
$$5 + 10 + 5(3) = \frac{5(3)(3+1)}{2}$$
$$30 = 30$$

$n = 4$:
$$5 + 10 + 15 + 5(4) = \frac{5(4)(4+1)}{2}$$
$$50 = 50$$

3. $n = 1$:
$$3(1) + 4 = \frac{1[3(1) + 11]}{2}$$
$$7 = 57$$

$n = 2$:
$$7 + 3(2) + 4 = \frac{2[3(2) + 11]}{2}$$
$$17 = 17$$

$n = 3$:
$$7 + 10 + 3(3) + 4 = \frac{3[3(3) + 11]}{2}$$
$$30 = 30$$

$n = 4$:
$$7 + 10 + 13 + 4(3) + 4 = \frac{4[3(4) + 11]}{2}$$
$$46 = 46$$

5. True for $n = 1$?
$$2(1) = 1(1 + 1)$$
$$2 = 1(2)$$
works for $n = 1$

Assume the formula works for $n = k$.
$$2 + 4 + 6 + \ldots + 2k = k(k + 1)$$

$$\boxed{2 + 4 + 6 + \ldots + 2k} + 2(k + 1) = \boxed{k(k + 1)} + 2(k + 1)$$
$$2 + 4 + 6 + \ldots + 2(k + 1) = k^2 + k + 2k + 2$$
$$= k^2 + 3k + 2$$
$$= (k + 1)(k + 2)$$

Since this is what results when $k + 1$ is used in the formula, we have shown that the formula works for $k + 1$ whenever it works for k.

7. True for $n = 1$?
$$4(1) - 1 = 1[2(1) + 1]$$
$$3 = 1(3) \qquad \text{works for } n = 1$$

Assume the formula works for $n = k$.
$$3 + 7 + 11 + \ldots + (4k - 1) = k(2k + 1)$$
$$\boxed{3 + 7 + 11 + \ldots + (4k - 1)} + [4(k+1) - 1] = \boxed{k(2k+1)} + [4(k+1) - 1]$$
$$3 + 7 + 11 + \ldots + [4(k+1) - 1] = 2k^2 + k + 4k + 4 - 1$$
$$= 2k^2 + 5k + 3$$
$$= (k+1)(2k+3)$$
$$= (k+1)[2(k+1) + 1]$$

Since this is what results when $k + 1$ is used in the formula, we have shown that the formula works for $k + 1$ whenever it works for k.

9. True for $n = 1$?
$$14 - 4(1) = 12(1) - 2(1)^2$$
$$10 = 12 - 2 \qquad \text{works for } n = 1$$

Assume the formula works for $n = k$.
$$10 + 6 + 2 + \ldots + (14 - 4k) = 12k - 2k^2$$
$$\boxed{10 + 6 + 2 + \ldots + (14 - 4k)} + [14 - 4(k+1)] = \boxed{12k - 2k^2} + [14 - 4(k+1)]$$
$$10 + 6 + 2 + \ldots + [14 - 4(k+1)] = 12k - 2k^2 + 14 - 4k - 4$$
$$= 12k + 12 - 2k^2 - 4k - 2$$
$$= 12(k+1) - 2(k+1)^2$$

Since this is what results when $k + 1$ is used in the formula, we have shown that the formula works for $k + 1$ whenever it works for k.

11. True for $n = 1$?

$$3(1) - 1 = \frac{1[3(1) + 1]}{2}$$

$$3 - 1 = \frac{1 \cdot 4}{2} \qquad \text{works for } n = 1$$

Assume the formula works for $n = k$.

$$2 + 5 + 8 + ... + (3k - 1) = \frac{k(3k + 1)}{2}$$

$$\boxed{2 + 5 + 8 + ... + (3k - 1)} + [3(k + 1) - 1] = \boxed{\frac{k(3k + 1)}{2}} + [3(k + 1) - 1]$$

$$2 + 5 + 8 + ... + [3(k + 1) - 1] = \frac{3k^2 + k}{2} + 3k + 3 - 1$$

$$= \frac{3k^2 + 7k + 4}{2}$$

$$= \frac{(k + 1)(3k + 4)}{2}$$

$$= \frac{(k + 1)[3(k + 1) + 1]}{2}$$

Since this is what results when $k + 1$ is used in the formula, we have shown that the formula works for $k + 1$ whenever it works for k.

13. True for $n = 1$?

$$1^2 = \frac{1(1 + 1)[2(1) + 1]}{6}$$

$$1 = \frac{1(2)(3)}{2} \qquad \text{works for } n = 1$$

Assume the formula works for $n = k$.

$$1^2 + 2^2 + 3^2 + ... + k^2 = \frac{k(k + 1)(2k + 1)}{6}$$

$$\boxed{1^2 + 2^2 + 3^2 + ... + k^2} + (k + 1)^2 = \boxed{\frac{k(k + 1)(2k + 1)}{6}} + (k + 1)^2$$

$$1^2 + 2^2 + 3^2 + ... + (k + 1)^2 = \frac{(2k^2 + k)(k + 1)}{6} + \frac{6(k + 1)(k + 1)}{6}$$

$$= \frac{(2k^2 + k + 6k + 6)(k + 1)}{6}$$

$$= \frac{(2k + 3)(k + 2)(k + 1)}{6}$$

$$= \frac{(k + 1)[(k + 1) + 1][2(k + 1) + 1]}{6}$$

Since this is what results when $k + 1$ is used in the formula, we have shown that the formula works for $k + 1$ whenever it works for k.

15. True for $n = 1$?
$$\frac{5}{3}(1) - \frac{4}{3} = 1\left(\frac{5}{6} \cdot 1 - \frac{1}{2}\right)$$
$$\frac{1}{3} = \frac{5}{6} - \frac{3}{6} \qquad \text{works for } n = 1$$

Assume the formula works for $n = k$.
$$\frac{1}{3} + 2 + \frac{11}{3} + \dots + \left(\frac{5}{3}k - \frac{4}{3}\right) = k\left(\frac{5}{6}k - \frac{1}{2}\right)$$
$$\boxed{\frac{1}{3} + 2 + \frac{11}{3} + \dots + \left(\frac{5}{3}k - \frac{4}{3}\right)} + \left[\frac{5}{3}(k+1) - \frac{4}{3}\right] = \boxed{k\left(\frac{5}{6}k - \frac{1}{2}\right)} + \left[\frac{5}{3}(k+1) - \frac{4}{3}\right]$$

$$\frac{1}{3} + 2 + \frac{11}{3} + \dots + \left[\frac{5}{3}(k+1) - \frac{4}{3}\right] = \frac{5}{6}k^2 - \frac{1}{2}k + \frac{5}{3}k + \frac{5}{3} - \frac{4}{3}$$
$$= \frac{5}{6}k^2 + \frac{7}{6}k + \frac{1}{3}$$
$$= (k+1)\left(\frac{5}{6}k + \frac{1}{3}\right)$$
$$= (k+1)\left[\frac{5}{6}(k+1) - \frac{1}{2}\right]$$

Since this is what results when $k + 1$ is used in the formula, we have shown that the formula works for $k + 1$ whenever it works for k.

17. True for $n = 1$?
$$\left(\frac{1}{2}\right)^1 = 1 - \left(\frac{1}{2}\right)^1$$
$$\frac{1}{2} = 1 - \frac{1}{2} \qquad \text{works for } n = 1$$

Assume the formula works for $n = k$.
$$\frac{1}{2} + \frac{1}{4} + \frac{1}{8} + \dots + \left(\frac{1}{2}\right)^k = 1 - \left(\frac{1}{2}\right)^k$$
$$\boxed{\frac{1}{2} + \frac{1}{4} + \frac{1}{8} + \dots + \left(\frac{1}{2}\right)^k} + \left(\frac{1}{2}\right)^{k+1} = \boxed{1 - \left(\frac{1}{2}\right)^k} + \left(\frac{1}{2}\right)^{k+1}$$
$$\frac{1}{2} + \frac{1}{4} + \frac{1}{8} + \dots + \left(\frac{1}{2}\right)^{k+1} = 1 - \frac{1}{2^k} + \frac{1}{2^{k+1}}$$
$$= 1 - \frac{2}{2^{k+1}} + \frac{1}{2^{k+1}}$$
$$= 1 - \frac{1}{2^{k+1}}$$
$$= 1 - \left(\frac{1}{2}\right)^{k+1}$$

Since this is what results when $k + 1$ is used in the formula, we have shown that the formula works for $k + 1$ whenever it works for k.

19. True for $n = 1$? $\qquad 2^{1-1} = 2^1 - 1$

$$2^0 = 2 - 1 \qquad \text{works for } n = 1$$

Assume the formula works for $n = k$.

$$2^0 + 2^1 + 2^2 + 2^3 + \ldots + 2^{k-1} = 2^k - 1$$

$$\boxed{2^0 + 2^1 + 2^2 + 2^3 + \ldots + 2^{k-1}} + 2^{(k+1)-1} = \boxed{2^k - 1} + 2^{(k+1)-1}$$

$$2^0 + 2^1 + 2^2 + 2^3 + \ldots + 2^{(k+1)-1} = 2^k + 2^k - 1$$

$$= 2 \cdot 2^k - 1$$

$$= 2^{k+1} - 1$$

Since this is what results when $k + 1$ is used in the formula, we have shown that the formula works for $k + 1$ whenever it works for k.

21. True for $n = 1$? $\qquad\qquad x - y$ is a factor of $x^1 - y^1$.

Assume that $x - y$ is a factor of $x^k - y^k$.

Is $x - y$ a factor of $x^{k+1} - y^{k+1}$?

$$x^{k+1} - y^{k+1} = x^{k+1} - xy^k + xy^k - y^{k+1}$$

$$= x\boxed{(x^k - y^k)} + y^k(x - y)$$

$$= x\boxed{(x-y)(\text{SOMETHING})} + y^k(x - y) \quad \{\text{by assumption above}\}$$

$$= (x - y)[x(\text{SOMETHING}) + y^k]$$

Thus, $x - y$ is a factor of $x^{k+1} - y^{k+1}$ when it is a factor of $x^k - y^k$.

23. True for $n = 3$? $\quad$ If a figure has three sides, then the sum $= (3 - 2)180° = 180°$. Assume that a k sided figure has a sum of angles equal to $(k - 2)180°$.
Take a polygon with $k + 1$ sides. Consider two adjacent sides (they have a common endpoint). Connect the endpoints the two adjacent sides **do not** have in common. You now have a k-sided polygon next to a triangle. The sum of the measures of the $(k + 1)$-sided polygon is then the sum of the sum of the measures of the k-sided polygon and the sum of the measures of the triangle.

$$\boxed{k+1 \text{ sum}} = \boxed{k \text{ sum}} + \boxed{\text{Triangle sum}}$$

$$\boxed{k+1 \text{ sum}} = (k-2)180° + 180°$$

$$= 180°k - 360° + 180°$$

$$= 180°k - 180°$$

$$= (k - 1)180°$$

$$= [(k + 1) - 2]180°$$

This is the sum the formula yields when $n = k + 1$, so the formula works for $k + 1$ whenever it works for k.

25. Assume the formula works for $n = k$.
$$1 + 2 + 3 + \dots + n = \tfrac{n}{2}(n+1) + 1$$

$$\boxed{1 + 2 + 3 + \dots + n} + (n+1) = \boxed{\tfrac{n}{2}(n+1) + 1} + (n+1)$$

$$1 + 2 + 3 + \dots + n + 1 = \left(\tfrac{n}{2} + 1\right)(n+1) + 1$$

$$= 2 \cdot \left(\tfrac{n}{2} + 1\right) \cdot \tfrac{1}{2}(n+1) + 1$$

$$= (n+2)\left(\tfrac{n+1}{2}\right) + 1$$

$$= \left(\tfrac{n+1}{2}\right)(n+2) + 1$$

Since this is what results when $k+1$ is used in the formula, we have shown that the formula works for $k+1$ whenever it works for k. However, the formula does not work for $n = 1$. Thus, the formula is not valid.

27. True for $n = 1$? $7^1 - 1 = 6$, which is divisible by 6.
Assume that $7^k - 1$ is divisible by 6.   Then $7^k - 1 = 6 \cdot x$, where $x \in \mathbb{R}$.
$7^{k+1} - 1 = 7^k \cdot 7 - 1 = (6x+1) \cdot 7 - 1 = 42x + 7 - 1 = 42x + 6 = 6(7x+1)$
Thus 7^{k+1} is divisible by 6 whenever 7^k is divisible by 6.

29. True for $n = 1$? $1 + r^1 = \dfrac{1 - r^2}{1 - r}$

$$1 + r = \frac{(1+r)(1-r)}{1-r} = 1 + r \quad \text{TRUE.}$$

Assume the formula works for $n = k$.
$$1 + r + r^2 + \dots + r^k = \frac{1 - r^{k+1}}{1 - r}$$

$$\boxed{1 + r + r^2 + \dots + r^k} + r^{k+1} = \boxed{\frac{1 - r^{k+1}}{1 - r}} + r^{k+1}$$

$$1 + r + r^2 + \dots + r^{k+1} = \frac{1 - r^{k+1}}{1 - r} + \frac{r^{k+1}(1-r)}{1 - r}$$

$$= \frac{1 - r^{k+1} + r^{k+1} - r^{k+2}}{1 - r}$$

$$= \frac{1 - r^{(k+1)+1}}{1 - r}$$

This is the sum the formula yields when $n = k+1$, so the formula works for $k+1$ whenever it works for k.

Exercise 9.6 (page 488)

1. $8 \cdot 6 \cdot 3 = 144$ lunches

3. $8 \cdot 10 \cdot 10 \cdot 10 \cdot 10 \cdot 10 = 8 \cdot 10^6$
$$= 8,000,000$$

5. Consider er to be a group that cannot be divided (let x represent er). Then the problem is to rearrange the letters *numbx*. This can be done $5 \cdot 4 \cdot 3 \cdot 2 \cdot 1 = 120$ ways. For each of these ways, you could rearrange the e and the r (to get re). Thus the total number of ways is $120 \cdot 2 = 240$.

7. The word must appear as $\boxed{} \, L \, U \, \boxed{} \, \boxed{}$, where one of the F's must appear in each box. There are $3! = 6$ ways of doing this.

9. $P(7,4) = \dfrac{7!}{(7-4)!} = \dfrac{7!}{3!} = 7 \cdot 6 \cdot 5 \cdot 4 = 840$

11. $C(7,4) = \dfrac{7!}{4!(7-4)!} = \dfrac{7!}{4!3!} = \dfrac{7 \cdot 6 \cdot 5 \cdot 4!}{4! \, 3!} = \dfrac{7 \cdot 6 \cdot 5}{3 \cdot 2} = 35$

13. $P(5,5) = \dfrac{5!}{(5-5)!} = \dfrac{5!}{0!} = 5! = 120$

15. $\dbinom{5}{4} = \dfrac{5!}{4!(5-4)!} = \dfrac{5 \cdot 4!}{4! \cdot 1!} = 5$

17. $\dbinom{5}{0} = \dfrac{5!}{0!(5-0)!} = \dfrac{5!}{1 \cdot 5!} = 1$

19. $P(5,4) \cdot C(5,3) = \dfrac{5!}{(5-4)!} \cdot \dfrac{5!}{3!(5-3)!} = \dfrac{5!}{1!} \cdot \dfrac{5 \cdot 4 \cdot 3!}{3! \cdot 2!} = 120 \cdot 10 = 1200$

21. $\dbinom{5}{3}\dbinom{4}{3}\dbinom{3}{3} = \dfrac{5!}{3!2!} \cdot \dfrac{4!}{3!1!} \cdot \dfrac{3!}{3!0!} = 10 \cdot 4 \cdot 1 = 40$

23. $\dbinom{68}{66} = \dfrac{68!}{66!2!} = \dfrac{68 \cdot 67 \cdot 66!}{66!2!} = \dfrac{68 \cdot 67}{2} = 2278$

25. $8! = 40,320$

27. Arrange the men, then the women.
$$5! \cdot 5! = 120 \cdot 120 = 14,400$$

29. Pick 3 from 30 numbers, with no two numbers the same.
$30 \cdot 29 \cdot 28 = 24,360$

31. $(8-1)! = 5040$

33. Think of the two who sit together as one big "person" at first. Then there are 5 "people" to arrange. $(5-1)! = 24$
You could switch the two who you have grouped: $2 \cdot 24 = 48$ ways.

35. As in #33, group each pair as a big "person" and arrange the 5 "people." $(5-1)! = 24$. Rearrange each pair: $2 \cdot 2 \cdot 24 = 96$ ways.

37. Order is not important here, so $\binom{10}{4} = \dfrac{10!}{4!6!} = \dfrac{10 \cdot 9 \cdot 8 \cdot 7}{4 \cdot 3 \cdot 2} = 210$ ways.

39. Order is important here, so $P(4,4) = 4! = 24$ ways.

41. 7 distinct letters, with order important: $P(7,7) = 7! = 5040$ ways.

43. Choose letters, then digits: $(25 \cdot 24) \cdot (9 \cdot 9 \cdot 8 \cdot 7) = 2{,}721{,}600$.

45. Assume that balls of the same color are not distinguishable and that order is not important. Imagine choosing the red and then the white.
$$\binom{6}{3}\binom{8}{3} = \frac{6!}{3!3!} \cdot \frac{8!}{3!5!} = \frac{6 \cdot 5 \cdot 4}{3 \cdot 2 \cdot 1} \cdot \frac{8 \cdot 7 \cdot 6}{3 \cdot 2 \cdot 1} = 20 \cdot 56 = 1120.$$

47. Assume that the people of each party are not distinguishable and that order is not important. Choose the Democrats first, then the Republicans.
$$\binom{12}{4}\binom{10}{3} = \frac{12!}{4!8!} \cdot \frac{10!}{3!7!} = \frac{12 \cdot 11 \cdot 10 \cdot 9}{4 \cdot 3 \cdot 2 \cdot 1} \cdot \frac{10 \cdot 9 \cdot 8}{3 \cdot 2 \cdot 1} = 59{,}400.$$

49. $\binom{17}{2} = \dfrac{17!}{2!15!} = \dfrac{17 \cdot 16 \cdot 15!}{2! \cdot 15!} = 126$. Then they could switch: $2(136) = 272$.

51. Each pair of points determines a line (order of points is not important). So find the number of ways of choosing 2 from 8: $\binom{8}{2} = \dfrac{8!}{2!6!} = \dfrac{8 \cdot 7 \cdot 6!}{2!6!} = 28$.

53. Find the number of ways of choosing 5 from 10 (order not important, since each can play any position): $\binom{10}{5} = \dfrac{10!}{5!5!} = \dfrac{10 \cdot 9 \cdot 8 \cdot 7 \cdot 6}{5 \cdot 4 \cdot 3 \cdot 2 \cdot 1} = 252$.

55. Choose 5 of the 30: $\binom{30}{5} = \dfrac{30!}{5!25!} = \dfrac{30 \cdot 29 \cdot 28 \cdot 27 \cdot 26}{5 \cdot 4 \cdot 3 \cdot 2 \cdot 1} = 142{,}506$ ways.

57. $C(12,10) = \dfrac{12!}{10!2!} = 66$ ways.

59.

```
                        1
                     1     1
                  1     2     1
               1     3     3     1
            1     4     6     4     1
         1     5     10    10    5     1
      1     6     15    20    15    6     1
   1     7     21    35    35    21    7     1
1     8     28    56    70   [56]   28    8     1
```

61. $c(n,n) = \frac{n!}{n!0!} = 1$

63. $\binom{n}{r} = \frac{n!}{r!(n-r)!}$

$$\binom{n}{n-r} = \frac{n!}{(n-r)!(n-(n-r))!}$$

$$= \frac{n!}{r!(n-r)!}$$

Exercise 9.7 (page 494)

1. $\{(1,H),(2,H),(3,H),(4,H),(5,H),(6,H),(1,T),(2,T),(3,T),(4,T),(5,T),(6,T)\}$

3. {A, B, C, D, E, F, G, H, I, J, K, L, M, N, O, P, Q, R, S, T, U, V, W, X, Y, Z}

5. $\frac{1}{6}$

7. $\frac{4}{6} = \frac{2}{3}$

9. $\frac{19}{42}$

11. $\frac{13}{42}$

13. $\frac{3}{8}$

15. $\frac{0}{8} = 0$

17. Possible rolls of 4: $(1, 3), (2, 2), (3, 1)$. Probability $= \frac{3}{36} = \frac{1}{12}$

19. $\frac{4}{52} \cdot \frac{4}{52} = \frac{1}{13} \cdot \frac{1}{13} = \frac{1}{169}$

21. $\frac{5}{12}$

23. There are $\binom{52}{13}$ total bridge hands. There are $4 \cdot \binom{13}{13}$ hands of only one suit.

probability $= \dfrac{4 \cdot \binom{13}{13}}{\binom{52}{13}} \approx 6.3 \times 10^{-12}$

25. 0 (only 4 aces in a deck)

27. $\frac{12}{52} = \frac{3}{13}$

29. $\dfrac{\binom{5}{5}}{\binom{6}{5}} = \frac{1}{6}$

31. # total rolls $= 6 \cdot 6 \cdot 6 = 216$;
rolls of 11: $\{(6,4,1),(6,3,2),(6,2,3),(6,1,4),(5,5,1),(5,4,2),(5,3,3),(5,2,4),$
$(5,1,5),(4,6,1),(4,5,2),(4,4,3),(4,3,4),(4,2,5),(4,1,6),(3,6,2),(3,5,3),$
$(3,4,4),(3,3,5),(3,2,6),(2,6,3),(2,5,4),(2,4,5),(2,3,6),(1,6,4),(1,5,5),(1,4,6)\}$

probability $= \frac{27}{216} = \frac{1}{8}$

33. $\dfrac{\binom{5}{3}}{2^5} = \dfrac{10}{32} = \dfrac{5}{16}$

35. $\{(P,P,P,P), (P,P,P,F), (P,P,F,P), (P,P,F,F), (P,F,P,P), (P,F,P,F),$
$(P,F,F,P), (P,F,F,F), (F,P,P,P), (F,P,P,F), (F,P,F,P), (F,P,F,F),$
$(F,F,P,P), (F,F,P,F), (F,F,F,P), (F,F,F,F)\}$

37. Use #35. $\dfrac{4}{16} = \dfrac{1}{4}$

39. Use #35. $\dfrac{4}{16} = \dfrac{1}{4}$

41. sum $= 1$

43. $\dfrac{32}{119}$

45. $\dfrac{\binom{8}{4}}{\binom{10}{4}} = \dfrac{70}{210} = \dfrac{1}{3}$

47. $P(A \cap B) = P(A) \cdot P(B|A)$
$\qquad\qquad = (0.3)(0.6) = 0.18$

49. $A =$ has luxury car
$B =$ has a personal computer
Given: $P(A) = 0.2,$
$\qquad\quad P(B|A) = 0.7$
$P(A \cap B) = (0.2)(0.7) = 0.14$

51. $A =$ watch news
$B =$ watch soap opera
Given: $P(A) = 0.75$
$\qquad\quad P(A \cap B) = 0.25$
$P(A \cap B) = P(A) \cdot P(B|A)$
$\qquad 0.25 = (0.75) \cdot P(B|A)$
$\qquad 0.33 \approx P(B|A)$
about 33%

53. No. $P(B)$ would have to equal 1.166, which is greater than 1.

Exercise 9.8 (page 500)

1. $\dfrac{26}{52} = \dfrac{1}{2}$

3. $\dfrac{26}{52} + \dfrac{4}{52} - \dfrac{2}{52} = \dfrac{28}{52} = \dfrac{7}{13}$

5. $\dfrac{4}{52} \cdot \dfrac{3}{51} = \dfrac{1}{13} \cdot \dfrac{1}{17} = \dfrac{1}{221}$

7. $\dfrac{13}{52} \cdot \dfrac{25}{51} = \dfrac{1}{4} \cdot \dfrac{25}{51} = \dfrac{25}{204}$

9. $\dfrac{6}{36} + \dfrac{5}{36} - \dfrac{0}{36} = \dfrac{11}{36}$

11. $\dfrac{3}{36} + \dfrac{18}{36} = \dfrac{21}{36} = \dfrac{7}{12}$

13. $\dfrac{7}{16} + \dfrac{3}{16} = \dfrac{10}{16} = \dfrac{5}{8}$

15. $\dfrac{7}{16} + \dfrac{6}{16} = \dfrac{13}{16}$

17. $\dfrac{7}{16} \cdot \dfrac{6}{16} = \dfrac{21}{128}$

19. $\dfrac{9}{16} \cdot \dfrac{8}{15} \cdot \dfrac{7}{14} = \dfrac{3}{20}$

21. $\dfrac{7}{7} \cdot \dfrac{1}{7} \cdot \dfrac{1}{7} = \dfrac{1}{49}$

23. $\dfrac{365}{365} \cdot \dfrac{364}{365} \cdot \dfrac{363}{365} \cdot \dfrac{362}{365} \cdot \dfrac{361}{365} \approx 0.973$

25. The probability will be 1 − the probability that neither solves the problem:

$$1 - \frac{3}{4} \cdot \frac{3}{5} = 1 - \frac{9}{20} = \frac{11}{20}$$

27. $\frac{1}{2} \cdot \frac{1}{3} \cdot \frac{3}{4} = \frac{1}{8}$

29. $\frac{1}{2} \cdot \frac{2}{3} \cdot \frac{1}{4} + \frac{1}{3} \cdot \frac{1}{2} \cdot \frac{1}{4} + \frac{3}{4} \cdot \frac{1}{2} \cdot \frac{2}{3} = \frac{9}{24} = \frac{3}{8}$

31. $\frac{1}{3} \cdot \frac{1}{2} \cdot \frac{5}{6} + \frac{2}{3} \cdot \frac{1}{2} \cdot \frac{5}{6} + \frac{2}{3} \cdot \frac{1}{2} \cdot \frac{1}{6} = \frac{17}{36}$

33. $0.05(0.60) = 0.03$

Exercise 9.9 (page 503)

1. $\frac{1}{6}$

3. $\dfrac{\frac{5}{6}}{\frac{1}{6}} = \frac{5}{1} = 5 \text{ to } 1$

5. $\dfrac{\frac{1}{2}}{\frac{1}{2}} = \frac{1}{1} = 1 \text{ to } 1$

7. $\frac{5}{36}$

9. $\dfrac{\frac{31}{36}}{\frac{5}{36}} = \frac{31}{5} = 31 \text{ to } 5$

11. $\dfrac{\frac{18}{36}}{\frac{18}{36}} = \frac{18}{18} = 1 \text{ to } 1$

13. $\dfrac{\frac{1}{13}}{\frac{12}{13}} = \frac{1}{12} = 1 \text{ to } 12$

15. $\dfrac{\frac{12}{52}}{\frac{40}{52}} = \frac{12}{40} = 3 \text{ to } 10$

17. $\frac{5}{5+2} = \frac{5}{7}$

19. 1 to 90

21. $\dfrac{\frac{15}{16}}{\frac{1}{16}} = \frac{15}{1} = 15 \text{ to } 1$

23. $\frac{1}{1+8} = \frac{1}{9}$

25. $E = \frac{1}{13} \cdot 5 + \frac{1}{13} \cdot 4 + \frac{11}{13} \cdot 0 = \frac{9}{13}$. No. The expected winnings are $\$\frac{9}{13}$.

27. $P(\text{three boys}) = \frac{1}{2} \cdot \frac{1}{2} \cdot \frac{1}{2} = \frac{1}{8}$. Odds against $= \dfrac{\frac{7}{8}}{\frac{1}{8}} = \frac{7}{1} = 7 \text{ to } 1$.

29. $P(5 \text{ heads}) = \dfrac{\binom{5}{5}}{2^5} = \frac{1}{32}$, $P(4 \text{ heads}) = \dfrac{\binom{5}{4}}{2^5} = \frac{5}{32}$, $P(3 \text{ heads}) = \dfrac{\binom{5}{3}}{2^5} = \frac{10}{32} = \frac{5}{16}$.

expectation $= \frac{1}{32} \cdot 5 + \frac{5}{32} \cdot 4 + \frac{10}{32} \cdot 3 = \frac{55}{32} \approx \$1.72$.

31. $\frac{1}{13} \cdot 1 + \frac{1}{13}(2 + 3 + 4 + 5 + 6 + 7 + 8 + 9 + 10) + \frac{3}{13}(10) = \frac{1}{13} + \frac{54}{13} + \frac{30}{13} = \frac{85}{13} \approx 6.54$.

33. The probability of getting a given question right by guessing is $\frac{1}{5}$. Find the number of ways of choosing 7 of the 8 to get right, then multiply by the probability of getting 7 right and 1 wrong.

$$\binom{8}{7} \cdot \left(\frac{1}{5}\right)^7 \cdot \frac{4}{5} = 8 \cdot \frac{1}{5^7} \cdot \frac{4}{5} = \frac{32}{5^8} = \frac{32}{390,625}$$

Chapter 9 Review Exercises (page 505)

1. $7! \cdot 0! \cdot 1! \cdot 3! = 5040 \cdot 1 \cdot 1 \cdot 6 = 30{,}240$

3. $(2a - b)^3 = (2a)^3 + \frac{3!}{1!2!}(2a)^2(-b) + \frac{3!}{2!1!}(2a)(-b)^2 + (-b)^3$

$$= 8a^3 - 12a^2b + 6ab^2 - b^3$$

5. The 4th term has b^3: $\frac{8!}{3!5!}a^5b^3 = 56a^5b^3$

7. The 7th term has $(-y)^6$: $\frac{9!}{6!3!}x^3(-y)^6 = 84x^3y^6$

9. $a_1 = 5$, $a_2 = 3(5) + 2 = 17$, $a_3 = 3(17) + 2 = 53$, $a_4 = 3(53) + 2 = 161$

11. $\displaystyle\sum_{k=1}^{4} 3k^2 = 3\sum_{k=1}^{4} k^2 = 3(1^2 + 2^2 + 3^2 + 4^2) = 3(30) = 90$

13. $\displaystyle\sum_{k=5}^{8} (k^3 + 3k^2) = \sum_{k=5}^{8} k^3 + 3\sum_{k=5}^{8} k^2$

$$= 5^3 + 6^3 + 7^3 + 8^3 + 3(5^2 + 6^2 + 7^2 + 8^2) = 1718$$

15. $a = 5, d = 4$
29th: $5 + (29 - 1)4 = 117$

17. $a = 6, d = -7$
15th: $6 + (15 - 1)(-7) = -92$

19. $a = 81$, $r = \frac{1}{3}$
11th: $ar^{10} = 81 \cdot \left(\frac{1}{3}\right)^{10} = \frac{1}{729}$

21. $a = 9$, $r = \frac{1}{2}$
15th: $ar^{14} = 9 \cdot \left(\frac{1}{2}\right)^{14} = \frac{9}{16,384}$

23. $a = 5$, $d = 4$, 40th term $= 141$

$$\text{sum} = \frac{40(5 + 161)}{2} = 3320$$

25. $a = 6, d = -7$, 40th term $= -267$

$$\text{sum} = \frac{40(6 - 267)}{2} = -5220$$

27. $a = 81, r = \frac{1}{3}$

$$\text{sum} = \frac{81 - 81 \cdot \left(\frac{1}{3}\right)^8}{1 - \frac{1}{3}} = \frac{3280}{27}$$

29. $a = 9, r = \frac{1}{2}$

$$\text{sum} = \frac{9 - 9 \cdot \left(\frac{1}{2}\right)^8}{1 - \frac{1}{2}} = \frac{2295}{128}$$

31. $a = \frac{1}{3}, \ r = \frac{1}{2}$

$$\text{sum} = \frac{\frac{1}{3}}{1 - \frac{1}{2}} = \frac{2}{3}$$

33. $a = 1, \ r = \frac{3}{2}$

There is no sum (r is greater than 1).

35. $\frac{3}{10} + \frac{3}{100} + \frac{3}{1000} + \cdots$

$a = \frac{3}{10}, \ r = \frac{1}{10}$

$$\text{sum} = \frac{\frac{3}{10}}{1 - \frac{1}{10}} = \frac{\frac{3}{10}}{\frac{9}{10}} = \frac{1}{3}$$

37. $\frac{17}{100} + \frac{17}{10,000} + \frac{17}{1,000,000} + \cdots$

$a = \frac{17}{100}, \ r = \frac{1}{100}$

$$\text{sum} = \frac{\frac{17}{100}}{1 - \frac{1}{100}} = \frac{\frac{17}{100}}{\frac{99}{100}} = \frac{17}{99}$$

39. $8 = 2 + 4d \ \Rightarrow \ d = \frac{3}{2}$

$2, \frac{7}{2}, 5, \frac{13}{2}, 8$

41. $8 = 2r^4 \ \Rightarrow \ r = \sqrt{2}$

$2, 2\sqrt{2}, 4, 4\sqrt{2}, 8$

43. $a = \frac{1}{3}, \ r = 3$

$$\text{sum} = \frac{\frac{1}{3} - \frac{1}{3} \cdot 3^8}{1 - 3} = \frac{3280}{3}$$

45. $64 = 4r^2 \ \Rightarrow \ r = 4: \quad 16$

47. in 10 yrs: $4000(1 + 0.05)^{10} = 6516$; 5 years ago: $4000(1 + 0.05)^{-5} = 3134$

49.

$$1^3 = \frac{1^2(1+1)^2}{4}$$

$$1 = \frac{1 \cdot 2^2}{4}$$

$$1 = 1$$

$$1^3 + 2^3 = \frac{2^2(2+1)^2}{4}$$

$$1 + 8 = \frac{4 \cdot 3^2}{4}$$

$$9 = 9$$

$$1^3 + 2^3 + 3^3 = \frac{3^2(3+1)^2}{4}$$

$$1 + 8 + 27 = \frac{9 \cdot 4^2}{4}$$

$$36 = 36$$

$$1^3 + 2^3 + 3^3 + 4^3 = \frac{4^2(4+1)^2}{4}$$

$$1 + 8 + 27 + 64 = \frac{16 \cdot 5^2}{4}$$

$$100 = 100$$

51. $P(8,5) = \dfrac{8!}{(8-5)!} = \dfrac{8!}{3!} = 6720$

53. $0! \cdot 1! = 1 \cdot 1 = 1$

55. $P(8,6) \cdot C(8,6) = \dfrac{8!}{2!} \cdot \dfrac{8!}{6!2!} = 20{,}160 \cdot 28 = 564{,}480$

57. $C(7,5) \cdot P(4,0) = \dfrac{7!}{5!2!} \cdot \dfrac{4!}{4!} = 21 \cdot 1 = 21$

59. $\dfrac{P(8,5)}{C(8,5)} = \dfrac{6720}{56} = 120$

61. $\dfrac{C(6,3)}{C(10,3)} = \dfrac{20}{120} = \dfrac{1}{6}$

63.

$$\begin{array}{cccccccc}
 & & & H & & & & & & & & & & T \\
 & & H & & & T & & & & & H & & & & T \\
 & H & & T & & H & & T & & & H & & T & & H & & T \\
H & & T & & H & & T & & H & & T & & H & & T & & H & & T
\end{array}$$

65. $\dfrac{\dbinom{4}{3}\dbinom{4}{2}}{\dbinom{52}{5}} = \dfrac{24}{2{,}598{,}960} = \dfrac{1}{108{,}290}$

67. Think of the two pairs as units. Arrange the 6 people and 2 units in the circle: $7! = 5040$. Multiply by 2 for each pair to switch the order of each pair: $5040 \cdot 2 \cdot 2 = 20{,}160$

69. $\dfrac{\dbinom{4}{4}\dbinom{4}{4}\dbinom{4}{4}\dbinom{4}{3}}{\dbinom{52}{13}} \approx \dfrac{4}{6.35021 \times 10^{11}} \approx 6.3 \times 10^{-12}$

71. $\dfrac{13}{52} + \dfrac{13}{52} - \dfrac{0}{52} = \dfrac{26}{52} = \dfrac{1}{2}$

73. $\dfrac{1}{\dbinom{52}{5}} = \dfrac{1}{2{,}598{,}960}$

75. Find $1 - P(4 \text{ heads}) = 1 - \left(\dfrac{1}{2}\right)^4 = \dfrac{15}{16}$

77. $P(4 \text{ girls}) = \left(\frac{1}{2}\right)^4 = \frac{1}{16}$; odds $= \dfrac{\frac{1}{16}}{\frac{15}{16}} = \frac{1}{15} = 1$ to 15

79. $P(\text{neither marry}) = \frac{1}{6} \cdot \frac{1}{4} = \frac{1}{24}$;

odds against neither becoming a husband = odds for neither marrying = 1 to 23.

81. $800(0.83) = 664$

Chapter 9 Test   (page 508)

1. $3! \cdot 0! \cdot 4! \cdot 1! = 6 \cdot 1 \cdot 24 \cdot 1 = 144$

3. 2nd term has $(2y)^1$: $\dfrac{5!}{1!4!}x^4(2y) = 10x^4y$

5. $\displaystyle\sum_{k=1}^{3} 4k + 1 = 4\sum_{k=1}^{3} k + \sum_{k=1}^{3} 1 = 4(1+2+3) + 1 + 1 + 1 = 27$

7. $a = 2, d = 3$; 10th term $= 2 + (10-1)3 = 29$; sum $= \dfrac{10(2+29)}{2} = 155$

9. $24 = 4 + (5-1)d \Rightarrow d = 5$:   4, 9, 14, 19, 24

11. $a = \frac{1}{4}, r = \frac{1}{2}$:   sum $= \dfrac{\frac{1}{4} - \frac{1}{4}\cdot\left(\frac{1}{2}\right)^{10}}{1 - \frac{1}{2}} = \dfrac{1023}{4}$

13. $c(0.75)^3 = 0.421875c$

15. True for $n = 1$?

$$1^3 = \frac{1^2(1+1)^2}{4}$$

$$1 = \frac{1 \cdot 2^2}{4} \qquad \text{works for } n = 1$$

Assume the formula works for $n = k$.

$$1^3 + 2^3 + 3^3 + \ldots + k^3 = \frac{k^2(k+1)^2}{4}$$

$$\boxed{1^3 + 2^3 + 3^3 + \ldots + k^3} + (k+1)^3 = \boxed{\frac{k^2(k+1)^2}{4}} + (k+1)^3$$

$$1^3 + 2^3 + 3^3 + \ldots + (k+1)^3 = \frac{k^2(k+1)^2}{4} + \frac{4(k+1)^3}{4}$$

$$= \frac{(k+1)^2[k^2 + 4(k+1)]}{4}$$

$$= \frac{(k+1)^2(k+2)^2}{4}$$

$$= \frac{(k+1)^2[(k+1)+1]^2}{4}$$

Since this is what results if $k+1$ is substituted for n in the formula, the formula works for $k + 1$ whenever it works for k.

17. $P(7,2) = \dfrac{7!}{5!} = 42$

19. $C(8,2) = \dfrac{8!}{2!6!} = 28$

21. $4! \cdot 4! = 24 \cdot 24 = 576$

23. $\dfrac{5!}{2!} = 60$

25. $\dfrac{1}{6}$

27. $\dfrac{\dbinom{13}{5}}{\dbinom{52}{5}} = \dfrac{1287}{2{,}598{,}960} = \dfrac{429}{866{,}320}$

29. $1 - 0.1 = 0.9$

31. $\dfrac{26}{52} + \dfrac{12}{52} - \dfrac{6}{52} = \dfrac{32}{52} = \dfrac{8}{13}$

33. $\dfrac{3}{1+3} = \dfrac{3}{4}$

Cumulative Review Exercises (page 509)

1. $2x + y = 8$
$x - 2y = -1$

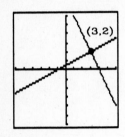

(3,2)

solution: $(3, 2)$

3. $5x = 3y + 12 \Rightarrow \boxed{3y = 5x - 12}$
$$2x - 3y = 3$$
$$2x - (5x - 12) = 3$$
$$2x - 5x + 12 = 3$$
$$-3x = -9 \Rightarrow x = 3$$

$$3y = 5x - 12$$
$$3y = 3 \Rightarrow y = 1$$

$$x = 3,\ y = 1$$

5.
$$\begin{bmatrix} 1 & 1 & 1 & 4 \\ 1 & 2 & -1 & 1 \\ 2 & -1 & 1 & 3 \end{bmatrix} \Rightarrow \begin{bmatrix} 1 & 1 & 1 & 4 \\ 0 & 1 & -2 & -3 \\ 0 & -3 & -1 & -5 \end{bmatrix} \Rightarrow \begin{bmatrix} 1 & 0 & 3 & 7 \\ 0 & 1 & -2 & -3 \\ 0 & 0 & -7 & -14 \end{bmatrix}$$

$$\begin{bmatrix} 1 & 0 & 3 & 7 \\ 0 & 1 & -2 & -3 \\ 0 & 0 & 1 & 2 \end{bmatrix} \Rightarrow \begin{bmatrix} 1 & 0 & 0 & 1 \\ 0 & 1 & 0 & 1 \\ 0 & 0 & 1 & 2 \end{bmatrix} \qquad \boxed{x = 1,\ y = 1,\ z = 2}$$

7. $\begin{bmatrix} 2 & 1 \\ 1 & 4 \end{bmatrix} + \begin{bmatrix} -1 & 2 \\ 2 & 3 \end{bmatrix} = \begin{bmatrix} 1 & 3 \\ 3 & 7 \end{bmatrix}$

9. $\begin{bmatrix} 2 & 1 \\ 1 & 4 \end{bmatrix} \begin{bmatrix} 2 & 0 & -1 \\ -1 & 2 & 2 \end{bmatrix} = \begin{bmatrix} 3 & 2 & 0 \\ -2 & 8 & 7 \end{bmatrix}$

11. $\begin{bmatrix} 2 & 6 & 1 & 0 \\ 2 & 4 & 0 & 1 \end{bmatrix} \Rightarrow \begin{bmatrix} 1 & 3 & 1/2 & 0 \\ 0 & -2 & -1 & 1 \end{bmatrix} \Rightarrow \begin{bmatrix} 1 & 3 & 1/2 & 0 \\ 0 & 1 & 1/2 & -1/2 \end{bmatrix}$

$$\begin{bmatrix} 1 & 0 & -1 & 3/2 \\ 0 & 1 & 1/2 & -1/2 \end{bmatrix} \quad \text{INVERSE:} \begin{bmatrix} -1 & 1.5 \\ 0.5 & -0.5 \end{bmatrix}$$

13. $\begin{vmatrix} -3 & 5 \\ 4 & 7 \end{vmatrix} = -3(7) - 5(4) = -21 - 20 - 41$

15. $x = \dfrac{\begin{vmatrix} 11 & 3 \\ 24 & 5 \end{vmatrix}}{\begin{vmatrix} 4 & 3 \\ -2 & 4 \end{vmatrix}}$

306

17. Find A and B such that:

$$\frac{-x+1}{(x+1)(x+2)} = \frac{A}{x+1} + \frac{B}{x+2}$$

$$-x+1 = A(x+2) + B(x+1)$$
$$-x+1 = Ax + 2A + Bx + B$$
$$-x+1 = (A+B)x + (2A+B)$$

$$A + B = -1$$
$$2A + B = 1$$

$$A = 2, B = -3 \;\Rightarrow\; \frac{-x+1}{(x+1)(x+2)} = \frac{2}{x+1} - \frac{3}{x+2}$$

19. $y < 2x + 6$

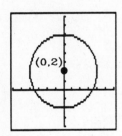

21. $x^2 + y^2 = 16$

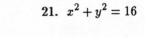

23. $$x^2 + y^2 - 4y = 12$$
$$x^2 + y^2 - 4y + 4 = 12 + 4$$
$$x^2 + (y-2)^2 = 16$$

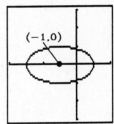

25. $$x^2 + 4y^2 + 2x = 3$$
$$x^2 + 2x + 1 + 4y^2 = 3 + 1$$
$$(x+1)^2 + 4y^2 = 4$$

$$\frac{(x+1)^2}{4} + \frac{y^2}{1} = 1$$

27. $\frac{x^2}{36} + \frac{y^2}{16} = 1$, or $\frac{y^2}{36} + \frac{x^2}{16} = 1$

29. $a = 2,\; c = 3 \Rightarrow b^2 = 5$

$$\frac{x^2}{4} - \frac{y^2}{5} = 1$$

31. 2nd term has $(2y)^1$: $\dfrac{8!}{1!7!}x^7(2y) = 16x^7y$

33. $\displaystyle\sum_{k=1}^{5} 2 = 5(2) = 10$

35. $a = -2$, $d = 3$

6th term: $-2 + (6-1)3 = 13$

$$\text{sum} = \frac{6(-2+13)}{2} = 33$$

37. $P(8,4) = \dfrac{8!}{4!} = 1680$

39. $C(12,10) = \dfrac{12!}{10!2!} = 66$

41. $4!6! = 24 \cdot 720 = 17{,}280$

43. $\dfrac{2}{36} = \dfrac{1}{18}$

45. $0.6(0.8) = 0.48$

47. True for $n = 1$?

$$3\cdot 1 + 1 = \frac{1(3\cdot 1 + 5)}{2}$$

$$4 = \frac{8}{2} \quad \text{works for } n = 1$$

Assume the formula works for $n = k$.

$$4 + 7 + 10 + \ldots + (3k+1) = \frac{k(3k+5)}{2}$$

$$\boxed{4 + 7 + 10 + \ldots + (3k+1)} + 3(k+1) + 1 = \boxed{\frac{k(3k+5)}{2}} + 3(k+1) + 1$$

$$4 + 7 + 10 + \ldots + 3(k+1) + 1 = \frac{k(3k+5)}{2} + 3k + 4$$

$$= \frac{3k^2 + 5k + 6k + 8}{2}$$

$$= \frac{(3k+8)(k+1)}{2}$$

$$= \frac{(k+1)[3(k+1)+5]}{2}$$

Since this is the result when $k+1$ is substituted for n in the formula, the formula works for $k+1$ whenever it works for k.